Optimization of Low-Carbon Energy Systems from Industrial to National Scale

Optimierung von CO_2-armen Energiesystemen vom industriellen bis zum nationalen Maßstab

Von der Fakultät für Maschinenwesen der
Rheinisch-Westfälischen Technischen Hochschule Aachen
zur Erlangung des akademischen Grades eines Doktors
der Ingenieurwissenschaften genehmigte Dissertation

vorgelegt von

Nils Julius Baumgärtner

Berichter: Univ.-Prof. Dr.-Ing. André Bardow
Apl. Prof. Dr.-Ing. Hans-Jürgen Koß
Assoc. Prof. Adam Brandt, Ph.D.

Tag der mündlichen Prüfung: 08. Oktober 2020

Diese Dissertation ist auf den Internetseiten der Universitätsbibliothek online verfügbar.

Aachener Beiträge zur Technischen Thermodynamik Band 29

Nils Julius Baumgärtner
Optimization of Low-Carbon Energy Systems from Industrial to National Scale
Optimierung von CO_2 -armen Energiesystemen vom industriellen bis zum nationalen Maßstab

ISBN: 978-3-95886-385-9

Bibliografische Information der Deutschen Bibliothek
Die Deutsche Bibliothek verzeichnet diese Publikation in der Deutschen Nationalbibliografie; detaillierte bibliografische Daten sind im Internet über http://dnb.ddb.de abrufbar.

Herstellung & Vertrieb:

1. Auflage 2020
© Wissenschaftsverlag Mainz GmbH - Aachen
Süsterfeldstr. 83, 52072 Aachen
Tel. 0241 / 87 34 34 00
www.Verlag-Mainz.de

ISSN: 2198-4832

Satz: nach Druckvorlage des Autors
Umschlaggestaltung: Druckerei Mainz

printed in Germany
D82 (Diss. RWTH Aachen University, 2020)

Danksagung

Die vorliegende Arbeit entstand im Rahmen meiner Tätigkeit als wissenschaftlicher Mitarbeiter am Lehrstuhl für Technische Thermodynamik der RWTH Aachen. Mein besonderer Dank gilt meinem Doktorvater André Bardow, du hast mich in allen meinen Plänen stets unterstützt. Ohne deine Betreuung, die anregenden Diskussionen und dein Vertrauen, wäre meine Zeit am Lehrstuhl, sowie diese Promotion, nicht möglich gewesen.
Weiterhin danke ich Hans-Jürgen Koß und Adam Brandt für die Übernahme des Koreferates im Rahmen meines Promotionsverfahrens, sowie Herrn Heinz Pitsch für die Übernahme des Prüfungsvorsitzes.

Allen aktiven und ehemaligen Kollegen des Lehrstuhls für Technische Thermodynamik und insbesondere der Energiesystemtechnik danke ich für die unvergesslichen Jahre, in denen ihr aus Kollegen Freunde geworden seid.
Ebenso möchte ich mich bei meiner ehemaligen Gruppenleiterin Maike Hennen dafür bedanken, dass dir meine Promotion und Probleme immer mindestens so wichtig waren wie deine eigenen. Björn Bahl danke ich für unsere intensiven Gespräche, zwei großartige Paper und den 'Kick-Start', den du meiner Promotion verliehen hast. Dir, Ludger Leenders, gilt mein Dank für die Zeit der gemeinsamen Gruppenleitung, wir waren stets ein starkes Team. Meinem Büro, im Besonderen Dinah Hollermann und Dominik Tillmanns, danke ich für die originellen Gespräche über Themen der Politik, Philosophie bis hin zur Chilizucht. Sarah Deutz, unsere Projekte haben mir ganz besonderen Spaß gemacht, danke für jede gemeinsame Betreuung, Diskussion und die vielen hundert Kaffees.
Außerdem danke ich meinen vielen außergewöhnlichen Studenten, ihr habt mich und ebenso meine Promotion maßgeblich geprägt. Ich habe mich besonders darüber gefreut, dass aus einigen von euch Kollegen und Freunde wurden.

Der Covestro AG möchte ich für ein spannendes Projekt und die damit einhergehende Finanzierung danken. Euch, Karen Perrey und Jochen Hammerschmidt, gilt mein besonderer Dank, es war mir stets eine Freude mit euch zusammenzuarbeiten. Es freut mich, dass wir auch weiterhin in Kontakt sind.

Von ganzem Herzen danke ich meinem Vater, der mir die Faszination für Natur und Technik nahegebracht hat und meiner Mutter, ohne deren Unterstützung meine akademische Laufbahn schon in der Schule gescheitert wäre.
Abschließend danke ich Dir Jana, für unser gemeinsames Leben in Aachen und dafür, dass du mir diese Promotion ermöglicht hast. Ich freue mich schon auf unseren nächsten Lebensabschnitt in Leverkusen!

Leverkusen, im November 2020 *Nils Baumgärtner*

'La vraie générosité envers l'avenir consiste à tout donner au présent'
'Real generosity toward the future lies in giving all to the present.'

From Mr. Albert Camus,
'L'Homme révolté 1951'

Contents

List of Figures

List of Tables

Notation

For readability, the notation of symbols only used in the appendix are introduced in the appendix itself.

Abbreviations and Acronyms

AbC	absorption chiller
AC	alternating current
A_{tfw}	acidification, terrestrial and freshwater
B	boiler
BAT	battery
B_{el}	electrode boiler
BEV	battery electric vehicle
B&C	branch-and-cut
CC	compression chiller
CHP	combined-heat-and-power engine
CI	climate impact
CO_2	carbon dioxide
DC	direct current
DeLoop	method for decomposition based long-term operational optimization
DSM	demand-side management
E_{fw}	eutrophication, freshwater
E_m	eutrophication marine
E_t	eutrophication terrestrial
ET_{fw}	ecotoxicity, freshwater
EU	European Union
ES	case study industrial energy system
FCV	fuel cell vehicle
G20	international group of 20 major economies
GCG	general-column-generation
GHG	greenhouse gas

GWP	global warming potential
H_2	hydrogen
HEX	heat exchanger
HP	heat pump
HT_c	human toxicity cancer
HT_{nc}	human toxicity non-cancer
INV	inverter
IR	ionizing radiation
LCA	life-cycle assessment
LCI	life-cycle inventory
LCIA	life-cycle impact assessment
LP	linear programming
LU	land use
MILP	mixed-integer linear programming
MINLP	mixed-integer non-linear programming
OD	ozone depletion
ORC	organic Rankine cycle
PCOF	photochemical ozone formation
PM	particulate matter
PS	case study pump system
PV	photovoltaic
R&A	relaxation and aggregation
$RiSES^4$	method for rigorous synthesis of energy supply and seasonal storage systems
RD_e	resource depletion, energy
RD_{mm}	resource depletion mineral and metal
SecMOD	sector-coupled energy system model
s.t.	subject to
ST	solar thermal
TS	thermal storage
UN	United Nations
USA	United States of America
vkm	vehicle kilometer
WIND	wind turbine
WS	water scarcity

Latin Symbols

A	coefficient matrices
AEF	average emission factor
$APVF$	annualized present value factor
b	right-hand-side vector
c	prices
$CAPEX$	capital expenditures
$C_{\mathrm{CO2}}^{\max}$	emission constraint
Δt	duration of a time step
ΔV	storage difference
e	emission factor
E	energy demand
EF	technology-specific emission factor
G	solar radiation
GWI	global warming impact
H	calorific value
i	interest rate
M	maintenance costs
MEF	marginal emission factor
N	number
NPV	net present value
$OPEX$	operational expenditures
P	input parameter
P	electricity flow
p	pressure
T	time
TAC	total annualized costs
u	speed
U	input capacity of energy-conversion units
V	output capacity of energy-conversion units
x	auxiliary variables
x_k	specific energy carrier costs
XEF	grid mix emission factor
y	auxiliary binary variables
z	binary assignment

Greek symbols

δ	binary on/off status and the current part-load performance
η	efficiency
ε	optimality gap
γ	binary existence
μ	specific GHG emissions
$\lfloor\ \rfloor$	underestimator
$\lceil\ \rceil$	overestimator

Superscripts

buy	buying
cool	cooling
el	electricity
heat	heating
i	investment
Imp	import
in	input
LB	lower bound
lt	low-temperature
m	maintenance
mar	marginal
max	maximal value
N	nominal value
pv	photovoltaic
o	operational
of	offshore
on	onshore
out	output
sell	selling
total	total volume flow
UB	upper bound
WS	wind speed

Subscripts

eq	equivalent
gas	gas consuming unit
grid	grid electricity
h	horizon
net	network connection
p	peak-power
T	transmission

Sets

j	auxiliary set containing 'network connection' and 'peak-power'
$k \in K$	set of energy carriers
$l \in L$	set of countries
$n \in C$	set of energy-conversion units
$p \in P$	set of typical periods
$pp \in PP$	set of power plants
$s \in S$	set of segments per period
$t \in T$	set of time steps
$t^{'} \in P$	set of time steps

Kurzfassung

Die Eindämmung des Klimawandels erfordert eine Reduzierung der Treibhausgasemissionen. Hauptverantwortlich für diese Emissionen ist der Energiesektor, der aktuell auf fossilen Brennstoffen basiert. Daher, müssen wir den Energiesektor so umgestalten, dass er auf CO_2-armen Technologien basiert. Dafür sind neue Energiesystemdesigns zusammen mit geeigneten Betriebsstrategien erforderlich. Diese neuen Designs und Betriebsstrategien lassen sich am besten durch mathematische Optimierung identifizieren. Allerdings führen CO_2-arme Technologien zu Herausforderungen bei der Lösung und korrekten Bewertung der Optimierungsprobleme.
CO_2-arme Technologien sind durch schwankende Verfügbarkeit bestimmt, die die Komplexität der Optimierung erhöht. Für komplexere Betriebsoptimierungen, entwickeln wir eine Zeitreihen-Zerlegungsmethode. Die Methode zerlegt das komplexe, zeitgekoppelte Betriebsproblem in Teilprobleme, liefert aber dennoch umsetzbare, nahoptimale Lösungen. Für komplexe Designauslegungen schlagen wir ein Lösungsverfahren basierend auf Zeitreihenaggregation vor. Das Lösungsverfahren teilt die Auslegung in zwei getrennte Probleme: ein aggregiertes, relaxiertes und ein aggregiertes, eingeschränktes Problem. Dadurch werden umsetzbare, nah-optimale Designs gefunden.
Zusätzlich erfordert die Transformation eine rigorose Bewertung von Treibhausgasemissionen und anderen Umweltkategorien. Die Bewertung von Treibhausgasemissionen durch industriellen Stromverbrauch ist schwierig, da das zugrunde liegende nationale Energiesystem nicht abgebildet wird. Daher entwickeln wir Methoden zur Berechnung von Treibhausgasemissionsfaktoren für Strom. Wir zeigen, dass diese Faktoren es industriellen Energiesystemen ermöglichen ihre Emissionen deutlich zu reduzieren. Auf nationaler Ebene muss eine Verlagerung von Problemen auf andere Umweltkategorien als den Klimawandel verhindert werden. Daher entwickeln wir eine nationale Energiesystemoptimierung mit Lebenszyklusanalyse, die 15 weitere Umweltkategorien berücksichtigt. Mit dem Modell optimieren wir einen Transformationspfad zu einem CO_2-arme Energiesystem, welcher zu Verbesserungen in vielen Kategorien, aber auch zu Verschlechterung in Umweltkategorien führt. Diese Kategorien müssen während der Transformation und bei der Entwicklung neuer Technologien berücksichtigt werden.
Insgesamt erleichtern die Methoden und Modelle in dieser Arbeit die Integration CO_2-arme Technologien in Energiesystemen.

Abstract

Climate change mitigation requires a reduction of greenhouse gas (GHG) emissions. The main emitter of GHG emissions is the energy sector, which today is based on fossil fuels. To mitigate climate change, we need to transform the energy systems to low-carbon technologies. For this purpose, new energy system designs are required along with appropriate operational strategies. In principle, these new designs and operational strategies can be identified best using mathematical optimization. However, low-carbon technologies impose challenges in solving and assessing the resulting optimization problems.

Low-carbon technologies are volatile, which increase the complexity of optimal synthesis and operation. To cope with the complexity of operational optimization, we develop a time-series decomposition method. The method decomposes the complex, time-coupled operational problem into smaller subproblems, while still providing feasible, near-optimal solutions. For the increased complexity in synthesis problems, we propose a method based on time-series aggregation. The method divides the original synthesis problem into two separate problems: one aggregated relaxed problem and another aggregated restricted problem, leading to feasible, near-optimal solutions.

In addition, the transformation process requires a rigorous assessment of greenhouse gas emissions and potential burden-shifting. In particular, the assessment of emissions due to electricity usage on the industrial scale is difficult, as the underlying national electricity system is not modeled. Therefore, we propose methods to compute industrial greenhouse gas emission factors for electricity. By exploiting these emission factors, industrial energy systems can significantly reduce their emissions. On the national scale, burden-shifting towards environmental impacts besides climate change needs to be prevented in the transformation. Hence, we develop a national energy system model and extend the optimization with life-cycle assessment considering 15 further environmental impacts. With the model, we compute a cost-optimal transformation pathway to a low-carbon energy system. The transformation leads to many co-benefits, but also to severe burden-shifting, which needs to be considered during the transformation process and in the development of new low-carbon technologies.

Overall, the methods and models in this thesis facilitate the integration of low-carbon technologies in energy systems.

Chapter 1

Introduction

The energy sector is the main contributor to anthropogenic greenhouse gas (GHG) emissions (IPCC, 2013). Climate change mitigation in line with the 1.5 °C goal requires a massive reduction of GHG emissions (IPCC, 2018). For the energy sector, this massive reduction of GHG emissions is the major challenge of the 21st century. Secretary-General of the United Nations (UN) Antònio Guterres said at the UN Climate Change Conference COP25:

> *'We simply have to stop digging and drilling, and take advantage of the vast possibilities offered by renewable energy and nature-based solutions.'*
>
> Antònio Guterres, UN Secretary-General,
> Madrid, December 2019 at the COP25

To master the challenge of reducing GHG emissions, we simply have to make use of the *'vast possibilities offered by renewable energy'*. For the energy sector, this means that we need to employ low-carbon technologies based on renewable energy. Energy systems supply final energy demands by converting primary and secondary energy. Current energy systems rely on fossil fuels as primary and secondary energy. Thus, employing low-carbon technologies based on renewable energy fundamentally transforms the energy system. This transformation affects all energy systems in all scales ranging from residential and industrial, to national and even global scales. Thus, the energy system needs to be transformed on all scales to reach the GHG emission targets.

The transformation of the energy system requires new designs and operation strategies suited for low-carbon technologies. Synthesis, i. e., design planning, and operation of energy systems based on mathematical optimization, will significantly reduce GHG emissions and costs of these new designs and operational strategies. Low-carbon technologies, however, impose challenges for the optimal synthesis and operation of energy systems:

Challenge 1: Long-term operation
Optimal operation has to handle increasingly volatile inputs, due to renewables, while planning horizons extend, due to a greater need for storage capacity.

Challenge 2: Rigorous synthesis with seasonal storage
Optimal synthesis needs to consider more time-dependent inputs due to renewable resources. Further, optimal synthesis needs to consider seasonal storage to cope with these renewable resources.

Challenge 3: Assessment of GHG emissions
Apart from economic objectives, the scope of optimal synthesis must include GHG emissions as well. The correct assessment of GHG emissions is challenging.

Challenge 4: Burden-shifting
Currently, GHG emissions and the resulting climate change are our main concern. However, the low-carbon transformation is likely to impose burden-shifting to other environmental issues. The scope of optimal synthesis must, therefore be extended to assess environmental impacts holistically.

In this work, we[1] extend the scope of energy system models to cope with these 4 challenges resulting from the integration of low-carbon technologies. This thesis focuses on the industrial and national scales; however, our proposed methods can easily be implemented on all the other scales.

1.1 Structure of this thesis

In Chapter 2, we review the scope of state-of-the-art energy system optimization and derive 4 current challenges for the optimization of energy system models imposed by low-carbon technologies. In Chapter 2.4, we state this thesis's contribution, which is to cope with the 4 challenges imposed by low-carbon technologies. Chapters 3-6 present the contributions to cope with the 4 challenges in detail:

Challenge 1: Long-term operation
In Chapter 3, we propose the method DeLoop to solve long-term operational optimization problems of energy systems with low-carbon technologies. Energy systems with low-carbon technologies require short-term flexible operation, while longer planning horizons become more important at the same time. Long planning horizons are necessary to handle emission limits, electricity network constraints, and storage systems.

[1]This thesis is written in the *pluralis modestiae*, first to avoid the excessive use of passive voice and second to emphasis that the research behind this thesis is done in collaboration with colleagues and my supervisor. The contribution of the author is pointed out at the beginning of each chapter.

Emission limits, network constraints, and storage systems introduce time-coupling constraints, which render the optimization more complex. To solve these complex optimization problems, DeLoop decomposes time-coupling constraints allowing for fast solutions with known solution quality while keeping a long planning horizon.

Challenge 2: Rigorous synthesis with seasonal storage
In Chapter 4, we propose the method RiSES^4 to exactly solve synthesis problems of energy systems with low-carbon technologies. To handle low-carbon technologies and storage systems, any time-dependent input data needs to be included in the synthesis problem. Thereby, the synthesis becomes challenging to solve, i. e., to find optimal and feasible solutions. RiSES^4 provides feasible solutions with known quality. To obtain a feasible solution and, thus, an upper bound, we use time-series aggregation and, subsequently, solve an operational problem. Lower bounds are obtained by two approaches: linear-programming relaxation and relaxation based on time-series aggregation. RiSES^4 renders the rigorous synthesis of low-carbon energy systems with seasonal storage possible and reduces computational time significantly.

Challenge 3: Assessment of GHG emissions
In Chapter 5, we show that using constant GHG grid emissions based on annual averages misquantifies emissions of energy systems. Therefore, we develop a synthesis model for low-carbon industrial energy systems with electricity grid connection. Subsequently, we propose 2 methods to compute time-dependent electricity grid emission factors. These grid emission factors serve as input to determine optimal economic and GHG emission trade-off curves. Our results show that time-dependent grid emissions enable climate-friendly demand-side management, significantly reducing emission.

Challenge 4: Burden-shifting
In Chapter 6, we show that the transformation towards a low-carbon national energy system leads to many co-benefits. However, at the same time, burdens are shifted to environmental categories other than climate change. To explore this burden-shifting, we first develop a synthesis model to optimize the transition path towards a low-carbon national energy system in Germany until 2050. We then extend the developed model by life-cycle assessment to consider environmental impacts holistically. For the transformation path, we compute environmental impacts. The results show certain environmental categories, such as metals and minerals resource depletion, which future national energy system models should consider.

Finally, in Chapter 7, we summarize the thesis and conclude. Furthermore, we outline possible future research topics.

Chapter 2

State-of-the-art in energy system optimization

This chapter provides an overview of the state-of-the-art in energy system optimization. We focus on challenges in optimization methods and the environmental scope of energy system models: both in the context of the expansion of low-carbon technologies. In Chapter 2.1, a general introduction is given to modeling and optimization of energy systems. In Chapter 2.2, we focus on industrial energy systems and in detail on their optimal operation (Chapter 2.2.1), optimal synthesis (Chapter 2.2.2), and their environmental scope (Chapter 2.2.3). In Chapter 2.3, we investigate the state-of-the-art of national energy system optimization. In Chapter 2.4, we state the contribution of this thesis.

2.1 Modeling and optimization of energy systems

Climate change mitigation in line with the 1.5 °C goal requires a massive reduction of greenhouse gas (GHG) emissions (IPCC, 2018). The mitigation, in combination with growing energy demands, forces a transition from fossil-based towards renewable-based energy systems. However, the integration of renewable-based low-carbon technologies into energy systems is challenging: Renewable generation is difficult to predict, and the availability is highly fluctuating. As a result, the power generation rarely correlates with the given energy demand (Haas et al., 2017).

Thus, we need to find new energy system designs and operational strategies to cope with challenges due to low-carbon technologies. Mathematical optimization is the most comprehensive approach to identify new energy system designs and operational strategies, as reviewed by Andiappan (2017). Mathematical optimization can find designs and operation schedules that best satisfy an objective, e. g., minimal cost or environmental impact (Demirhan et al., 2019). Mathematical optimization of energy system design is usually performed through a superstructure-based approach (Gong and You, 2015) and (Mencarelli et al., 2020): The superstructure represents all possible technologies and configurations of energy systems from which the optimal combination can be selected.

In general, mathematical optimization of energy system design is a two-stage problem consisting of a synthesis stage and an operational stage (Lin et al., 2016). On the synthesis stage, decisions on the type and size of units are made, determining the energy system design. On the operational stage, for each time step, the on/off status and the load allocation are determined for each energy supply and storage unit. Synthesis optimization requires the simultaneous optimization of design and operation. Thus, synthesis optimization also includes operational optimization, whereas operational optimization is solved based on a fixed design.

In general, synthesis of energy systems requires a formulation as a mixed-integer nonlinear programming problem (MINLP) (Bruno et al., 1998) and (Goderbauer et al., 2016). Discrete decisions are required to consider existence of units on the design stage and their on/off status on the operation stage. Typically, non-linear part-load performance and investment-cost curves have to be considered for each unit in the synthesis of energy systems (Bruno et al., 1998). These non-linearities are often linearized by piecewise linearization (Floudas, 1995; Misener et al., 2009) and (Voll et al., 2013). Further, for extensive models such as national energy systems, discrete decisions are often modeled in a continuous manner to avoid binary variables (Lopion et al., 2018). Thus, synthesis problems of national energy systems are typically formulated as linear programming (LP) problems (Lopion et al., 2018). In contrast, synthesis problems of industrial energy systems are often modeled as mixed-integer linear programming (MILP) problems (Elia and Floudas, 2014).

In the next section, we focus on industrial energy systems models and their optimization.

2.2 Industrial energy systems

Industrial energy systems, also called utility systems, consist of energy-conversion technologies based on-site, which are used to supply energy to industrial production facilities such as chemical, pharmaceutical, and food production. Thus, supply infrastructure is closely located to final energy demands (Mancarella, 2014). Thereby, on-site energy-conversion technologies reduce energy losses due to transportation and non-optimal operation of centralized conversion technologies (Voll, 2014). Therefore, on-site energy systems are increasingly widespread in the industrial sector and also in the residential sectors.

We review 50 available industrial and residential energy system models to determine common types of model formulation, energy demands, and energy-conversion units, see Appendix A. As stated in Chapter 2.1, industrial energy system models would require a formulation as MINLP. However, MINLP problems are complex to solve and are often solved only to local optimality. Instead, MILP problems enable fast convergence to the global optimum of the linearized problem (Floudas, 1995). Thus, 38 of the 50 reviewed models for industrial energy systems are formulated as MILP problems by linearization. Only a few approaches solve a non-linear MINLP problem directly, or by genetic, or successive optimization methods.
Industrial energy systems usually supply multiple final energy demands, such as cooling, electricity, heating, and steam. All reviewed energy system models supply electricity, heat and/or steam. 19 models consider cooling demand.
To supply these final energy demands, industrial energy systems convert primary and secondary energy (Frangopoulos, 2018). Industrial energy systems are typically connected to the national energy systems to exchange primary and secondary energy such as gas and electricity. For the conversion, industrial energy systems are often complex systems employing multiple conversion technologies and storage systems. The most common heat-providing units are boilers and combined-heat-and-power engines. The most common cooling-providing units are compression chillers and absorption chillers, and the most common renewable units are solar-thermal panels and photovoltaics.
The multiple types of energy demands and the different types of energy-conversion units, render industrial energy system very complex. Optimization of these complex systems leads to significant economic and ecological benefits (Di Somma et al., 2015).

In this thesis, we focus on the energy system only, i. e., on industrial sites' supply side. Thus, we neglect efficiency measures such as demand-side management by the energy consumer as, e. g., proposed by Bahl et al. (2017b), or integrated optimization of energy and production systems as proposed by Leenders et al. (2019). Further, we assume that the heat exchanger network is part of the demand side, and thus not part of the optimization problem.

In the next 3 subsections, we focus on the operational optimization, synthesis optimization, and environmental assessment of GHG emissions for industrial energy systems.

2.2.1 Long-term operational optimization of energy systems

Typically, optimization problems of industrial energy systems are formulated as mixed-integer linear programming (MILP) problems, cf. Chapter 2.1. For small-scale MILP problems, state-of-the-art solvers can provide solutions quickly (Grossmann, 2012). The computational effort of solving operational problems, however, increases polynomial with the considered problem size. Thus, the complexity of operational problems of energy systems is weakly NP-hard (unless P = NP) (Goderbauer et al., 2019).
For the operational optimization of energy systems, the size of the MILP problem depends on the number of considered time steps. Thus, the high number of time steps in long-term operational optimization leads to large-scale MILP problems.
A common solution approach to solve these large-scale MILP problems is to split the time horizon into shorter time intervals (direct decomposition). The decomposed periods can be solved independently and often at acceptable time. However, direct decomposition is only possible if no time-coupling constraints or variables are present (Bradley et al., 1977). For industrial energy systems, including low-carbon technologies, time-coupling constraints and variables are very common, e. g., by peak-power prices, network-connection fees, cogeneration subsidies, and storage systems (Bischi et al., 2019). To solve even large-scale MILP operational problems with time-coupling constraints, various solution methods exist, Table 2.1.

Table 2.1: Examples of solution methods for long-term operational optimizations with time-coupling constraints and variables.

Method	Examples	Strengths (+) / Limits (−)
Model simplifications	Piacentino and Cardona (2008), Yokoyama (2013)	+ easily applicable − problem specific
Non-rigorous	Kazarlis et al. (1996), Park et al. (2000), Kavvadias and Maroulis (2010), Renaldi and Friedrich (2017), Bischi et al. (2019)	+ allows parallelization + accurate model − no quality measure
Decomposition	Yokoyama and Ito (1996), Al-Agtash and Su (1998), Rong et al. (2008), Nasrolahpour et al. (2016), Wang et al. (2016)	+ allows parallelization + accurate model + provide quality measure − slow convergence − complex formulation

Model simplifications solve such long-term operational problems by simplifying component models to reduce the complexity of the optimization problem. For example, Piacentino and Cardona (2008) relax binary variables, Palmintier and Webster (2014) lump multiple components into one to reduce binary variables, and Yokoyama (2013) converts complex non-linear functions into easier linear functions. These simplifications are easy to implement and render the long-term optimization problem solvable, even when considering time-coupling constraints and variables. However, model simplifications limit the physical accuracy of optimization models, which may lead to suboptimal or even infeasible solutions in practice.

To prevent suboptimal or infeasible solutions, sufficiently accurate models are required. For these more accurate optimization models, non-rigorous methods can to solve the resulting complex optimization problems. Genetic algorithms have been used for unit commitment (Kazarlis et al., 1996), for capacity expansion (Park et al., 2000), and for trigeneration (power, heat, and cooling) (Kavvadias and Maroulis, 2010). For operational optimization considering cogeneration subsidies, Bischi et al. (2019) use a rolling horizon approach, including long-term foresight by typical weeks. Renaldi and Friedrich (2017) optimize the operation of short- and long-term storage systems by using 2 time grids, one with short and the other with long time intervals per time step. The advantage of these non-rigorous methods is that they normally yield good solutions for difficult problems and often allow fast computation, e. g., by parallelization. However, in general, it is not possible to evaluate the quality of the proposed solution (Hanne and Dornberger, 2017).

As an alternative, decomposition approaches can be used to obtain solutions of known solution quality. Decomposition methods have been applied to the optimization of energy systems with coupling constraints and variables. Problems with coupling constraints have been decomposed by Dantzig-Wolfe decomposition (Yokoyama and Ito, 1996) and Lagrangian relaxation (Al-Agtash and Su, 1998) and (Rong et al., 2008). To decompose problems with coupling variables, Benders' decomposition is suitable (Nasrolahpour et al., 2016). Modified decomposition methods can cope with both time-coupling constraints and variables in MILP optimizations. To solve problems with both coupling constraints and variables, Wang et al. (2016) apply a combination of Lagrangian relaxation and Benders' decomposition.
Although much progress has been made in automated Dantzig-Wolfe decomposition, it is still hard to automatically identify an appropriate decomposition. Further, solution times remain long for many problems (Bergner et al., 2015). Lagrangian methods require good multipliers for fast convergence. However, the selection of Lagrangian multipliers is difficult, which has lead to the development of a large variety of methods (Conejo et al., 2006). Benders' decomposition often shows slow convergence due to many time-consuming iterations. Although methods to improve convergence of Benders' decomposition have been proposed, its formulation and implementation remain complex (Rahmaniani et al., 2017).

In summary, no solution method exists for long-term operational optimizations, including low-carbon technologies, which combines all strengths of the approaches: is easily applicable, uses a physically accurate model, allows for parallelization, and shows fast convergence. Thus, operational optimization usually neglects long time horizons (Demirhan et al., 2019). Hence, we identify the consideration of long time horizons in operational optimization as first challenge of energy system optimization.

2.2.2 Rigorous synthesis optimization of energy systems

Synthesis problems of energy systems are even more complex to solve than operational problems, as in synthesis problems, both operation and design need to be optimized simultaneously, see Chapter 2.1. Synthesis problems of energy systems are generally strongly NP-hard (Goderbauer et al., 2019), and thus, cannot be solved efficiently (unless P = NP). In particular, this implies that the computational effort increases exponentially with the problem size.

As for operational optimization, practical application has shown that advanced solvers can solve small-scale MILP instances of synthesis problems at acceptable time (Grossmann, 2012). However, synthesis problems of energy systems consider time scales ranging from minutes up to several years of planning (Floudas et al., 2016). As a result, large time series have to be considered in synthesis problems. The operation stage usually depends on multiple of these large time series, e. g., demand profiles, electricity prices, and availability of renewable resources for low-carbon technologies (Pfenninger et al., 2014). Consequently, the resulting synthesis problems are large-scale MILPs, which are computationally challenging even for the most advanced solvers and are often not solvable within reasonable computational time or memory limits.
In particular, the integration of energy storage increases the complexity of the synthesis problem, because energy storage systems require a large and highly resolved time series (Adhau et al., 2014) and (Gabrielli et al., 2018) and lead to time-coupling constraints (Rong et al., 2008). However, short-term storage systems must be considered in the synthesis of energy systems with low-carbon technologies, as Zerrahn et al. (2018) show by a literature review. Cebulla et al. (2017) conclude that energy systems with high shares of renewable energy have to cope with longer periods of low energy generation, thus, creating a need for long-term storage to balance seasonal variations in the availability of renewables. Keck et al. (2019) quantify that (seasonal) storage systems are able to reduce system costs of low-carbon energy systems by 13-22 %. Thus, seasonal energy storage is necessary to balance longer periods of low renewable generation and significantly reduce system costs. Hence, both short-term and long-term storage systems are needed for the successful synthesis of low-carbon energy systems (Cebulla et al., 2018).

To still solve the resulting synthesis problems, various solution methods based on the problem structure and aggregation exist, as reviewed by Bahl et al. (2018b).
The two-stage character of the synthesis problem forms the basis of two-stage algorithms as developed by Fazlollahi et al. (2014) and Elsido et al. (2017): First-stage decisions on the investment of units are computed by evolutionary algorithms, while rigorous optimization methods determine the second-stage decisions for the operation. However, an evolutionary algorithm does not provide rigorous information on the solution quality. Yokoyama et al. (2015) therefore, propose a decomposition method utilizing the two-stage character of synthesis problems, which allows to evaluate the solution quality. In the proposed decomposition method of Yokoyama et al. (2015), the first-stage decisions are solved with relaxed second-stage decisions. Following, the second-stage decisions are solved with fixed first-stage decisions. However, as Yokoyama et al. (2015) conclude, the implementation of this method is complicated as

the input data needs to be prepared manually. In more recent work of Yokoyama et al. (2018b), the method is further enhanced by parallel computing, and, in Yokoyama et al. (2019), the method is extended by a clustering approach reducing computational effort of the first stage. Still, the methods are unable to cope with seasonal storage systems.

Besides exploiting the problem structure, the size of the synthesis problem can be reduced by model aggregation to solve large-scale problems (Grossmann, 2012). Model aggregation methods are discussed in Mancarella (2014). Mancarella (2014) distinguishes between the aggregation of spatial distribution, the number of energy flows, and multiple energy-conversion units into 1 component. Besides these aggregations methods, often, time-series aggregation is applied in synthesis problems for the design of chemical processes and energy systems, as reviewed by Maravelias and Sung (2009) and categorized by Nahmmacher et al. (2016). However, state-of-the-art time-series aggregation methods often consider independent typical days, e. g., Bahl et al. (2018a), and thus cannot include seasonal storage into the synthesis optimization. To consider seasonal storage with time-series aggregation, Rager (2015) and Samsatli et al. (2016) select typical days for each month of the year or per season, resulting in coupled typical days. However, this approach does not represent the original time series accurately, as diversity of days within a month or season might not be captured by only one typical day (Kotzur et al., 2018). Therefore, Renaldi and Friedrich (2017), Kotzur et al. (2018), Gabrielli et al. (2018) as well as van der Heijde et al. (2019) developed synthesis methods for energy, systems including seasonal storage systems based on time-series aggregation. Renaldi and Friedrich (2017) introduce a second time grid, typical periods are coupled by Gabrielli et al. (2018); Kotzur et al. (2018) and van der Heijde et al. (2019). All their methods enable integration of seasonal storage systems into energy systems synthesis in reasonable computational time. However, these synthesis methods only solve a reduced synthesis problem. Thus, the solution does not correspond to the solution of the original synthesis problem employing the full time series. As a result, the solution quality of the reduced synthesis problem for the full time series is unknown and the resulting design might even be infeasible.

To prove feasibility, Kannengießer et al. (2019) developed a method based on time-series aggregation and the two-stage character of the synthesis problem. In the method, feasibility is proven by a second-stage optimization with the full time series. However, the solution quality remains unknown. Hoffmann et al. (2020) review time-series aggregation methods and conclude that determining the solution quality is still an open research question. Also, Buchholz et al. (2019) state that the trade-off between solution quality and computational difficulty from aggregation makes validation techniques key. In particular, the synthesis optimization of storage technologies has been shown to be more sensitive to time-series aggregation than other generation technologies (Frew and Jacobson, 2016).

Thus, to obtain a feasible solution of large-scale problems with known quality, exact solution strategies are needed (Grossmann, 2012). Pöstges and Weber (2019) propose a time-series aggregation method, which leads to feasible solutions with known solution quality only for small-scale problems. For large-scale synthesis problems, Bahl et al. (2018a) and Bruche and Tsatsaronis (2019) proposed exact decomposition method based on time-series aggregation with known solution quality. The methods exploit both the problem structure and model aggregation.
However, the proposed decomposition methods are not applicable to time-coupling constraints as occurring in seasonal storage. However, as shown above, for low-carbon technologies, seasonal storage systems are crucial. Still, no solution method exists to rigorously solve synthesis problems of low-carbon energy systems with seasonal storage systems. We identify this missing solution method for seasonal storage systems to synthesize energy systems as the second challenge of energy system optimization.

2.2.3 Environmental scope of industrial energy system models

The required substantial reduction of GHG emissions is a major challenge for the industry (Bazzanella and Ausfelder, 2017). For instance, the European chemical industry emits about 128.2 Mt of GHGs per year, mostly due to its energy demand (Bazzanella and Ausfelder, 2017). Hence, efficient energy supply in industry is key to reduce industrial GHG emissions and thus to achieve the ambitious GHG reductions.

In industry, energy is mainly supplied by on-site energy systems combined with a connection to the public electricity and gas grid. Thus, reduction of GHG emissions requires the design of on-site, low-carbon energy systems with grid connection. The overall GHG emissions of low-carbon energy systems depend strongly on the emissions associated with grid electricity, the so-called Scope II emissions according to GHG Protocol Initiative (2004). Today, grid emissions are usually accounted for by annually-averaged emission factors, but emissions are, in fact, time-dependent (Kopsakangas-Savolainen et al., 2017). This time-dependency mainly results from the balance of varying demand and supply. The time-dependency increases with a larger share of generation from wind and solar, which not only leads to increased variation in electricity prices, but also in grid emissions (Kelley et al., 2019).

Here, also, an opportunity arises, since temporal shifting of the electricity demand by demand-side management can reduce not only costs but also GHG emissions associated with electricity purchase (Burre et al., 2020). Kopsakangas-Savolainen et al. (2017) demonstrate that operational optimization considering hourly-based grid emissions can reduce emissions from electricity consumption. Kelley et al. (2019) optimize production schedules for an air separation unit, showing the potential to reduce emissions by considering time-dependent grid emissions. Following the ideas presented in this thesis, Pina et al. (2020) use multi-objective optimization for the synthesis of a tri-generation system on the residential scale that incorporates time-dependent grid emissions. Thus, fluctuating grid emissions enable climate-friendly demand-side management. In contrast, studies employing constant grid emissions cannot exploit the full emission reduction potential of shifting consumption (Kono et al., 2017) and resulting actions are often even misleading (Hawkes, 2010).

Despite the importance of time-dependent grid emission factors, no general accepted factors are public and there are many inconsistent methods to compute them. Besides, no synthesis optimization model for industrial energy systems investigates the impact of time-dependent grid emissions on the potential to reduce GHG emissions. We identify missing methods to compute emission factors and their integration in the synthesis of energy systems as the third challenge of energy system optimization.

2.3 National energy system optimization

In the previous sections, we reviewed industrial energy systems optimization in the context of the integration of low-carbon technologies. However, to reach the GHG reduction targets, it is necessary to transform the energy system at all scales, also on the national level. Reduction measures on the national scale mainly focus on the electricity sector through the integration of low-carbon technologies. However, electricity production accounts only for approximately 40 % of GHG emissions in the G20 countries (Olivier et al., 2017), 25 % in EU-28 (Eurostat, 2018), and 38 % in Germany (Salb et al., 2018). Thus, GHG emissions can only be massively reduced by also targeting other high-emission sectors, such as heating and transportation (Ruhnau et al., 2019).

In low-carbon energy systems, the sectors electricity, heat, and transportation are interconnected via electricity. This sector coupling makes the development of a decarbonization strategy challenging (DeCarolis et al., 2017). To capture this challenge, national energy system models can support the development of decarbonization strategies (Mirakyan and de Guio, 2015).

The decarbonization strategies created with national energy system models envision a major transformation by the vast expansion of low-carbon technologies. Although these low-carbon technologies perform well regarding direct GHG emissions, they lead to indirect GHG emissions due to construction (McDowall et al., 2018) and might also induce other environmental impacts, such as ecotoxicity and resource depletion (Rauner and Budzinski, 2017) and (Algunaibet and Guillén-Gosálbez, 2019). Thus, employing low-carbon technologies for GHG reductions may shift burden to other environmental impacts (Algunaibet and Guillén-Gosálbez, 2019). With reduced GHG emissions, these other environmental impacts will gain more importance during the transformation of the energy system (Rauner and Budzinski, 2017).

To consider all environmental impacts, energy system models have recently been extended by life-cycle assessments (LCA) (Ringkjøb et al., 2018). For example, Turconi et al. (2014) investigate the future environmental impacts of the Danish electricity sector for different scenarios of heat and power cogeneration and biomass. Kouloumpis et al. (2015) assessed environmental impacts for 16 scenarios for the UK electricity sector up to the year 2070. Berrill et al. (2016) present an LCA of 44 European electricity scenarios, from conventional energy system using carbon capture and storage and systems based mostly on renewable generation technologies. García-Gusano et al. (2016) calculate environmental impacts of a business-as-usual scenario for the electricity sector of Norway up to the year 2050. Rauner and Budzinski (2017) used multi-objective optimization to include environmental impacts in an expansion problem of the German electricity system. Algunaibet and Guillén-Gosálbez (2019) used multi-objective optimization to quantify the burden-shifting potential of GHG emission reductions on other environmental impacts of the electricity system of the USA. These energy system models provide a comprehensive view on environmental impacts by including LCA, but focus on the electricity sector only. Thus, cross-sectoral interconnectivity is neglect.

Volkart et al. (2017) combine a cross-sectoral energy system model with LCA to evaluate 3 scenarios for the Swiss energy transition. The used cross-sectoral energy system model, however, considers no spatial resolution, i. e., no electricity grid, and only low time-resolution with 6 time steps. Recently, Blanco et al. (2020) performed ex-post LCA of the JRC-EU-TIMES model including multiple sectors, however, the LCA is not part of the optimization itself. Also worth mentioning, is the work of (Gençer et al., 2020), in which they developed the SESAME-tool to evaluate and compare energy transition pathways, however, SESAME does not include optimization. And the work of Li et al. (2020), in which they developed a new modeling approach to optimize the carbon loop of biomass and carbon capture, use and storage for the carbon neutrality of Switzerland in 2050. However,Li et al. (2020) focus on climate change and do not include an holistic life-cycle assessment.

Lopion et al. (2018) review energy system models and conclude that 4 elements are required to provide holistic insight for national GHG-reduction strategies: (1) cross-sectoral technologies (Lopion et al., 2018) and (Pregger et al., 2013); (2) high temporal resolution (Lopion et al., 2018; Mallapragada et al., 2018) and (Victoria et al., 2019); (3) high spatial resolution (Lopion et al., 2018; Haller et al., 2012) and (Victoria et al., 2019); and (4) evaluating environmental impacts holistically–for which LCA is best suited (Lopion et al., 2018) and (Ringkjøb et al., 2018). Most of the energy models reviewed by Lopion et al. (2018) aim only for cost-optimal GHG reduction strategies, but do not consider other environmental impacts (Spittler et al., 2019), e. g., REMIX (Gils et al., 2017), TIMES (Loulou et al., 2016), OSeMOSYS (Howells et al., 2011).

A model is required that includes all 4 elements to derive optimal GHG emission reduction pathways considering other environmental impacts. We identify the integration of an holistic assessment via LCA into a sector-coupled energy system with high temporal and spatial resolution as the forth challenge for energy system optimization.

2.4 Contribution of this thesis

Synthesis and operation of energy systems is assisted by well-established state-of-the-art energy system models and optimization methods. However, our review of energy system models reveals 4 challenges when dealing with low-carbon technologies:

Challenge 1: Long-term operation

Challenge 2: Rigorous synthesis with seasonal storage

Challenge 3: Assessment of time-dependent GHG emissions

Challenge 4: Burden-shifting in national energy systems

This thesis extends the scope of energy system models to handle these 4 challenges. The contribution of this thesis is presented in Chapters 3-6 as illustrated in Fig. 2.1.

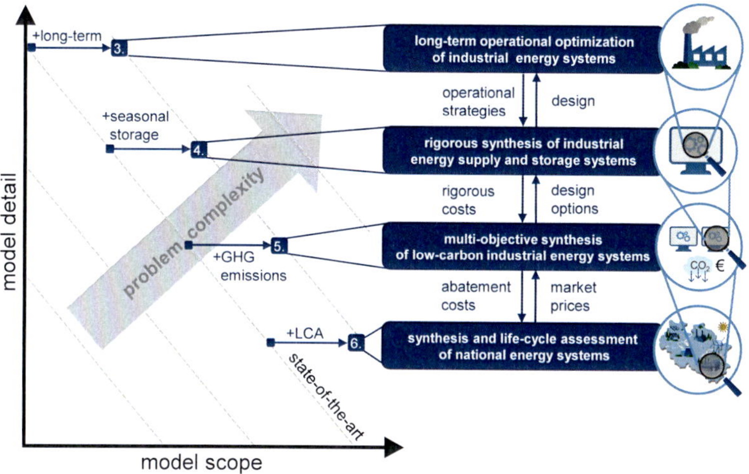

Figure 2.1: Contribution of this thesis: expanding the scope of energy system models. The numbers show the chapter in which the scope expansions are described. LCA: life-cycle assessment, GHG: greenhouse gas

In Fig. 2.1, we define *problem complexity* according to Krämer and Engell (2018). Thus, the term *problem complexity* does not imply a formal quantitative relationship, but indicates the level of computational effort to find the optimal solution for the investigated problem. Problem complexity increases with higher model scope and higher model detail. The level of model detail represents the degree of accuracy in which the

technologies are modeled, e. g., if part-load is considered or not. Model scope describes the range of information which is provided as results, e. g., if the design is optimized or if the model needs a design as input. In this thesis, we investigate state-of-the-art energy systems at different model detail and expand their model scope. This expansion of model scope increases the complexity of the optimization.
In Fig. 2.1 on the right-hand site, we show that information between different scopes of energy systems can be exchanged. By exchanging information, energy system models might take advantage of lower or higher levels of model scope without increasing model detail or model scope. However, the information required is very case specific. Thus, in this work, we focus on general methods for each model scope individually. In detail, we tackle each of the 4 challenges as follows.

Chapter 3, Challenge 1: Long-term operation
Long-term operational optimization of energy systems with low-carbon technologies results in complex, time-coupled problems. These problems can directly decomposed into smaller subproblems, in the absence of time-coupling constraints and variables. However, time-coupling is common in energy systems with low-carbon technologies, e. g., due to (seasonal) storage systems and peak-power prices. Current solution methods are not suited to decompose these time-coupled problems. To expand the scope of these current solution methods, we propose the method DeLoop for the Decomposition-based Long-term operational optimization of energy systems with time-coupling.

Chapter 4, Challenge 2: Rigorous synthesis with seasonal storage
The synthesis of energy systems requires the simultaneous optimization of the design and operation, which results in complex, large-scale problems. Problem complexity increases further by time-coupling constraints, e. g., due to seasonal storage, which are crucial for low-carbon energy systems. Existing rigorous solution methods are not suited to handle time-coupling constraints due to seasonal storage. To fill this gap, we propose the method RiSES[4] that allows for the Rigorous Synthesis of Energy Supply and Seasonal Storage Systems with low-carbon technologies.

Chapter 5, Challenge 3: Assessment of GHG emissions
In state-of-the-art energy system models, electricity from the grid is usually assumed to have annually averaged GHG emission factors. However, GHG emissions from electricity production are in fact time-dependent. These time-dependency offers the potential to reduce emissions by temporal shifting of electricity demand. In Chapter 5, we investigate the impact of time-dependent grid emissions on the synthesis of

energy systems with low-carbon technologies. For this purpose, we extend the scope of energy system models by time-dependent grid emission factors and determine economic and GHG emissions trade-off curves.

Chapter 6, Challenge 4: Burden-shifting in national energy systems
State-of-the-art energy system models for national decarbonization strategies focus on GHG emissions. However, an exclusive focus on GHG emissions carries the risk of burden-shifting to other environmental impacts. In Chapter 6, we present a national energy system model comprising the sectors of electricity, heat, and private transportation. To account for potential burden-shifting, we expand the scope of the presented model by LCA. We provide detailed insights into the development of environmental impacts of a highly interconnected low-carbon energy system up to 2050.

All presented methods contribute to the integration of low-carbon technologies in energy systems of industrial to national scale. In summary, we extend the scope of state-of-the-art energy system optimization by cutting-edge solution methods and environmental assessment methods; both tailored for low-carbon technologies. Thus, we improve the tools to integrate low-carbon technologies into the energy system and facilitate reductions of GHG emissions. Our work thus supports the Secretary-General of the United Nations (UN) Antònio Guterres's statement at the UN Climate Action Summit:

'The climate emergency is a race we are losing, but it is a race we can win. We have the tools: technology is on our side.'

Antònio Guterres UN Secretary-General,
New York, September 2019 at the UN Climate Action Summit

Chapter 3

Long-term operational optimization of energy systems

In this chapter, we extend the scope of operational optimization problems to consider long-term effects. Thus, we overcome Challenge 1, cf. Chapter 2.4. We investigate an industrial energy system with a given design and an economic objective function. This level of model scope allows for a high degree of model detail, cf. Fig. 2.1. To extend the scope of operational optimization methods, we propose a time-series Decomposition-based solution method for Long-term operational optimization problems (**DeLoop**).

DeLoop copes with long-term operation of energy systems with low-carbon technologies, including both time-coupling constraints and variables. DeLoop combines the strengths of the approaches reviewed in Chapter 2.2.1: DeLoop is easily applicable, uses a physically accurate model, allows for parallelization, and shows fast convergence.

In Chapter 3.1, we state a generic long-term operational optimization problem with time-coupling. DeLoop is presented in detail in Chapter 3.2. In Chapter 3.3, we apply DeLoop to a real-world industrial operational problem and validate the results by a computational study. We conclude our findings in Chapter 3.4.

Major parts of this chapter are reproduced by permission of Elsevier from:

Baumgärtner, N., Shu, D., Bahl, B., Hennen, M., Hollermann, D., and Bardow, A. (2020). DeLoop: Decomposition-based Long-term Operational Optimization of energy systems with time-coupling constraints. *Energy*, 198, 117272.

Contribution report: Writing the draft, principal author, conceptual development of the method, calculation, and evaluation of results.

3.1 Time-coupling constraints and variables in operational optimization

In this section, we present a comprehensive model for the optimization of low-carbon industrial energy systems.
In this chapter, we use the low-carbon industrial energy system model as operational optimization model. The model is also used in Chapter 4 as single objective synthesis model and in Chapter 5 as multi-objective synthesis model. To avoid unnecessary repetitions, we refer to this chapter when possible. Thus, in the other chapters, we state only changes in model detail and model scope.

3.1.1 Model for low-carbon industrial energy systems

In Chapter 2.2, we review available models for industrial energy systems with focus on model type, supplied energy demands, and energy conversion-technologies. Based on the reviewed state-of-the-art, we model a low-carbon industrial energy system.
We follow most authors and formulate the low-carbon industrial energy system model as piecewise-linearized MILP problem.
The energy systems reviewed from literature supply a combination of electricity, cooling, heating, and steam. In our model, we consider all four types of energy demands. In literature, many types of energy-conversion technologies are considered. We include most energy-conversion units considered in the literature in our low-carbon industrial energy system model. Some specific energy-conversion units are not considered in the model: Energy-conversion units considering different steam levels (turbines and steam compressors) are neglected as only one steam level is present in the case study. For short-term electricity storage, we consider batteries, and thus neglect the expensive combination of electrolysis, hydrogen storage and fuel cells. However, these energy-conversion units could be added to the model if desired.
In Table 3.1, we summarize all energy-conversion units considered in the proposed low-carbon industrial energy system model. Furthermore, for each energy-conversion unit, we list in- and outputs, Table 3.1. In each industrial case study, Chapters 3.3, 4.3, and 5.3, we state which components of the energy system model are used in the (super-)structure.

Table 3.1: All possible energy-conversion units n considered in the (super-)structure C of the proposed low-carbon industrial energy system model. Additionally, the inputs ($\dot{U}_n$) and outputs ($\dot{V}_n$) for each energy- conversion unit are shown.

units ($n \in \mathcal{C}$)	**input** ($\dot{U}_n$)	**output** ($\dot{V}_n$)
absorption chiller (AbC)	heat	cooling
battery (BAT)	electricity	electricity
boiler (B)	gas	steam
combined-heat-and-power engine (CHP)	gas	electricity, heat
compression chiller (CC)	electricity	cooling
electrode boiler (B_{el})	electricity	steam
heat exchanger (HEX)	steam	heat
heat pump (HP)	electricity	heat
inverter (INV)	electricity	electricity
organic Rankine cycle (ORC)	steam	heat, electricity
photovoltaic (PV)	radiation	electricity
solar thermal (ST)	radiation	heat
thermal storage (TS)	heat/cooling	heat/cooling
wind turbine (WIND)	wind	electricity

In this chapter, we use the model to state an operational problem for energy systems as a mixed-integer linear programming (MILP) problem. The stated problem is based on the previous work of our group (Voll et al., 2013) and has been extended to consider time-series by Bahl et al. (2017a) and storage units by Bahl et al. (2018b).

We extend the model by time-coupling constraints: emission limits, seasonal storage, network-connection fees, peak-power prices. Commonly, these time-coupling constraints extend over one year of operation. However, the proposed method can handle time-coupling constraints on arbitrary time horizons. Furthermore, the heat demand is separated into a low-temperature heat demand and a steam demand. Additionally, we include more energy-conversion units based on the literature review. In this section, for readability, we present a simplified version of the problem. The detailed MILP model is described in Appendix A.

We optimize the operational expenditures $OPEX$ of the energy system Eq. (3.1)

$$\min_{\dot{V}_{n,t},\delta_{n,t},\dot{V}^{\max}_{\mathrm{grid}},\dot{V}_{\mathrm{grid},t},V_{n,t},x,y} \quad OPEX$$

with

$$OPEX := \tag{3.1}$$
$$\overbrace{\sum_{t\in\mathcal{T}}\left(\Delta \mathrm{t}_t \sum_{n\in\mathcal{C}} \mathrm{c}^{\mathrm{o}}_{n,t}\cdot\frac{\dot{V}_{n,t}}{\eta_n}\right)}^{fuel/electricity\ costs} + \overbrace{\sum_{n\in\mathcal{C}} M^{\mathrm{N}}_n}^{maintenance\ costs} \underbrace{+\, c^{(\mathrm{low/high})}_{\mathrm{net}}\cdot\sum_{t\in\mathcal{T}}\Delta \mathrm{t}_t \dot{V}_{\mathrm{grid},t}}_{network-connection\ fees} + \underbrace{\mathrm{c}^{(\mathrm{low/high})}_{p}\cdot\dot{V}^{\max}_{\mathrm{grid}}}_{peak-power\ costs}$$

The operational expenditures, Eq. (3.1), typically sum fuel and electricity costs, maintenance costs M^{N}_n for all components n, network-connection fees, and peak-power costs. Fuel and electricity costs are the sum of the output power $\dot{V}_{n,t}$ divided by the efficiency η_n of every component $n \in \mathcal{C}$ for every time step t and multiplied by the specific operation cost $\mathrm{c}^{\mathrm{o}}_{n,t}$ and the duration Δt_t of a time step, Eq. (3.1). The operating costs $\mathrm{c}^{\mathrm{o}}_{n,t}$ for gas-consuming components is the gas price, whereas the operating costs for electricity-consuming components is the electricity price, each in €/kWh. The operating costs are zero for renewable technologies, steam and heat-consuming technologies. The gas and electricity prices account for the amounts of energy used. The maintenance costs M for a unit n are constant for a given time horizon and given in Appendix A. The network-connection fees result from the electricity consumption $\sum_{t\in\mathcal{T}}\dot{V}_{\mathrm{grid},t}$ multiplied by network-connection fees $c^{(\mathrm{low/high})}_{\mathrm{net}}$ that depend on the duration of utilization. The peak power $\dot{V}^{\max}_{\mathrm{grid}}$ multiplied by a peak-power price c_{p} results in additional operational expenditures $OPEX$, Eq. (3.1). The peak-power price c_{p} also depends on the duration of utilization.

The optimization problem is subject to constraints, which are presented in the following.

$$\sum_{n\in\mathcal{C}\setminus\mathcal{C}_{\text{stor}}} \dot{V}_{n,t} + \dot{V}_{\text{grid},t} + \sum_{n\in\mathcal{C}_{\text{stor}}} (\dot{V}^{\text{out}}_{n,t} - \dot{V}^{\text{in}}_{n,t}) = \dot{E}_t, \qquad \forall t\in\mathcal{T}, \tag{3.2}$$

$$A_1\dot{V}_{n,t} + \widetilde{A}_1\delta_{n,t} \leq b_1, \qquad \forall t\in\mathcal{T}, \forall n\in\mathcal{C}, \tag{3.3}$$

$$A_2 x + \widetilde{A}_2 y \leq b_2, \qquad \forall t\in\mathcal{T}, \forall n\in\mathcal{C}, \tag{3.4}$$

$$V_{n,t} + \Delta t_t \cdot (\dot{V}^{\text{in}}_{n,t} - \dot{V}^{\text{out}}_{n,t}) = V_{n,t+1}, \qquad \forall t\in\mathcal{T}, \forall n\in\mathcal{C}_{\text{stor}}, \tag{3.5}$$

$$|\dot{V}_{\text{grid},t}| \leq \dot{V}^{\max}_{\text{grid}}, \qquad \forall t\in\mathcal{T}, \tag{3.6}$$

$$c_{\text{j}} = \begin{cases} c^{\text{high}}_{\text{j}} & \text{, if } \frac{\sum_{t\in\mathcal{T}} \dot{V}_{\text{grid},t}\cdot\Delta t_t}{\dot{V}^{\max}_{\text{grid}}} \leq 2500\text{ h} \\ c^{\text{low}}_{\text{j}} & \text{, otherwise} \end{cases} \qquad \text{j}\in\{\text{net,p}\}, \tag{3.7}$$

$$\sum_{t\in\mathcal{T}} \Delta \text{t}_t \cdot \left[\left(\dot{V}_{\text{grid},t}\cdot c^{\text{CO2}}_{t,el}\right) + \sum_{n\in\mathcal{C}_{gas}} \left(\frac{\dot{V}_{n,t}}{\eta_n}\cdot c^{\text{CO2}}_{t,gas}\right)\right] \leq C^{\max}_{\text{CO2}}, \tag{3.8}$$

$$\delta_{n,t}, y\in\{0,1\}; x\in\mathbb{R}^a; \dot{V}_{n,t}\in\mathbb{R}; V_{n,t}, \dot{V}^{\max}_{\text{grid}}\in\mathbb{R}^+ \qquad \forall t\in\mathcal{T}, \forall n\in\mathcal{C}.$$

Eq. (3.2) represents the energy balance. The energy balance enforces that the component's output power $\dot{V}_{n,t}$, the grid electricity $\dot{V}_{\text{grid},t}$ and the net energy output of the storage units $\dot{V}^{\text{out}}_{n,t} - \dot{V}^{\text{in}}_{n,t}$ meet the energy demand $\dot{E}_t$ for every time step t. Eq. (3.3) is the generic representation of further (in)equalities with the coefficient matrices A_1, $\widetilde{A}_1$, and the vector b_1. These (in)equalities determine the component's binary on/off status $\delta_{n,t}$ and the current part-load performance. Surrogate Eq. (3.4) summarizes additional constraints such as constraints for describing unit behavior, e. g., the output limitation of photovoltaics, or the choice to either charge or discharge storage systems. The detailed MILP model of the low-carbon industrial energy system, including part-load behavior, minimal loads, and linearizations, is presented in Appendix A.

Long-term operational optimization is challenging due to both time-coupling constraints and variables which occur in the following equations: In the storage energy balance Eq. (3.5), the net energy input of the storage units $\dot{V}^{\text{in}}_{n,t} - \dot{V}^{\text{out}}_{n,t}$ couples the current storage level $V_{n,t}$ to the future storage level $V_{n,t+1}$. Thereby, the storage balance is a coupling equation for the entire time series.

Eq. (3.6) ensures that the peak power $\dot{V}^{\max}_{\text{grid}}$ is the maximum power exchanged with the grid. Thus, $\dot{V}^{\max}_{\text{grid}}$ is a time-coupling variable for the entire time series. The network-connection fees, as well as the peak-power price depend, on the duration of utilization, e. g., in Germany (StromNEV, 2005). Eq. (3.7) determines the price $c^{(\text{low/high})}_{\text{i}}$ as a function of the duration of utilization. Binary variables model the

non-linear dependence of prices on the duration of utilization. The multiplication of these binary variables and continuous decision variables lead to bilinear terms, cf. Appendix A. The linearization of the bilinear terms results in both time-coupling constraints and variables. Eq. (3.8) sets an upper limit for greenhouse gas emissions $C_{\mathrm{CO2}}^{\max}$ from electricity usage and all gas-consuming units $\mathbb{C}_{gas}$. This emission limit represents a coupling constraint for the entire time series.
Overall, Eq. (3.1-3.8) represents a long-term operational problem that includes long-term time-coupling constraints and variables. In the next section, we propose the time-series decomposition-based solution algorithm DeLoop to solve such long-term operational problems.

3.2 Time-series decomposition for long-term operational optimization

The proposed time-series decomposition-based solution algorithm, DeLoop, generates feasible solutions of long-term operational optimization problems with known solution quality. In this section, we first present the idea underlying DeLoop and then the details of the time-series decomposition.

3.2.1 Solution method

DeLoop provides lower and upper bounds for the operational optimization problem, Eq. (3.1-3.8). DeLoop computes these lower and upper bounds in a parallel computing mode, Fig. 3.1. After each calculation of a new upper bound, we calculate the optimality gap ε and check if the user-defined optimality gap ε_{DeLoop} is satisfied. The optimality gap ε is the relative gap between the best possible objective (lower bound) and the best-found objective (upper bound):

$$\varepsilon := \frac{OPEX^{\text{upper bound}} - OPEX^{\text{lower bound}}}{OPEX^{\text{upper bound}}} \leq \varepsilon_{DeLoop}. \tag{3.9}$$

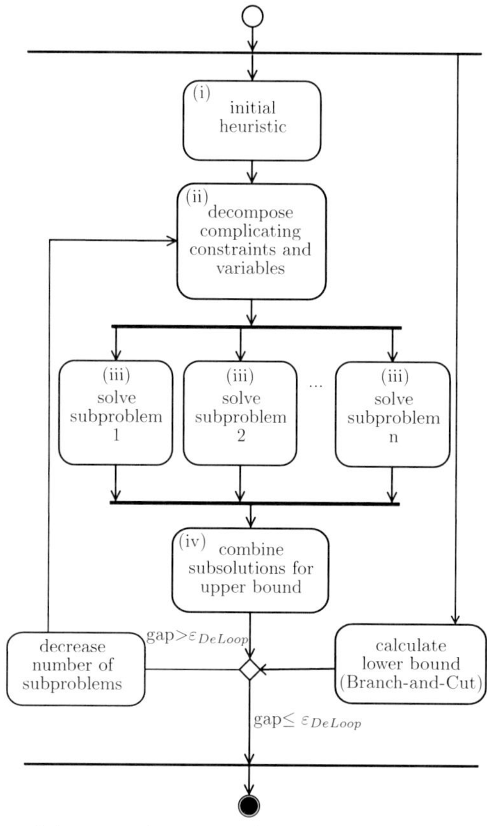

Figure 3.1: Overview of the proposed time-series decomposition method DeLoop to create feasible solutions of operational optimization problems with known solution quality

Upper bounds are obtained from feasible solutions. These feasible solutions are computed by decomposing the original operational problem into smaller subproblems. The upper bound is tight, i. e., near-optimal, if the impact of time-coupling constraints and variables is small. The tightness of the upper bound is commonly unknown prior to optimization. The upper bounds are tightened if the desired optimality gap ε_{DeLoop} is not satisfied. DeLoop tightens the upper bounds by iteratively decreasing the number of subproblems. Thereby, DeLoop improves the representation of time coupling within the decomposition. In general, the original long-term operation problem can be decomposed into any number of subproblems (less than or equal to the size of the time series T). Each of the subproblems represents a subsequence of the full time series. However, the subproblems should preferably have a similar number of time steps to achieve similar solution times when solved in parallel. For simplicity, we only allow integer divisors of the time steps T as possible numbers of subproblems, which leads to subproblems of the same size. The algorithm for upper bounds is described in Chapter 3.2.2.

For the lower bounds, we relax all binary variables ($\delta_{n,t}$, $y \in \{0,1\}$) of the operational problem, Eq. (3.1-3.8), thereby converting the complex MILP into an LP which can be solved efficiently. This solution of the relaxed problem serves as the first lower bound and is the root node for the Branch-and-Cut procedure. The lower bound is tight if the impact of binary decision variables is small. The tightness of the lower bound is commonly unknown prior to optimization. Subsequently, to tighten the lower bound, the Branch-and-Cut procedure starts by branching on binary variables and cutting off branches that cannot improve the solution (IBM, 2016).

After each calculated upper bound, DeLoop compares the current upper bound to the current lower bound to calculate the optimality gap ε. The iterative calculation of the lower and upper bounds stops when an optimality gap ε_{DeLoop} is satisfied. As a post-processing step, the solution quality can be further enhanced by warm-starting the original problem with the final solution of DeLoop.

3.2.2 Time-series decomposition for the upper bound

DeLoop obtains feasible solutions (upper bounds) of the original long-term operational problem (Eq. (3.1-3.8)) in 4 steps, Fig. 3.1:

(i) Initial heuristic to initialize the number of subproblems
(ii) Decomposition of time-coupling constraints and variables
(iii) Optimizing subproblems in parallel computing mode
(iv) Combining subsolutions to upper bound

In step (i), we select an initial number of subproblems. Subsequently, in step (ii), DeLoop decomposes the long-term operational problem, including its complicating constraints and variables into subproblems. The decomposition into subproblems reduces the overall complexity of the operational problem. Thus, the subproblems can be solved efficiently in parallel computing mode in step (iii). In step (iv), the solutions of all subproblems are combined into a feasible solution of the original operational problems, resulting in an upper bound. In the following, we present the details of steps (i-iv) of the proposed time-series decomposition method DeLoop.

Step (i): Initial heuristic

Initialization of the decomposition method DeLoop requires an initial number of subproblems. In principle, one could always start with the maximum number of subproblems. In practice, we found it more efficient to identify an initial number of subproblems, leading to minimal computational time for the first solution. To identify a good estimate for this initial number, a heuristic is employed. In our experience, the choice of the initial number only affects the speed of convergence but not the final optimal solution. The algorithm ensures that the final solution always satisfies the desired optimality gap. Increasing the number of subproblems decreases the size of the individual subproblems and, therefore, also the calculation time per subproblem. However, more subproblems have to be solved in parallel by a limited number of cores, which might again increase the overall computation time. Thus, the calculation time is minimal for a particular number of subproblems.

To efficiently identify a good number of subproblems, leading to low computational time to generate a first feasible solution, we employ a heuristic that tests decreasing numbers of subproblems, starting with the maximum number of subproblems. For each tested decomposition, we decompose the problem into subproblems but solve only 1 of the resulting subproblems. The time for the decomposition and the solution time of the 1 subproblem are recorded and extrapolated to the full number of subproblems. DeLoop decreases the number of subproblems until the extrapolated time increases. Thereby, DeLoop estimates the number of subproblems, resulting in minimal calculation time at low computational costs. The identified number of subproblems is selected as the initial number of subproblems.

In the case study, we rigorously tested the initial heuristic by not only solving 1 but all subproblems. We found that the heuristic identifies the number of subproblems leading to the minimal solution time in 3 out of 6 cases and very close to the minimal solution time in the other cases.

Step (ii): Decomposition of time-coupling constraints and variables

First, we decompose all complicating constraints and variables of the subproblems. The decomposition is generic and conducted automatically based on the type of time-coupling constraint or variable in 4 steps:

(1) For storage-like constraints, Eq. (3.5), we fix start and end values of the storage-level variable for each subproblem. To ensure feasibility, the end value of each subproblem has to be consistent with the start value of the subsequent subproblem. In the initial decomposition, the variables representing the start and end values are fixed to 1 identical but arbitrary value for all subproblems. In the subsequent decompositions, the fixed start and end values are changed to the

values resulting from the preceding optimization. By this adaptation, DeLoop iteratively improves the solution. We illustrate this iterative improvement for 5 and 4 subproblems in Fig. 3.2.

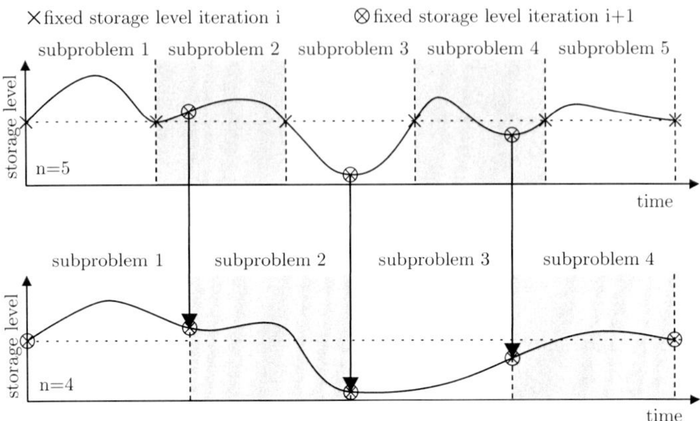

Figure 3.2: In DeLoop, start and end values for storage-like constraints are fixed in each subproblem. In the subsequent decomposition with fewer subproblems (n=4), these start and end values are taken from the previous optimization results with more subproblems (n=5). Storage levels are piecewise linear in the actual model.

(2) Independent variables in every subproblem substitute time-coupling variables like Eq. (3.6-3.7). Thus, for example, network-connection fees and the peak-power prices are calculated for each subproblem separately. Thereby, the resulting subproblems become independent of each other.

(3) For time-coupling constraints, like the emission limits in Eq. (3.8), a fraction of the limit is allocated to each subproblem. However, equally distributed limits may lead to poor overall convergence or even infeasible subproblems. Therefore, DeLoop computes individual limits for each subproblem. To compute individual subproblem limits, e. g., of emissions, we first identify the minimal possible emissions by an independent minimization of the emissions within each subproblem. Then, the total limit is divided proportionally according to the minimal possible emissions of each subproblem.

(4) To improve performance, we extend each subproblem by aggregated time steps. The aggregated time steps represent the missing parts of the full time series in each subproblem in an averaged manner. Furthermore, critical time steps can be added to each subproblem. Critical time steps are case-study specific and often unknown beforehand. For the optimization of the network-connection fees and

the peak-power price, the peak-power demand represents a critical time step. Thus, we add a critical time step for the expected peak-power demand. Here, the expected peak-power demand is approximated based on the power demand, the maximal power produced, and the minimal power consumed by all units. For a detailed description of the identification of the peak power demand, see Appendix D. Thus, in each subproblem, the corresponding time interval of the long-term operational optimization is solved with the full accuracy of the time series, and the rest with low accuracy. For the low-accuracy representation, we use 3 aggregated time steps for each subproblem: (1) one averaged time step for all time steps before the time interval corresponding to the subproblem, (2) one averaged time step between the subproblem and the critical time step, and (3) one averaged time step after the critical time step. Thus, 3 aggregated time steps are added to each subproblem, except for the first and last subproblem, where only 2 aggregated time steps are added. In Fig. 3.3, we illustrate this subproblem extension for 5 subproblems and the power demand.

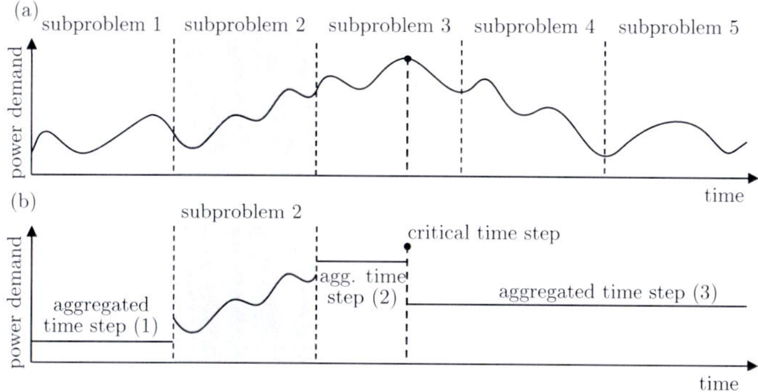

Figure 3.3: Illustration of the subproblem extension for 5 subproblems for the power demand time series. a) The full time series is decomposed into five intervals, each belonging to a subproblem. The time step with the peak-power demand is potentially a critical time step and highlighted by a dot. b) The time series of subproblem 2 is extended by the critical time step and by aggregated time steps represented by average values, respectively.

Step (iii): Optimizing subproblems in parallel computing mode

DeLoop decomposes the original problem into smaller and independent subproblems that can be quickly solved by parallelization. To further enhance computational speed, the solution of the independent minimization of minimal emissions (step (ii)(3)) is used to warm-start the optimization of each subproblem.

Step (iv): Combining subsolutions to upper bound

In step (iv), the solutions of the subproblems are combined into a feasible solution of the original operational problem. For this purpose, the parts of the subsolutions with low accuracy are discarded, i. e., the added aggregated and critical time steps (step (ii)(4)). Subsequently, we merge all subsolutions: for each time interval, the subsolution is used that has solved this time interval with full accuracy. After these adaptations, the combination of all subproblems is a feasible solution. The combination yields a feasible solution as all time-coupling constraints and variables needed for feasibility have been decomposed and fixed to the same values at the boundaries of the subproblems, as described in step (ii). The time-coupling variables, such as the peak-power demand $\dot{V}_{\text{grid}}^{\text{max}}$, are then recalculated by taking all merged subproblems. The recalculation contains only fixed parameters; thus, no further optimization is required. As the time-coupling constraints and variables have been fixed and the objective function is recalculated, the solution is an upper bound for the original operational problem.

In the next section, we apply DeLoop to a complex long-term operational optimization problem with time-coupling constraints and variables.

3.3 Case study: Long-term operation

To validate the proposed time-series decomposition method, we apply DeLoop to a long-term operational MILP problem presented in Chapter 3.1 and Appendix A. We optimize an industrial energy system with a given design of 3 boilers, 5 compression chillers, 4 absorption chillers, 7 heat exchangers, 3 inverters, 1 combined heat and power engine, 1 photovoltaic system, 1 battery, and 2 storage tanks, one for hot and one for cold water, Fig. 3.4. A table with all units' capacities can be found in Appendix A. All results are generated using 4 Intel-Xeon CPUs at 3 GHz and 64 GB RAM. All MILP problems are solved using CPLEX 12.6.3.0 (IBM, 2016), and the model is written in General Algebraic Modeling System 24.7.3 (GAMS Development Corporation, 2019). For this case study, 4 cores are employed to solve subproblems in parallel. Time series with hourly resolution are available for steam, low-temperature heating, cooling, and electricity demand, Fig. 3.5a. The renewable-based units introduce additional time series to the synthesis problem. We consider measured time series data for solar radiation and wind speed (Schneider and Ketzler, 2014), Fig. 3.5b. Moreover, time-varying electricity purchase and selling prices are incorporated in the model, Fig. 3.5c. The day-ahead electricity price from epexspot (2017) is used and scaled to a typical industrial level of $16\,cent/kWh$ buying and $10\,cent/kWh$ selling prices.

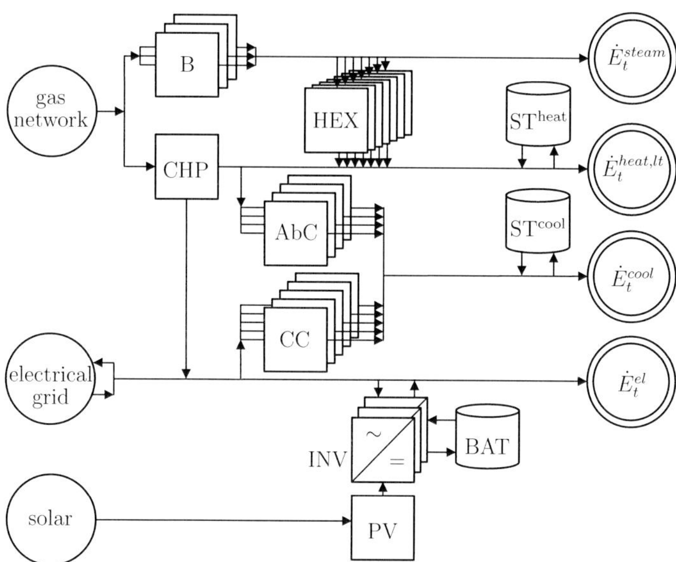

Figure 3.4: Visualization of the design of the energy system used in the case study, consisting of boilers (B), heat exchangers (HEX), a combined heat and power unit (CHP), absorption chillers (AbC), compression chillers (CC), inverters (INV), a photovoltaic system (PV), a heat storage unit ($\mathrm{ST}^{\mathrm{heat}}$), a cold storage unit ($\mathrm{ST}^{\mathrm{cool}}$), and a battery (BAT). A table with all capacities can be found in Appendix A.

To calculate efficiencies dependent on ambient-temperature, see Appendix A, we include data for the ambient temperature (Schneider and Ketzler, 2014), Fig. 3.5d. We consider one year of operation, because time-coupling constraints and variables for industrial energy systems often span one year. To render the case study computational tractable for the benchmark calculations, we here consider the time-series with a two-hourly resolution.

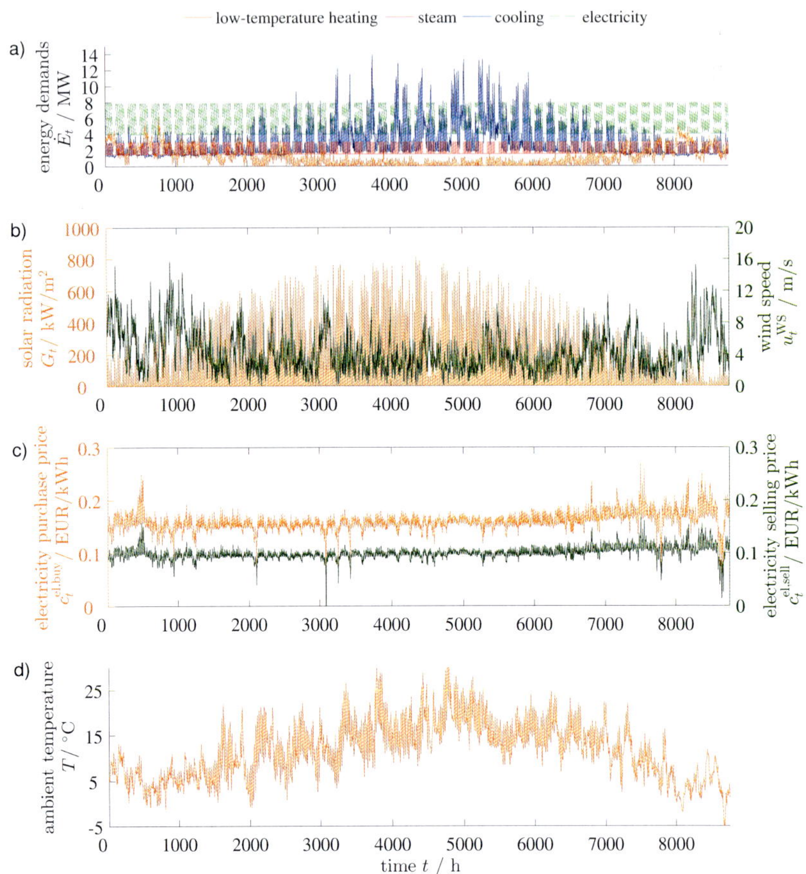

Figure 3.5: Full time series for the energy system optimization: a) energy demands, b) wind speed and solar irradiation, c) electricity prices for purchase and selling, and d) ambient temperature.

The original operational problem after the presolve contains $6 \cdot 10^5$ equations and $5 \cdot 10^5$ variables (incl. $2 \cdot 10^5$ binaries) with $18 \cdot 10^5$ nonzero elements. The long-term operational problem is highly coupled due to the storage and battery systems, an annual emission limit, peak-power prices, and network-connection fees, Eq. (3.1-3.8).

As benchmark, we intended to solve the original operational problem with the General-Column-Generation (GCG) solver (Gamrath and Lübbecke, 2010) and (Gleixner et al., 2018). The GCG solver decomposes coupled problems into independent subproblems and a connective master problem by a Dantzig-Wolfe decomposition and solves the decomposed problems via the Branch-Cut-and-Price algorithm. However, the GCG solver did not provide any decomposition besides the trivial decomposition in only 1 subproblem. Thus, as benchmark, we solve this 1 subproblem—ergo the original long-term operational problem—directly with CPLEX 12.6.3.0 (IBM, 2016). In CPLEX, we employ the deterministic parallel mode such that all methods use the same number of computational cores. To validate the computational results, we repeat the calculation for 5 instances generated by statistical noise using Latin hypercube sampling on the data ($\pm 5\,\%$) (Mckay et al., 2000).

DeLoop finds a feasible solution within a $2\,\%$ optimality gap in 5,656 s on average, Fig. 3.6. The average computational time of the benchmark takes 32 times longer (182,152 s) to obtain solutions of equal quality. On average, after 786 s, the proposed time-series decomposition method generates the first feasible solution with known solution quality In contrast, on average, the benchmark takes 55 times longer (43,573 s) to provide the first feasible solution. Thus, the proposed decomposition method outperforms the benchmark in all instances by more than an order of magnitude.

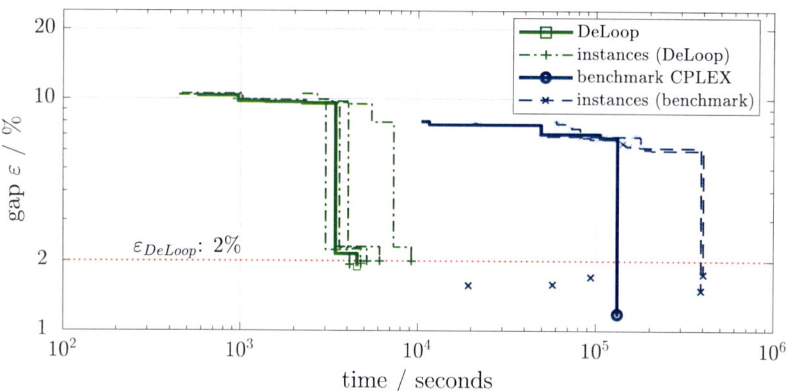

Figure 3.6: Gap ε of the proposed decomposition method DeLoop and the benchmark CPLEX as function of the solution time. A marker indicates final optimality gap ε and solution time of all calculations. The required optimality gap ε_{DeLoop} of $2\,\%$ is marked in dotted red.

The solution method DeLoop is further analyzed by the lower bound and the upper bound of the operational expenditures *OPEX* of the original instance as a function of the solution time, Fig. 3.7.

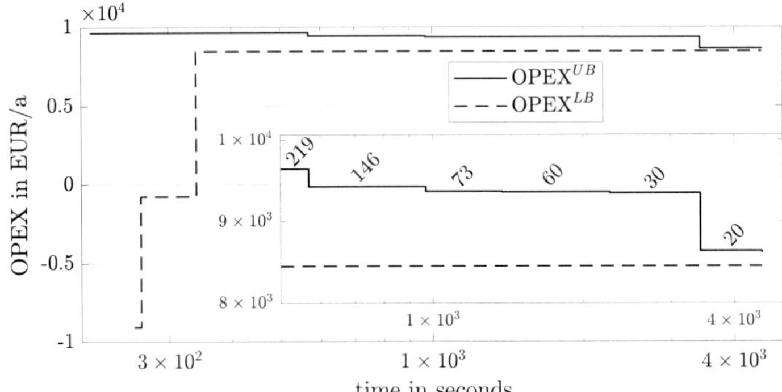

Figure 3.7: Operational expenditures *OPEX* of the lower and upper bounds of DeLoop as function of the solution time. The figure includes a zoom on the objective *OPEX* for better visualization of the improvement of the upper bound. The diagonal numbers are the number of considered subproblems in each iteration.
Note: The gap ε is evaluated after each iteration of the upper bound; thus, for the first feasible solution of the original problem after 209 s, no solution quality is determined. Therefore, the first gap ε is computed for the second feasible solution found, cf. Fig. 3.6.

The linear-programming relaxation provides the first bound very quickly after 255 s. The first lower bound is the solution of the linear programming relaxation of the MILP, i. e., the root node of the Branch-and-Cut procedure. In this relaxed problem, negative operational costs are possible; thus, the industrial energy system earns money, although all energy demands are supplied. Within 340 s, the lower bound of the operational expenditures *OPEX* rises twice sharply to reach a value of just 0.01 % below its final value. After 340 s, the lower bound is already a very tight relaxation of the original long-term operational problem.

For the upper bound, the initial heuristic (step (i)) composes the original problem into 219 subproblems. The solution of these 219 subproblems results in the first upper bound after 209 s. In subsequent iterations, the solution quality slightly improves as the number of subproblems is iteratively decreased. The improvement is due to subproblems being larger, which allows the peak-power demands and costs to be reduced and allows the distribution of the total emission limit among the subproblems to be improved. At 3425 s, with a decomposition into 20 subproblems, a large improvement in solution quality is achieved by reducing the network-connection fees. At this point, a gap of 3 % is reached, which, for this case study, is set as the starting gap for the post-processing warm-start procedure. The warm start satisfies the desired optimality gap of 2 % after a total computational time of 4,548 s. Considering fewer but longer subproblems enables DeLoop to represent long-term effects better. By the iterative solution approach, DeLoop identifies the maximal time span necessary for the subproblems to consider all long-term effects accurately. The improvement of the upper bound in the considered case study shows that accurate consideration of long-term effects reduces operating costs by about 10 %.

The investigated case study shows that DeLoop is very time-efficient for handling time-coupling constraints and variables in long-term operational optimization problems. Although time-coupling is very common, not all of the presented time-coupling constraints and variables are always present at once in operational optimization problems. Therefore, we investigate the performance of DeLoop for less severe cases of time-coupling. For comparability, we also use CPLEX as benchmark. Again, to validate the computational results, we repeat the calculation for 5 additional instances with ±5 % variation on the data.

We analyze the solution time for the original case study (*original*) and 3 further cases of less severe time-coupling, where we exclude different time-coupling equations from the original long-term operational problem: *no emission limits*, *no network fee*, and *neither* emission limits nor network fees, Fig. 3.8. All cases still include time-coupling due to the storage and battery systems.

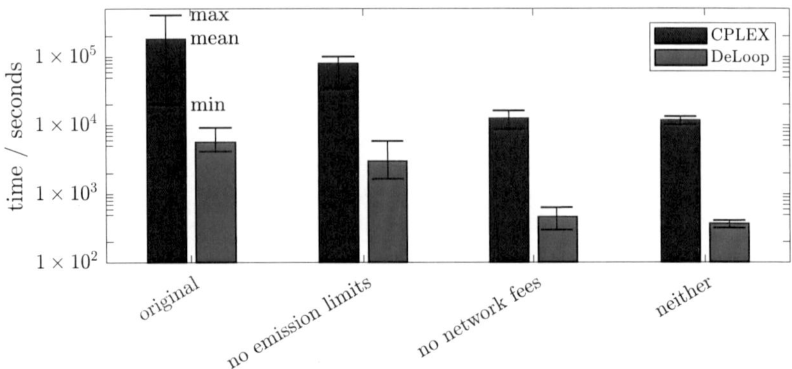

Figure 3.8: Solution time of the proposed decomposition method DeLoop and the benchmark with CPLEX for the investigated combinations of complicating constraints and variables: *original* includes all complicating constraints and variables (Eq. (3.1-3.8)); thus, *original* is the unaltered case study; *no emission limits* does not include the time-coupling constraint related to the emission limit (Eq. (3.1-3.7)); *no network fee* does not include the time-coupling constraints and variables related to the network-connection fees (Eq. (3.1-3.5 and 3.8)); *neither* includes neither the time-coupling constraints and variables related to the emission limit nor related to the network-connection fees (Eq. (3.1-3.5)). The top line of the bar represents the maximal solution time needed of the 6 investigated instances (marked by *max* at the very left bar). The bottom line represents the minimal (*min*) and the middle line the average solution time (*mean*) of the 6 instances.

In all 6 instances of each of the investigated 3 cases with less severe time-coupling, DeLoop outperforms the benchmark, here on average, by a factor of 28. In the case of *no emission limits*, DeLoop requires 2992 s on average to solve the long-term operational problem, whereas the benchmark takes 80353 s. For the cases of *no network fees* and *neither*, DeLoop needs about 400 s to find a solution satisfying the optimality gap, whereas the benchmark takes 30 times as long (≈12000 s).

The investigation shows: combining multiple time-coupling constraints and variables leads to very challenging optimization problems; the network-connection fees increase the solution time of the problem more than the emission limit as the network-connection fees result in a combination of both time-coupling constraints and variables; operational problems considering storage systems alone as time-coupling variable already result in long computational times.

Overall, the computational study shows that DeLoop enhances computational speed for long-term operational problems over a broad range of time-coupling.

3.4 Summary and conclusion

Long-term operational optimization of energy systems with low-carbon technologies is challenging. The complexity of the operational problem strongly increases due to time-coupling constraints and variables that are common in industrial applications. Additionally, time-coupling constraints prohibit the direct decomposition of long-term operational problems. As a consequence, long-term operational optimization is often not solvable within reasonable computational time or memory limits (Challenge 1, Chapter 2.4).

To solve this challenge, we propose a time-series decomposition method (DeLoop), providing feasible solutions with known solution quality. The method decomposes the original long-term operational problem with time-coupling constraints into smaller subproblems. These subproblems can be quickly solved in parallel computing mode. Subsequently, DeLoop recombines the solutions of the subproblems into a feasible solution of the original long-term operational problem while still representing a rigorous decomposition. The method is generally applicable to long-term operational problems considering time-coupling constraints and variables.

Here, DeLoop is applied to an operational problem of an industrial energy system with low-carbon technologies with a high degree of model detail. The operational problem exhibits both time-coupling constraints and variables. Time-coupling is due to storage systems, emission limits, peak-power prices, and network-connection fees. First, the case study results illustrate the importance of long-term operational planning, as significant cost reductions of about 10 % can be achieved. Second, the proposed method DeLoop provides fast convergence, outperforming a commercial solver in a large computational study on average by a factor of 32. Third, DeLoop also outperforms a commercial solver for less severely time-coupled problems on average by a factor of 28, showing the broad applicability of DeLoop. DeLoop is very time-efficient for handling time-coupling. Thereby, DeLoop renders real-world long-term operational problems solvable. Thus, DeLoop enables significant cost reductions for these real-world applications.

In this chapter, the model scope is the operational optimization of energy systems with a fixed design. We extended the model scope successfully with the method DeLoop to consider long-term time-coupling constraints and variables. In the next chapter, we extend the model scope to investigate not only the operation but also the synthesis of energy systems.

Chapter 4

Rigorous synthesis of energy supply and storage systems

In this chapter, we extend the scope of synthesis optimization problems to consider time-variant inputs and storage systems. Thus, we solve Challenge 2, cf. Chapter 2.4. We determine the cost-optimal design of an industrial energy system. This level of model scope still allows for a high model detail, cf. Fig. 2.1. However, in comparison to Chapter 3, we assume constant part-load behavior and peak-power prices. To extend the scope of synthesis optimization methods, we propose the Rigorous Synthesis method of Energy Supply and Seasonal Storage Systems (**RiSES⁴**). RiSES^4 surpasses the limitation of approaches reviewed in Chapter 2.2.2. RiSES^4 addresses the chronology of time steps by using typical periods. RiSES^4 simultaneously under- and overestimates all time-dependent input parameters. To further improve the computational time, RiSES^4 employs 2 competitive lower bounds and parallel computing. In Chapter 4.1, we extend the operational problem of Chapter 3.1 to a generic synthesis problem, accounting for volatile inputs and time-coupling. In Chapter 4.2, we present RiSES^4 in detail. In Chapter 4.3, we apply RiSES^4 to 2 real-world industrial synthesis problems and validate the results by large computational studies. Finally, we present our conclusion in Chapter 4.4.

Major parts of this chapter are reproduced by permission of Elsevier and ECOS from:

Baumgärtner, N., Bahl, B., Hennen, M., and Bardow, A. (2019). RiSES3: Rigorous Synthesis of Energy Supply and Storage Systems via time-series relaxation and aggregation. *Computers & Chemical Engineering*, 127, 127-139.

Baumgärtner, N., Temme, F., Bahl, B., Hennen, M., and Bardow, A. (2019). RiSES4: Rigorous Synthesis of Energy Supply Systems with Seasonal Storage by relaxation and time-series aggregation to typical periods. *Proceedings of ECOS 2019 - the 32ND International conference on Efficiency, Cost, Optimization, Simulation and environmental impact of energy systems*, June 23-28, 2019, Wroclaw, Poland

Contribution report: Writing the draft, principal author, development and implementation of the method, calculation and evaluation of results.

4.1 Generic industrial synthesis problem

In this section, we extend the scope of the long-term operational problem stated in Chapter 3.1 (Eq. 3.1-3.8) to a synthesis problem. For this purpose, first, the objective is extended. For synthesis problems, as economic criterion, the net present value (NPV) is considered the most appropriate criterion for synthesis problems (Pintarič and Kravanja, 2015). The net present value leads to designs balancing capital and operational expenditures. As the net present value is in general negative for energy systems, we prefer to change the sign of the objective function and divide all terms by the annualized present value factor $APVF$. These changes lead to the total annualized costs TAC, Eq. (4.1). The objective function consists of 2 parts representing the two-stage character of the synthesis problem, the annualized capital and operational expenditures, $CAPEX$ and $OPEX$:

$$\min_{\dot{V}_{n,t},\delta_{n,t},\dot{V}_n^{\mathrm{N}},\gamma_n,\dot{V}_{\mathrm{grid}}^{\mathrm{max}},\dot{V}_{\mathrm{grid},t},x,y} \quad TAC = CAPEX + OPEX. \tag{4.1}$$

The investment costs $CAPEX$, Eq. (4.2), are calculated by multiplying the nominal (superscript ^{N}) capacity $\dot{V}_n^{\mathrm{N}}$ of each component by the specific investment (superscript ^{i}) costs $\mathrm{c}_n^{\mathrm{i}}$ for all units $n \in \mathcal{C}$ and dividing by the annualization factor $APVF$. The annualized present value factor is $APVF = \frac{(i+1)^{T_h}-1}{i(i+1)^{T_h}}$. The interest rate is $i = 8\,\%$ and the time horizon $T_h = 15a$.

$$CAPEX = \frac{1}{APVF} \cdot \sum_{n\in\mathcal{C}} \mathrm{c}_n^{\mathrm{i}} \dot{V}_n^{\mathrm{N}}. \tag{4.2}$$

For the operational expenditures $OPEX$, Eq. (4.3), fuel and electricity costs are the same as in the operational optimization problem in Chapter 3.3. The maintenance costs are calculated by multiplying the nominal capacity of each component $\dot{V}_n^{\mathrm{N}}$ by the specific maintenance (superscript ^{m}) factor $\mathrm{c}_n^{\mathrm{m}}$. In addition, industrial users usually pay for the maximum power used. Thus, we also consider a peak-power price (subscript $_{\mathrm{p}}$) in the operational costs $OPEX$ by the electrical peak power $\dot{V}_{\mathrm{grid}}^{\mathrm{max}}$ multiplied by a specific grid fee c_{p} in €/kW^{max}, which is constant in contrast to Chapter 3. Also, the network-connection fee is neglected to simplify the model detail.

$$OPEX := \overbrace{\sum_{t\in\mathcal{T}} \left(\Delta \mathrm{t}_t \sum_{n\in\mathcal{C}} \mathrm{c}_{n,t}^{\mathrm{o}} \cdot \frac{\dot{V}_{n,t}}{\eta_n} \right)}^{fuel/electricity\ costs} + \overbrace{\sum_{n\in\mathcal{C}} \mathrm{c}_n^{\mathrm{m}} \dot{V}_n^{\mathrm{N}}}^{maintenance\ costs} + \underbrace{c_p \cdot \dot{V}_{\mathrm{grid}}^{\mathrm{max}}}_{peak-power\ costs} \tag{4.3}$$

OPEX directly depend on the set of time steps $\mathcal{T}$, while *CAPEX* only depend on one-time investment decisions about the capacity $\dot{V}_n^{\mathrm{N}}$. The total annualized costs *TAC* are minimized subject to constraints modeling the characteristics of the energy system. The constraints regarding the operation of the energy system are the same here as for the operational optimization. Thus, Eq. (3.2-3.6) is also part of the synthesis problem, whereas the long-term operational constraints for network-connection fees (Eq. (3.7)) and emissions limits (Eq. (3.8)) are neglected. Additionally, to render the synthesis problem computationally tractable, we fix part-load efficiencies to the nominal efficiency. Thus, only Eq. (4.4) is added as an additional constraint:

$$\begin{aligned}
\text{s.t.}\quad & A_3\dot{V}_n^{\mathrm{N}} + \widetilde{A}_3\gamma_n \leq b_3, && \forall n \in \mathcal{C}, \\
& \dot{V}_n^{\mathrm{N}} \in \mathbb{R}^+, \gamma_n \in \{0,1\}, && \forall n \in \mathcal{C}, \\
& \Delta\dot{V}_{n,t} \in \mathbb{R}, && \forall t \in \mathcal{T}, \forall n \in \mathcal{C}_{\mathrm{stor}}, \\
& \dot{V}_{\mathrm{grid}}^{\mathrm{max}} \in \mathbb{R}^+, \dot{V}_{\mathrm{grid,t}} \in \mathbb{R}, x \in \mathbb{R}, y \in \{0,1\}.
\end{aligned} \tag{4.4}$$

Eq. (4.4) summarizes (in)equalities, with the coefficient matrices A_3, $\widetilde{A}_3$ and the vector b_3. Eq. (4.4) determines the binary existence γ_n of units and the nominal power $\dot{V}_n^{\mathrm{N}}$. Additionally, Eq. (3.4) here also includes bilinear sizing constraints $\dot{V}_{n,t} \leq \gamma_n\dot{V}_n^{\mathrm{N}}$ which result from the linearization of the non-linear investment curves. The bilinear term $\gamma_n\dot{V}_n^{\mathrm{N}}$ is linearized using Glover's reformulation (Glover, 1975), which introduces additional variables that are included in the vector x. For each time step t, the equations that have to be stated include the output power $\dot{V}_{n,t}$, the storage variables, and the on/off status $\delta_{n,t}$. These equations thus depend on the size of the time series $\mathcal{T}$. For large time series $\mathcal{T}$, the original synthesis problem is often not solvable in reasonable computational time or within memory limits.
To solve large-scale synthesis problems, we propose the rigorous synthesis method RiSES[4] in the next section.

4.2 The RiSES[4] method

RiSES[4] aims to solve MILP synthesis problems that depend on large-scale time series. RiSES[4] is based on 3 parallel branches, Fig. 4.1, to calculate upper bounds and 2 competitive lower bounds. In the master branch, upper bounds are obtained from feasible solutions. These feasible solutions are computed using time-series aggregation and restriction. The upper bounds are tightened by iteratively increasing the time resolution. The final time resolution is reached once a convergence criterion is fulfilled. The algorithm of the master branch is described in Chapter 4.2.1.

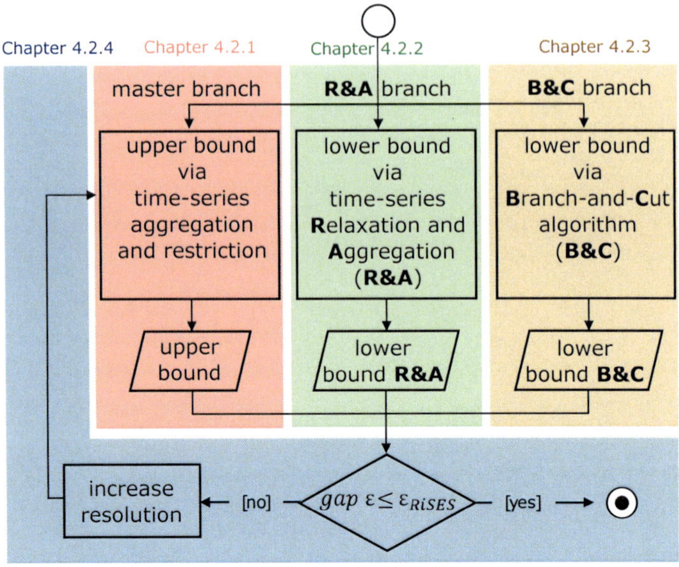

Figure 4.1: RiSES[4]: Rigorous Synthesis of Energy Supply and Storage Systems via time-series relaxation and aggregation. Lower and upper bounds are computed and compared to satisfy an optimality gap $\varepsilon_{\mathrm{RiSES}}$.

RiSES[4] employs 2 competitive relaxation methods to compute lower bounds: the **R&A** branch is based on time-series **R**elaxation and **A**ggregation of input parameters; the **B&C** branch is based on linear-programming relaxation implemented as the **B**ranch-and-**C**ut procedure. In the RiSES[4] method, the tighter of these 2 relaxation, i. e., the one with the worse objective, is used as lower bound to calculate the optimality gap ε. In the **R&A** branch (see Fig. 4.1), lower bounds are obtained via time-series aggregation and simultaneous relaxation. These lower bounds are tightened by increasing the time resolution. The algorithm of the **R&A** branch is described in Chapter 4.2.2. In the **B&C** branch, lower bounds are obtained via linear-programming relaxation implemented as Branch-and-Cut procedure of a state-of-the-art MILP solver. The branching procedure tightens these lower bounds. The algorithm of the **B&C** branch is briefly described in Chapter 4.2.3.

After each iteration of the master branch, RiSES[4] compares the current upper bound with the available lower bounds of the parallel branches **R&A** and **B&C** to calculate the optimality gap ε, Chapter 4.2.4. The method stops when the resulting gap ε satisfies the desired optimality gap $\varepsilon_{\text{RiSES}}$. If the optimality gap $\varepsilon_{\text{RiSES}}$ is not satisfied, the time resolution of the time-series aggregation of the master branch is increased, Chapter 4.2.4.

4.2.1 Master branch to calculate upper bounds

In the master branch, feasible solutions (upper bounds) of the original synthesis problem are calculated based on 3 steps, Fig. 4.2.

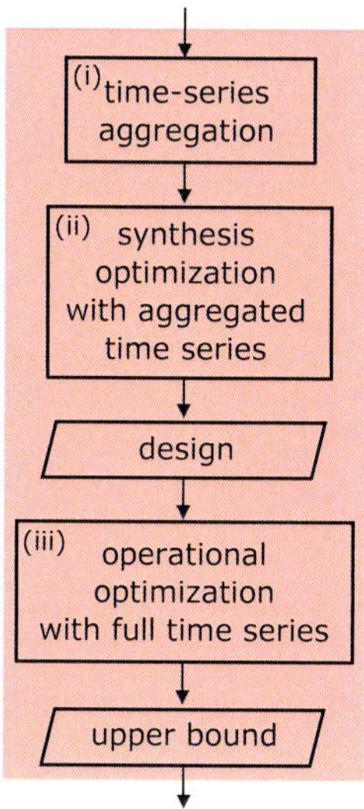

Figure 4.2: RiSES[4]: The master branch of RiSES[4] consists of 3 steps: (i) time-series aggregation, (ii) synthesis optimization with aggregated time series yielding a design candidate. This design is fixed for the (iii) operational optimization with the full time series to obtain an upper bound.

In step (i), time-series aggregation reduces the complexity of the synthesis problem. Thereby, the synthesis problem with the aggregated time series can be solved efficiently in step (ii). The solution of the aggregated problem results in a design candidate for the energy system. The feasibility of the design candidate is tested in the operation optimization for the full time series in step (iii). The operation optimization of a feasible design candidate for the full time-series results in an upper bound TAC^{UB} for the original synthesis problem.

In step (i), we employ the time-series aggregation method based on Bahl et al. (2018b), which has been extended for seasonal storage by coupled typical periods as in Kotzur et al. (2018). First, the method identifies the length of typical periods by looking for periodic patterns in the time-series data using autocorrelation, as in Box and Jenkins (2015), by the 'autocorr' function of MathWorks (2017). Second, the identified period length is used to split the original time series into periods. Third, the periods are aggregated to typical periods $\mathcal{P}$ based on k-means clustering. Within each typical period p, the time steps t are further aggregated to segments s. To aggregate segments, we adapt the k-means idea and calculate the average of a set of randomly chosen consecutive time steps. We use the average value with the lowest Euclidean distance to the original time steps for the entire aggregated segment. The employed time-series aggregation method thus aggregates in 2 dimensions: the number of typical periods N_p and the number of segments N_s per typical period, Fig. 4.3. The employed time-series aggregation is based on k-means clustering using the 'kmeans' function of MathWorks (2017), which employs the k-means++ algorithm. Recently, Teichgräber and Brandt (2019) compared clustering methods based on k-medoids, k-means, dynamic time warping based clustering, and k-shape. They show that dynamic time warping and k-shape based clustering slightly improve the results for operational problems, including storage systems. Luo et al. (2019) also improve a k-means-based clustering algorithm based on parametric bootstrap and Zatti et al. (2019) propose an optimization-based approach to select clusters. Both methods leading to slightly more accurate clusters. However, all observed differences of clustering algorithms are minor, as Schütz et al. (2018) also showed.

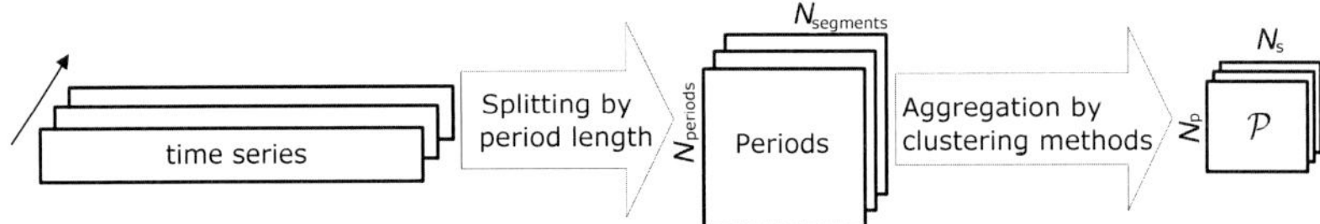

Figure 4.3: Aggregation to typical periods. All time series are split into periods according to the identified typical period length. Each period $N_{periods}$ still consists of the original number of segments $N_{segments}$. Afterward, the time-series is aggregated by clustering methods to N_p typical periods p with N_s aggregated segments s in each typical period. (Adapted from Bahl et al. (2018b)).

Time-series aggregation by typical periods maintains the chronology within each typical period and thereby enables to consider time-coupling constraints and storage within each typical period (intra-period). However, as non-consecutive periods are clustered, only intra-period storage is directly possible. To model seasonal storage, an inter-period storage difference has to be considered. This inter-period storage difference ΔV_p is defined as the difference of the storage level from the beginning to the end of each typical period p. Fig. 4.4 schematically shows the inter-period storage difference ΔV_p for 3 typical periods.

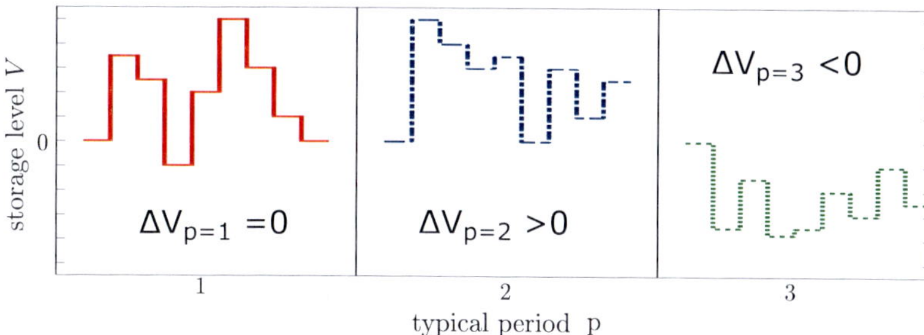

Figure 4.4: Schematic representation of the intra-period storage difference ΔV_p for 3 typical periods

To consider these inter-period storage differences for seasonal storage, a second inter-period time grid couples the typical periods (Kotzur et al., 2018). To couple the typical periods in the second time grid, we assign each original unclustered period to its corresponding typical period. Thus, the second time grid contains the information on the chronological order of the typical periods. Table 4.1 shows exemplary this assignment for 3 typical periods and 365 unclustered periods. By this chronological order, the inter-period storage level differences are then also ordered.

Table 4.1: Look-up table to assign each unclustered period to the corresponding typical period p. The columns assign unclustered periods to typical periods, leading to ordered typical periods in the second row.

unclustered period	1	2	3	4	5	6	...	362	363	364	365
typical period p	2	3	3	2	1	1	...	2	2	3	1

The superposition of this inter-period storage level differences ΔV_p and the intra-period storage level within each typical period results in the actual storage level, Fig. 4.5. Considering the actual storage level from the superposition enables RiSES^4 to design seasonal storage systems in the aggregated synthesis problem. Thereby, RiSES^4 overcomes this shortcoming of the method previously proposed by Bahl et al. (2018a), Chapter 2.2.2.

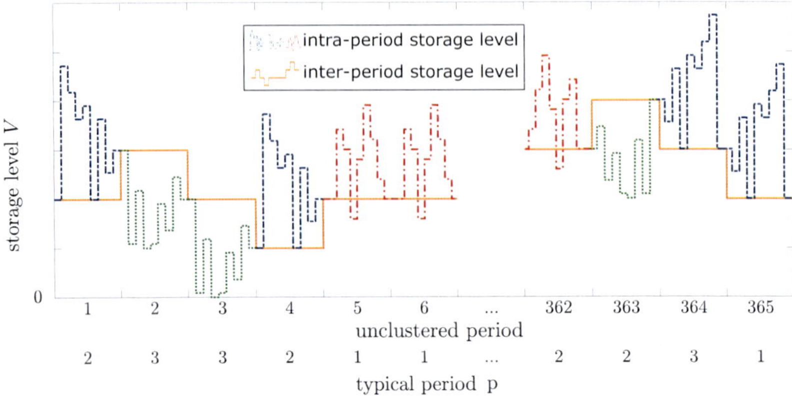

Figure 4.5: Schematic representation of the superposition of the intra-period and inter-period storage difference

In step (ii), the typical periods $\mathcal{P}$ from step (i) are used instead of the full time series T, resulting in an aggregated synthesis problem. The time-series aggregation reduces complexity by orders of magnitudes as the aggregated synthesis problem needs to be solved for fewer aggregated time steps $N_p \times N_s = |\mathcal{P}| \ll |T|$ only. The solution of the aggregated synthesis optimization yields a design candidate for the energy system. The design candidate is the solution of the first-stage decisions of the original synthesis problem, i. e., the selection and sizing of units.

In step (iii), the design candidate of the energy system is employed to obtain the second-stage decisions. For this purpose, the first-stage decisions of the design candidate are fixed in the original synthesis problem. By fixing the first-stage decisions in the original synthesis problem, the problem reduces to an operational optimization problem with the full time series. Fixing the first-stage decisions reduces the problem complexity and thus allows efficient solving of the resulting operation optimization problem.
However, the feasibility of the design candidate cannot be guaranteed for the full time series yet. To increase the probability of a feasible design candidate, we add peak demands in step (ii) to the aggregated time series as single time steps with a duration of zero and no chronological order (Bahl et al., 2017a). If the resulting design is still infeasible for the original demand time series in step (iii), we increase the time resolution of the time-series aggregation. This strategy proved to be efficient for our case

study. Alternatively, the aggregated time series could be expanded by feasibility time steps, as in Bahl et al. (2018b). If the resulting design is feasible, the solution of the operation problem leads to an upper bound TAC^{UB} of the original synthesis problem. In RiSES^4, the solution of the operational optimization problem yields simultaneously an upper bound for the original synthesis problem, as the operational problem is a restricted problem of the original synthesis problem, since design variables are fixed. To determine the optimality gap ε, a lower bound is calculated based on 2 parallel branches described in the following Chapter 4.2.2 and Chapter 4.2.3.

4.2.2 R&A branch to calculate lower bounds

In the time-series **R**elaxation and **A**ggregation (**R&A**) branch, lower bounds are obtained via time-series aggregation and simultaneous under- and overestimation of the time-dependent input parameters P_t. The **R&A** branch consists of 3 steps, Fig. 4.6.

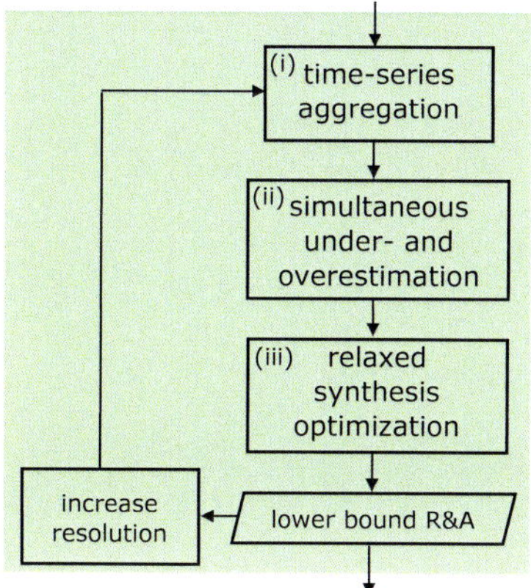

Figure 4.6: **R&A** branch of RiSES^4 consists of 3 steps: (i) time-series aggregation, (ii) simultaneous under- and overestimation of time-dependent input parameters and (iii) relaxed synthesis optimization.

In step (i), the complexity of the synthesis problem is reduced by time-series aggregation. In step (ii), we simultaneously under- and overestimate all time-dependent input parameters to generate a relaxed synthesis problem. In step (iii), the resulting aggregated and relaxed synthesis problem is solved. The lower bound $TAC^{\mathrm{LB}_{R\&A}}$ of the aggregated relaxed synthesis problem is a lower bound of the original synthesis optimization, as shown below.

In step (i), we employ the same time-series aggregation methods as in the master branch, Chapter 4.2.1, resulting in typical periods p with aggregated segments s within each typical period.

In step (ii), the time-dependent input parameters P_t of these typical periods are simultaneously under- and overestimated. Hereby, we overcome the shortcoming of the previously proposed method by Bahl et al. (2018a), Chapter 2.2.2, where only time-dependent demand data could be considered. Simultaneous under- and overestimation allow for considering of any time-dependent input parameter in the synthesis problem, besides demand data, e. g., prices and renewable resources. Mathematical details of the simultaneous under- and overestimation are described in the following. The time-series aggregation finds an assignment $z_{t,p,s}$ of each unclustered period to a typical period p and each time step t within a typical period to a segment s. Thus, each time step t of the original time series is assigned to a specific segment s within a period p. Next, time-series aggregation would represent all input parameters of the assigned time steps t with one value for the segment s. Here, instead of clustering the input parameters, we identify an underestimator $\lfloor P_{p,s} \rfloor$ and an overestimator $\lceil P_{p,s} \rceil$ for each segment s in each typical period p, Fig. 4.7. The underestimator $\lfloor P_{p,s} \rfloor$ is the smallest value of all time steps assigned to a segment of a typical period,

$$\lfloor P_{p,s} \rfloor \leq z_{t,p,s} P_t, \qquad \forall t \in \mathcal{T}, \forall p \in \mathcal{P}, \forall s \in \mathcal{S}. \tag{4.5}$$

For the overestimator $\lceil P_{p,s} \rceil$ the same holds true vice versa with the largest value of all time steps within a segment s,

$$\lceil P_{p,s} \rceil \geq z_{t,p,s} P_t, \qquad \forall t \in \mathcal{T}, \forall p \in \mathcal{P}, \forall s \in \mathcal{S}. \tag{4.6}$$

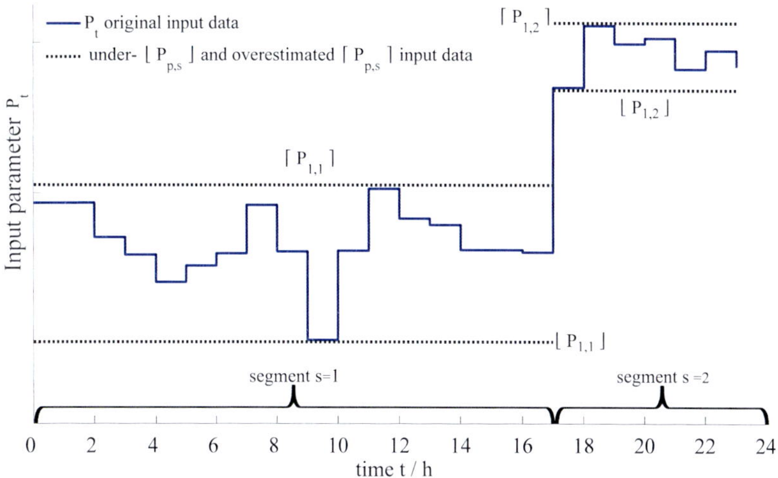

Figure 4.7: Schematic representation of the simultaneous under- and overestimation of time-dependent input parameters P_t, for 1 typical period p with 2 segments s.

In step (iii), the under- and overestimated input parameters $\lfloor P_{p,s} \rfloor$, $\lceil P_{p,s} \rceil$ are subsequently used to solve a relaxed aggregated synthesis optimization problem, Eq. (4.7-4.11):

$$\min_{\dot{V}_{n,t'},\delta_{n,t'},\dot{V}_n^{\mathrm{N}},\gamma_n,\dot{V}_{\mathrm{grid}}^{\max},\dot{V}_{\mathrm{grid},t'},x,y} \quad TAC^{\mathrm{LB}_{R\&A}} = \dots$$

$$\overbrace{\frac{1}{APVF}\sum_{n\in\mathcal{C}} \mathrm{c}_n^{\mathrm{i}}\dot{V}_n^{\mathrm{N}}}^{CAPEX} + \overbrace{\sum_{t'\in\mathcal{P}}\left(\Delta \mathrm{t}_{t'} \sum_{n\in\mathcal{C}\setminus\mathcal{C}_{\mathrm{stor}}} [\mathrm{c}_{n,t'}^{\mathrm{o}}]\frac{\dot{V}_{n,t'}}{\eta_n}\right) + \sum_{n\in\mathcal{C}} \mathrm{c}_n^{\mathrm{m}}\dot{V}_n^{\mathrm{N}} + \mathrm{c}_p \cdot \dot{V}_{\mathrm{grid}}^{\max}}^{OPEX} \tag{4.7}$$

s.t.

$$\lfloor \dot{E}_{t'} \rfloor \leq \sum_{n \in \mathcal{C} \setminus \mathcal{C}_{\text{stor}}} \dot{V}_{n,t'} + \dot{V}_{\text{grid},t'} + \sum_{n \in \mathcal{C}_{\text{stor}}} (\dot{V}^{\text{out}}_{n,t'} - \dot{V}^{\text{in}}_{n,t'}) \leq \lceil \dot{E}_{t'} \rceil, \qquad \forall t' \in \mathcal{P}, \tag{4.8}$$

$$A_1 \dot{V}_{n,t'} + \widetilde{A}_1 \delta_{n,t'} \leq b_1, \qquad \forall t' \in \mathcal{P}, \forall n \in \mathcal{C}, \tag{4.9}$$

$$A_2 x + \widetilde{A}_2 y \leq b_2, \qquad \forall n \in \mathcal{C}, \tag{4.10}$$

$$A_3 \dot{V}^{\text{N}}_n + \widetilde{A}_3 \gamma_n \leq b_3, \qquad \forall t' \in \mathcal{P}, \forall n \in \mathcal{C}, \tag{4.11}$$

$$V_{n,t'} + \Delta \text{t}_{t'} \cdot \sum_{n \in \mathcal{C}_{\text{stor}}} (\dot{V}^{\text{in}}_{n,t'} - \dot{V}^{\text{out}}_{n,t'}) = V_{n,t'+1}, \qquad \forall t' \in \mathcal{P}, \forall n \in \mathcal{C}_{\text{stor}} \tag{4.12}$$

$$|\dot{V}_{\text{grid},t'}| \leq \dot{V}^{\text{max}}_{\text{grid}}, \qquad \forall t' \in \mathcal{P}, \tag{4.13}$$

$$\dot{V}^{\text{N}}_n \in \mathbb{R}^+, \gamma_n \in \{0,1\}, \qquad \forall n \in \mathcal{C},$$

$$\dot{V}_{n,t'} \in \mathbb{R}^+, \delta_{n,t'} \in \{0,1\}, \qquad \forall t' \in \mathcal{P}, \forall n \in \mathcal{C},$$

$$\dot{V}^{\text{max}}_{\text{grid}} \in \mathbb{R}^+, \dot{V}_{\text{grid},t'} \in \mathbb{R}, x \in \mathbb{R}^a, y \in \{0,1\}.$$

Here, we highlight the differences between the relaxed synthesis problem, Eq. (4.7-4.13), and the original synthesis problem. The set of time steps T is reduced by the time-series aggregation to $\mathcal{P}$ with $N_p \times N_s = |\mathcal{P}| \ll |T|$, Chapter 4.2.1. In Eq. (4.7), the specific operation cost $\text{c}^{\text{o}}_{n,t'}$ as part of the OPEX are relaxed by the underestimator $\lfloor \text{c}^{\text{o}}_{n,t'} \rfloor$ for positive, e. g., buying gas, and by the overestimator $\lceil \text{c}^{\text{o}}_{n,t'} \rceil$ for negative operational cost, e. g., selling electricity; here represented by $[\text{c}^{\text{o}}_{n,t'}]$. In Eq. (4.8), the sum of the units output power $\dot{V}_{n,t'}$ and the net energy flow of the storage units $\Delta \dot{V}_{n,t'}$ are relaxed and bound by the energy demand under- $\lfloor \dot{E}_{t'} \rfloor$ and overestimator $\lceil \dot{E}_{t'} \rceil$ at every time step t' of the aggregated periods. Thus, over- and underproduction within the bounds are possible. All other equations of the aggregated synthesis problem are identical to the original synthesis problem. The relaxed synthesis problem, Eq. (4.7-4.13), can be solved efficiently due to the reduced number of aggregated periods $\mathcal{P}$. After each calculation of a lower bound, we increase the time resolution in step (i) and restart the **R&A** branch. To increase time resolution, we employ the same heuristic as for the master branch, described in Chapter 4.2.4. If the master branch finishes an iteration, the best lower bound $TAC^{\text{LB}_{R\&A}}$ is provided as candidate for the lower bound in RiSES[4]. Depending on the problem studied, linear-programming relaxation might serve as a tighter relaxation of the original synthesis problem. Thus, we employ a second competitive lower bound based on the Branch-and-Cut procedure.

4.2.3 B&C branch to calculate lower bounds

In the **B**ranch-and-**C**ut (**B&C**) branch, lower bounds are obtained via the Branch-and-Cut procedure commonly included in state-of-the-art solvers. In these solvers, first, all binary variables ($\delta_{n,t}, \gamma_n, y \in [0, 1]$) of the original synthesis problem are relaxed. Thereby, the complex MILP is converted into an LP, which can be solved efficiently. The solution of the relaxed problem serves as first lower bound candidate of the **B&C** branch. Subsequently, the Branch-and-Cut procedure starts by branching on binary variables and cutting off branches that cannot perform better than the best known integer solution (IBM, 2016). In general, the **B&C** branch could further be improved by a warm-start procedure based on feasible solutions of the master branch. Thereby, the **B&C** branch could provide both lower and upper bounds of the original synthesis problem. We tested the warm-start strategy in preliminary calculations and did not observe a large impact on solution time but did not tune the method yet. In the following, the warm-start procedure of the **B&C** branch is not used. If the master branch finishes an iteration, the best current lower bound $TAC^{\text{LB}_{B\&C}}$ is provided as lower bound candidate in RiSES^4.
Using the lower bound of Chapter 4.2.2 and Chapter 4.2.3 and upper bound of Chapter 4.2.1, an optimality gap ε can be calculated.

4.2.4 Optimality gap ε and increase of time resolution

The best lower bound candidate TAC^{LB} resulting from the **R&A** branch, Chapter 4.2.2, or the **B&C** branch, Chapter 4.2.3, is compared with the best upper bound TAC^{UB} from the master branch, Chapter 4.2.1, and checked if the desired optimality gap $\varepsilon_{\text{RiSES}}$ is satisfied:

$$\varepsilon := \frac{TAC^{\text{UB}} - TAC^{\text{LB}}}{TAC^{\text{UB}}} \leq \varepsilon_{\text{RiSES}}. \tag{4.14}$$

RiSES^4 solves internal MILP optimization problems in the master branch, Chapter 4.2.1, and the **R&A** branch, Chapter 4.2.2. These MILP problems require the definition of an internal optimality gap $\varepsilon_{internal}$. These internal optimality gaps might add up and prevent the convergence of RiSES^4 below the gap ε, Eq. (4.14). Thus, to ensure convergence of RiSES^4, an internal optimality gap sufficiently smaller than ε_{RiSES} must be chosen.
If the optimality gap ε does not satisfy the required optimality gap $\varepsilon_{\text{RiSES}}$, we iteratively increase the time resolution until the optimality gap $\varepsilon_{\text{RiSES}}$ is satisfied. To choose whether to increase the number of periods N_p or the number of segments N_s

in the master branch, we employ a heuristic based on finite backward differences of the solution quality. The heuristic selects the larger backward difference as most promising direction to increase the resolution of the aggregation (Bahl et al., 2018b). Details of the employed heuristic are stated in Appendix E.1. In a previous paper by Bahl et al. (2018b), they studied only the upper bound but computed a full grid of choices for both the number of periods N_p and the number of segments N_s. The results showed that the heuristic selection strategy usually leads to good solutions that are either optimal or very close to the optimum. The iterative increase of the time resolution for the time-series aggregation stops, as soon as $\varepsilon \leq \varepsilon_{\text{RiSES}}$, Eq. (4.14), is satisfied, yielding a feasible solution of the original synthesis problem with known solution quality.
In the next section, we apply RiSES^4 to 2 real-world industrial synthesis problems.

4.3 Real-world industrial case studies

To validate the proposed rigorous synthesis method, we apply RiSES^4 to 2 real-world synthesis problems. In Chapter 4.3.1, we study a low-carbon industrial energy system providing electricity, low-temperature heat, steam, and cooling for an industrial site. In Chapter 4.3.2, we study a pump system providing cooling water to a chemical site.

4.3.1 Industrial energy system (ES)

In this section, we apply RiSES^4 to the MILP synthesis problem presented in Chapter 4.1 and Appendix A. Here, we use a superstructure with 3 units of each energy-conversion technology (absorption chiller, boiler, CHP engine, compression chiller, electrode boiler, heat exchanger, and heat pump), additional roof-top PV, an inverter station, and a wind turbine. As storage systems, the superstructure includes a battery system and 1 storage tank for hot and 1 for cold water. The model allows continuous sizing of all units and storage systems. For all units, we assume constant part-load efficiencies.

In this case study, the energy demand for steam, low temperature heat, cooling, electricity, and electricity purchase and selling prices are provided for 1 year, Fig. 3.5. To render the case study computational tractable for the benchmark, we consider here the time-series with a two-hourly resolution. We refer to this original data as 'original instance.' Five further instances are generated by statistical noise using Latin hypercube sampling (Mckay et al., 2000) with a variation by the time series of $\pm 5\,\%$.

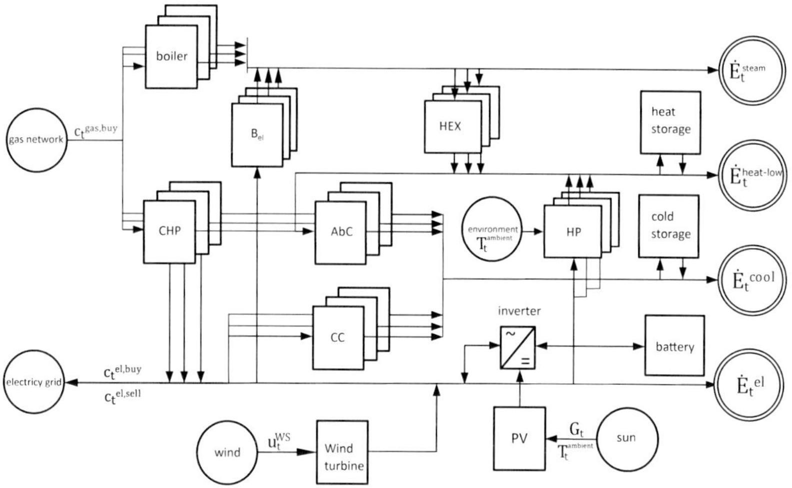

Figure 4.8: Superstructure of the case study: absorption chillers (AbC), battery (BAT), boilers (B), combined-heat-and-power engines (CHP), compression chillers (CC), electrode boilers (B_{el}), heat exchanger (HEX), heat pumps (HP), inverter stations, photovoltaic (PV). Redundant units of utilities are shown. The model considers time-varying energy demands ($\dot{E}_{cool,t}, \dot{E}_{heat-low,t}, \dot{E}_{steam,t}, \dot{E}_t^{el}$), electricity prices ($c_t^{el,buy}, c_t^{el,sell}$), ambient temperature ($T_t^{ambient}$) and resources of renewable energies (G_t, u_t^{WS}).

After presolve, the original synthesis problem with full time series contains $9 \cdot 10^5$ equations and $4 \cdot 10^5$ variables ($1.4 \cdot 10^5$ binaries) with $24 \cdot 10^5$ nonzero elements. As benchmark, we solve this original synthesis problem directly with CPLEX (IBM, 2016).

In CPLEX, we employ the deterministic parallel mode such that all methods use the same number of computational cores. The benchmark and the RiSES[4] calculations are performed using 4 Intel-Xeon CPUs with 3.0 GHz and 64GB RAM. All MILP problems are solved using the CPLEX version 12.6.3.0. The optimality gap ε_{RiSES} of RiSES[4] is set to 2 %. An internal optimality gap smaller than ε_{RiSES} must be chosen to enable convergence, Chapter 4.2.4. Here, we intuitively choose $\frac{\varepsilon_{RiSES}}{4} = 0.5\,\%$ as the internal MILP optimality gaps $\varepsilon_{internal}$ of the master branch, Chapter 4.2.1, and the **R&A** branch, Chapter 4.2.2. The time limit of the synthesis optimizations in these branches is set to 3 hours. The time limit of the operational optimizations in Chapter 4.2.1 is set to 20 minutes.

The time-series aggregation method identifies the period length, Chapter 4.2.1. The highest values for normalized sum of the autocorrelation are found for time lags of 24 hours and 168 hours (= 1 week). Since shorter periods are expected to lead to smaller optimization problems, we picked the period length 24 hours.

The proposed rigorous synthesis method RiSES[4] satisfies the optimality gap $\varepsilon_{RiSES} = 2\,\%$ in 9 minutes (524 s), Fig. 4.10. The optimality gap ε_{RiSES} can be satisfied using 13 aggregated time steps resulting in a synthesis problem with only 6698 equations, 2113 variables (493 binaries), and 19,000 nonzero elements. Thus, the complexity is reduced by about 2 orders of magnitude. For illustration, the method is not stopped once the optimality gap ε_{RiSES} is reached but continued to a time limit of 3 hours. Considering more periods and segments further increases the accuracy slightly, Fig. 4.10. The master branch's highest resolution employs 22 periods consisting of 11 segments each and reaches an optimality gap $\varepsilon = 0.89\,\%$. For an example, the aggregated time series with 13 aggregated time steps is shown in Fig. 4.9 and the highest time resolution with 246 aggregated time steps is shown in Appendix E.2, Fig. E.1. Although the aggregated demands and prices, in particular cooling and low-temperature demand, differ from the original time series, excellent designs are obtained.

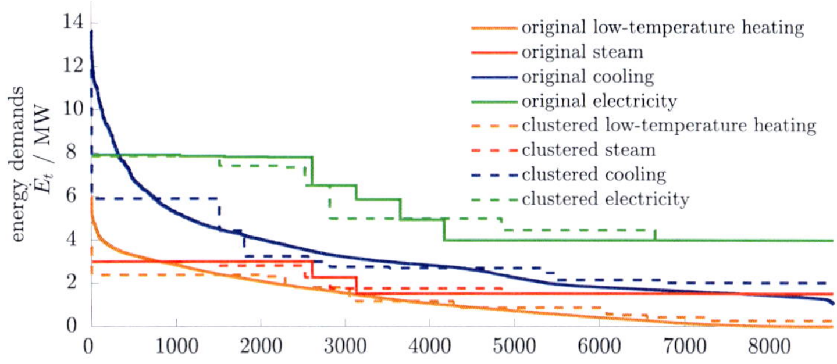

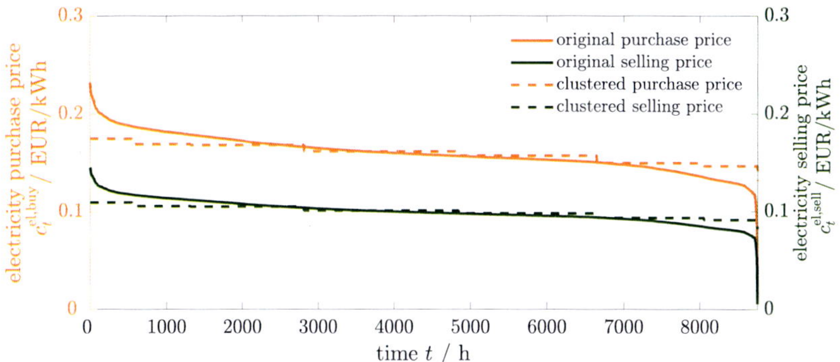

Figure 4.9: Sorted time series of case study ***ES*** for low temperature heat, steam, cooling, and electricity demands for 1 year with a time step length of $\Delta t_t = 2\,\mathrm{h}$. Additionally, the used aggregated time series with 13 time steps (P=13) for the first design of RiSES[4] satisfying the optimality gap ε is shown. Time series for wind speed, solar radiation, and temperature are omitted for simplicity.

The computational time for the upper bound in the master branch consists of 3 parts: (i) time-series aggregation, (ii) aggregated synthesis optimization, and (iii) operational optimization. The computational time for time-series aggregation and the aggregated synthesis optimization increases with increasing time resolution: for a small number of time steps, it only takes a few seconds, which increases up to several minutes for large numbers of time steps (4-900 s). In contrast, the operational optimization is independent of the time resolution of the time-series aggregation, because the full time series is used. The operational optimization takes, on average, about 130 s (20-700 s).

In summary, RiSES[4] obtains a feasible design candidate satisfying the optimality gap $\varepsilon_{RiSES} = 2\,\%$ within a few seconds. However, the feasibility proof by the operational optimization and the optimality proof by the lower bounds takes longer.

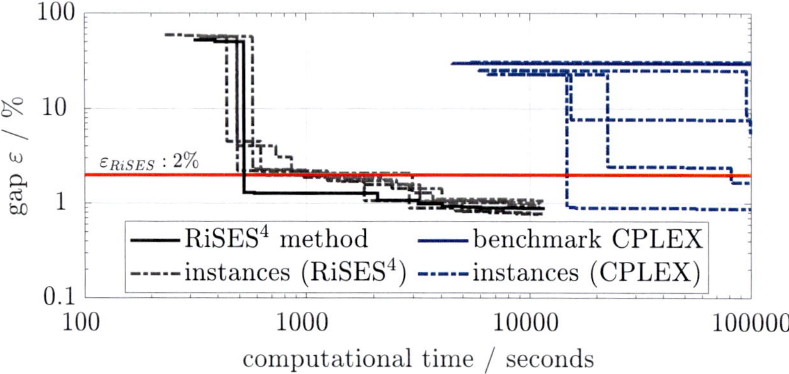

Figure 4.10: Optimality gap ε of the proposed rigorous synthesis method RiSES[4] and the benchmark CPLEX as function of the solution time for the low-carbon industrial energy system of case study ***ES***. The required optimality gap ε of $2\,\%$ is marked in thick red.

The benchmark solves the original synthesis problem directly with CPLEX. The benchmark provides the first feasible solution after 75 minutes (4489 s) and reaches the time limit of 10^5 seconds (~28 hours), still with a relative gap of $29.9\,\%$, Fig. 4.10.

For all instances, the performance of RiSES[4] is very similar, Fig. 4.10. RiSES[4] outperforms the benchmark in all instances, always satisfying the desired optimality gap ε_{RiSES} in under 1 hour. The performance of the benchmark differs: in all instances, feasible solutions were found within the time limit of 10^5 seconds; however, only in 2 instances, a solution satisfying the required optimality gap ε_{RiSES} is found; whereas in the 3 other instances, the optimality gap ε remains between 5 and $31\,\%$. Thus, RiSES[4] always satisfies the required optimality gap $\varepsilon_{RiSES} = 2\,\%$ before the benchmark provides any feasible solution at all. Further, RiSES[4] reaches the required optimality gap on average at least 65 times faster than the benchmark CPLEX.

Fig. 4.11 shows the lower bounds $TAC^{\mathrm{LB}_{R\&A}}$ and $TAC^{\mathrm{LB}_{B\&C}}$ together with the upper bound TAC^{UB} for the original instance. Additionally, the optimality gap ε of RiSES[4] is plotted as function of the solution time.

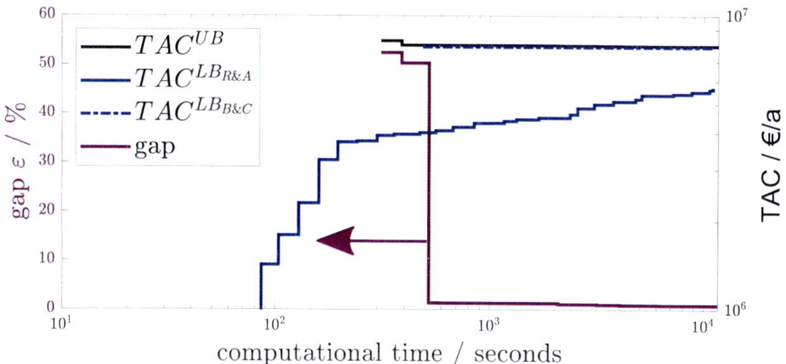

Figure 4.11: Total annualized costs TAC of the lower and upper bounds of RiSES[4] as function of the solution time for case study ***ES*** (right axis). Additionally, the optimality gap ε of RiSES[4] is plotted (left axis).

Branch **R&A** provides the first lower bound $TAC^{\mathrm{LB}_{R\&A}}$ very fast, based on the proposed simultaneous under- and overestimation. The first lower bounds are found within a few seconds. However, lower bound $TAC^{\mathrm{LB}_{R\&A}}$ converges slowly over many iterations. After 3 hours, optimality could not be proven by lower bound $TAC^{\mathrm{LB}_{R\&A}}$. However, the convergence of the **R&A** branch could be tuned by faster increasing time resolution of the heuristic, Appendix E.1.

In contrast, the linear-programming relaxation in the **B&C** branch provides the first bound after 524 seconds. The first lower bound $TAC^{\mathrm{LB}_{B\&C}}$ is the solution of the linear programming relaxation of the MILP. Thus, all binary variables are relaxed and the resulting linear program is solved. This $TAC^{\mathrm{LB}_{B\&C}}$ only improves slightly with increasing solution time. However, the first lower bound $TAC^{\mathrm{LB}_{B\&C}}$ is already a very tight relaxation of the original synthesis problem, Eq. (4.7-4.11). Thus, the **B&C** branch provides fast optimality proof in this case study. RiSES[4] further reduces the remaining gap ε between the first upper bound TAC^{UB} and the lower bound $TAC^{\mathrm{LB}_{B\&C}}$. In this case study, the slight improvement is based on improved design candidates from the master branch, as an increasing number of aggregated time steps is considered.

Table 4.2 shows the first RiSES[4] design candidate satisfying the optimality gap, the best design candidate within the time limit of 10^4 seconds, and, as a benchmark, the best design found from CPLEX within the time limit of 10^5 seconds, all for the original instances. In general, very different design candidates can have almost the same

Table 4.2: Optimization results for case study ***ES***, by RiSES^4 with 13 aggregated time steps ($\mathcal{P}$ = 13), the best found solution within the time limit with 246 aggregated time steps ($\mathcal{P}$ = 246), and the benchmark with CPLEX. Nominal capacity of each installed unit, costs TAC, gap ε, and computational time of the solution. Units of which no capacity is built are not included in the table.

unit /MW	**RiSES^4** $\mathcal{P}$ = 13	**RiSES^4** $\mathcal{P}$ = 246	**CPLEX**
Boiler	7.5 / 5.5 / –	7.5 / – / –	7.3 / – / –
CHP engine	3.2 / 3.2 / –	3.2 / 3.2 / –	– / – / –
Compression chiller	1.3 / 1.3 / –	3.7 / 2.6 / –	6.5 / 6.0 /–
Absorption chiller	5.9 / 5.1 / –	3.7 / 3.6 / –	– / – /–
Heat exchanger	5.0 / 5.0 / –	4.5 / – / –	5.0 / – /–
$\text{Storage}_{\text{heating}}$ /MW h	–	2.0	113.8
$\text{Storage}_{\text{cooling}}$ /MW h	–	–	11.7
TAC /10^6 €	8.03	7.91	11.2
Gap ε /%	1.3	0.89	29.9
Time /s	524	10^4	10^5

solution quality, such that different designs are likely to appear for different solution methods (Voll et al., 2015). The plurality of near-optimal solutions is discussed in detail by Hennen et al. (2017).

Here, the general set-ups of both designs by RiSES^4 differ only in some units. The major design changes take place for the cooling units and the boiler capacity. For the cooling units, the compression chillers are enlarged from the first RiSES^4 design ($\mathcal{P}$ = 13) compared to the best RiSES^4 design ($\mathcal{P}$ = 246) and the absorption chillers are scaled down. At the same time, 1 boiler is removed from the design. Further, a small storage unit is added to the design. In RiSES^4, rigorous synthesis of seasonal storage systems is possible, Chapter 4.2.1. Storage systems are built in almost every design candidate but not in the first design candidate satisfying the solution quality (RiSES^4 $\mathcal{P}$ = 13). However, the solution quality is already 1.3 % for the first design candidate (RiSES^4 $\mathcal{P}$ = 13), showing that small storage systems or even designs without any storage systems are sufficient for excellent quality.

In the benchmark problem with CPLEX, only a solution quality of 29.9 % is achieved within the time limit of 10^5 seconds. Compared to the RiSES^4 design candidates, no CHP engines are build and cooling is only supplied by electric-driven compression chillers. Additionally, large storage systems for cooling and heating are added, which in summary is more expensive.

RiSES^4 shows very good convergence and yields designs with excellent quality for all instances of case study ***ES***. RiSES^4 outperforms the benchmark with CPLEX in all instances in terms of solution time on average at least by a factor of 65. Additionally, RiSES^4 always provides solutions satisfying the desired optimality gap $\varepsilon_{RiSES} = 2\,\%$ before the benchmark even found a first feasible solution. To further validate RiSES^4, we apply the method to another real-world synthesis problem.

4.3.2 Pump system (PS)

In the case study ***PS***, the real-world synthesis problem is a pump system providing a chemical site with cooling water presented in Bahl et al. (2018a), Fig. 4.12. The pump system supplies several customers via a cooling network. The required total volume flow $\dot{V}_t^{total}$ of cooling water has to be supplied at a minimum pressure gradient of $p_{\min}$ 6 bar, Fig. 4.13. The pressure loss of the piping network and the cooling towers is included in the pressure difference $p_{\min}$. Thus, the system boundary is set around the pump system itself, Fig. 4.12. The synthesis problem minimizes the total annualized cost by identifying the appropriate design, i. e., the type and sizing of each pump. In this case study, the key decision is the technology choice between variable-speed pumps with a frequency converter and fixed-speed pumps without the efficiency loss of the frequency converter. However, fixed-speed pumps require throttling for non-nominal operation. For each pump type, different sizes are available described by the nominal volume flow $\dot{V}_N$. A maximum number of six pumps is considered in the superstructure.

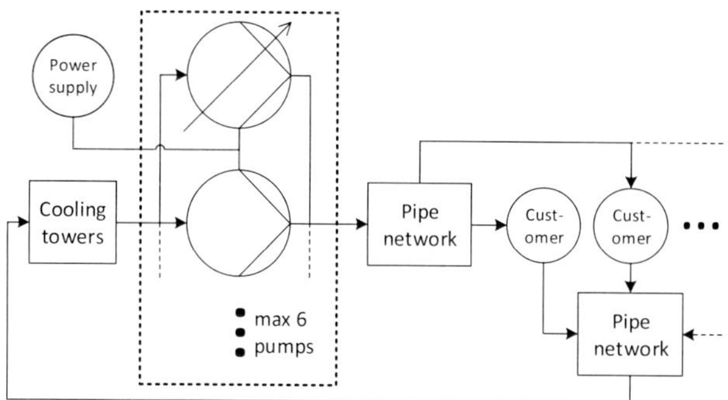

Figure 4.12: Superstructure of case study ***PS***, (dotted rectangle), which is connected to several customers and cooling towers via a pipe network. All pumps are connected to the external electricity grid (Bahl et al., 2018a).

The synthesis problem of case study (PS) differs significantly from the first case study (ES). The case study (PS) represents a design problem without storage systems, which shows the broad applicability of RiSES[4]. The complete MILP formulation (energy balances, performance curves, etc.) is stated in Bahl et al. (2018a). For completeness of this thesis, we restate the model formulation in Appendix B. The original instance consists of 2 years with daily demand data for the volume flow rate, Fig. 4.13.

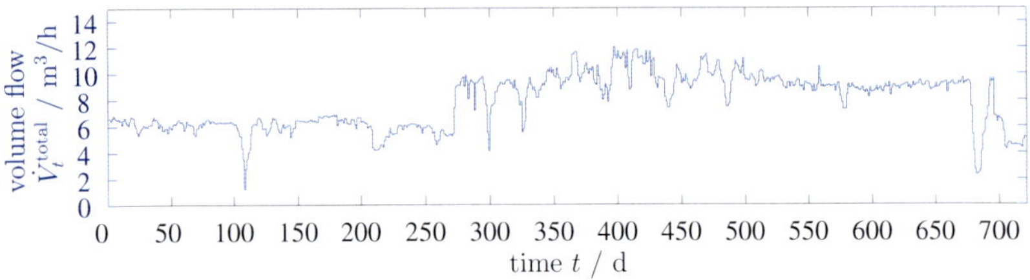

Figure 4.13: Time series of the volume flow rate $\dot{V}_t^{total}$ for case study ***PS***. The time series consists of 2 years with a time-step length of 1 day.

The original synthesis problem with full time series after presolve contains $6 \cdot 10^5$ equations and $4.6 \cdot 10^5$ variables ($2 \cdot 10^5$ binaries) with $24 \cdot 10^5$ nonzero elements. All calculations are performed with the same software and hardware settings as in case-study ***ES***, Chapter 4.3.1. As for case study ***ES***, to validate the computational results, we repeat the calculation of the original time series for 5 generated instances by statistical noise using Latin hypercube sampling with a variation of the time series of $\pm 5\,\%$ (Mckay et al., 2000). Chronology of time steps is only needed if time-coupling constraints are considered in the synthesis problem, e. g., due to storage systems or ramping constraints. In the case study ***PS***, no storage systems or ramping constraints are included; thus, chronology of time steps is not considered as in Bahl et al. (2018a). Since no segmentation is necessary for time-series aggregation, we increase the number of considered time steps in each iteration by 5.

The proposed rigorous synthesis method RiSES[4] satisfies the required optimality gap $\varepsilon_{RiSES} = 2\,\%$ in under 2 hours (5455 s), Fig. 4.15. The required optimality gap $\varepsilon_{RiSES} = 2\,\%$ can be satisfied with only 4 aggregated time steps in the master branch resulting in a synthesis problem with only 2933 equations, 1905 variables (968 binaries) and 9908 nonzero elements, reducing the problem size by about 2 orders of magnitude compared to the original synthesis problem.
In contrast to case study ***ES***, the **R&A** branch provides the optimality proof in case

study ***PS***. The optimality proof requires 44 aggregated time steps in the **R&A** branch, resulting in a synthesis problem with 32,573 equations, 21,406 variables (10,421 binaries) and 110,556 nonzero elements. Still, the complexity of the aggregated synthesis problem is reduced by about 1 order of magnitude compared to the original synthesis problem.

For illustration, the method is not stopped as the optimality gap ε is reached but continued for 3 hours. The optimality gap ε still decreases remarkable with increasing solution time, in contrast to case study ***ES*** (cf. Fig. 4.10 and Fig. 4.15). The optimality proof is always provided by lower bound $TAC^{\mathrm{LB}_{R\&A}}$, which shows good convergence here.

As an example for the time-series aggregation, we show the aggregated time series with 4 aggregated time steps in Fig. 4.14 and the highest time resolution used in the master branch with 39 aggregated time steps in Appendix E.2, Fig. E.2. The aggregated volume flow differs significantly from the original time series. The aggregated time series consists only of 3 peak values and the mean value of the volume flow. Still, this aggregation is sufficient for a design with excellent solution quality.

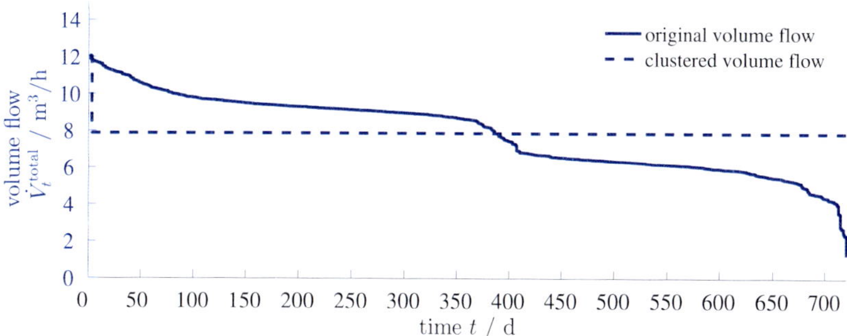

Figure 4.14: Sorted time series of the volume flow rate $\dot{V}_t^{total}$ for case study ***PS***. The time series consists of 2 years with a time step length of 1 day. Additionally, the used aggregated time series with 4 time steps (P=4) for the first design of RiSES[4] satisfying the optimality gap ε is shown.

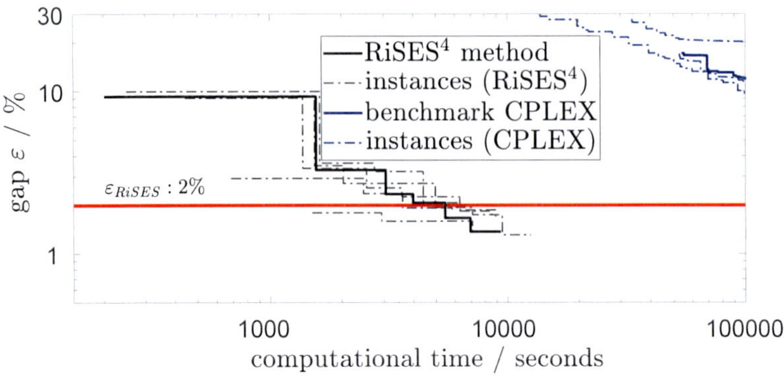

Figure 4.15: Gap ε of the proposed rigorous synthesis method RiSES^4 and the benchmark CPLEX as function of the solution time for case study ***PS***. The required optimality gap ε of 2 % is marked in thick red.

As a benchmark, we solve the original synthesis problem directly with CPLEX 12.6.3.0. The benchmark provides the first feasible solution after 15 hours (54,497 s) and reaches the time limit of 10^5 seconds with a remaining relative gap $\varepsilon = 11.9\,\%$. The performance is very similar for all instances of our method, Fig. 4.15. RiSES^4 outperforms the benchmark in all instances, always satisfying the desired gap within less than 2 hours. In contrast, the benchmark using CPLEX does not find any feasible solution in the same time for any instance.

Fig. 4.16 shows the lower TAC^{LB} and upper bounds TAC^{UB} of the original instance of the master and the parallel branches **R&A** and **B&C**. Additionally, the optimality gap ε of RiSES^4 is plotted as function of the solution time. As seen for case study ***ES***, the **R&A** branch based on simultaneous under- and over-estimation provides first lower bounds within a few seconds. However, in contrast to case study ***ES***, the lower bound $TAC^{\text{LB}_{R\&A}}$ shows good convergence towards the optimal solution. The **B&C** branch provides the first lower bound after 355 seconds. The lower bound $TAC^{\text{LB}_{B\&C}}$ also improves with increasing solution time, but is always inferior to lower bound $TAC^{\text{LB}_{R\&A}}$. In contrast to case study ***ES***, the optimality of the upper bound could not have been proven within the time limit with the lower bound $TAC^{\text{LB}_{B\&C}}$ from the linear-programming relaxation. Thus, depending on the problem studied, the linear programming relaxation or the relaxation of input parameters serve as a tighter relaxation of the original synthesis problem. The master branch provides the first

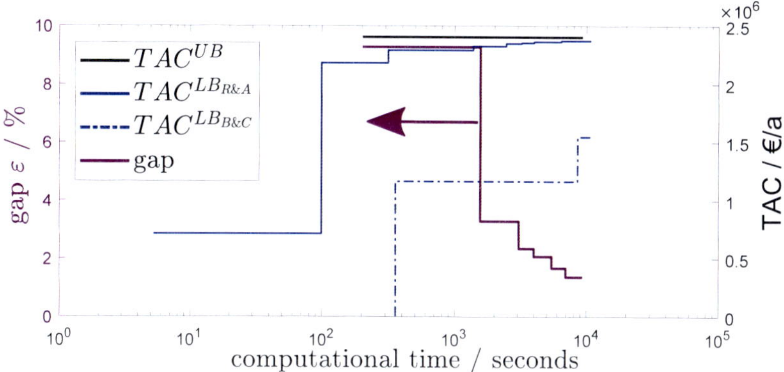

Figure 4.16: Total annualized costs TAC of the lower and upper bounds of RiSES[4] as function of the solution time of case study ***PS*** (right axis). Additionally, a zoom of the optimality gap ε of RiSES[4] is plotted (left axis).

upper bound after 250 seconds. The upper bound is not improved with increasing time resolution. Thus, the aggregated time series of 1 time step together with peak demands already results in an excellent design candidate as proven by $TAC^{\mathrm{LB}_{B\&C}}$.
For case study ***PS***, as for case study ***ES***, very few time steps are necessary to find a feasible design candidate satisfying the required optimality gap $\varepsilon_{RiSES} = 2\,\%$. The design specifies the type of pump (variable-speed or fixed-speed) and the nominal volume flow rate $\dot{V}_N$. In Table 4.3, we compare the design candidate of RiSES[4] satisfying the required optimality gap $\varepsilon_{RiSES} = 2\,\%$ with 4 aggregated time step, the last design candidate within the time limit of 10^4 seconds, and the best design found from the benchmark with CPLEX within the time limit of 10^5 seconds.

The benchmark with CPLEX provides a design candidate with only 1 fixed-speed pump and 5 flexible variable-speed pumps. This is a feasible design; however, 10 % more expensive than the design candidates provided by RiSES[4] with only 1 variable-speed pump each. Variable-speed pumps improve flexibility, but their efficiency is lower than for the fixed-speed pumps due to their frequency converters. The flexibility by 1 variable-speed pump seems to be sufficient for case study ***PS***, as shown by the cheaper design obtained by RiSES[4].

Table 4.3: Optimization results for case study $\boldsymbol{PS}$, by RiSES^4 with 4 aggregated time steps ($\mathcal{P} = 4$), the best found solution within the time limit with 39 aggregated time steps ($\mathcal{P} = 39$), and the benchmark with CPLEX. Nominal capacity of each installed pump, costs TAC, gap ε, and computational time of the solution.

pumps/$10^3 \frac{m^3}{h}$	**RiSES^4** $\mathcal{P} = 4$	**RiSES^4** $\mathcal{P} = 39$	**CPLEX**
Fixed-speed	3/3/2/–/–/–	3/3/2/2/–/–	3/–/–/–/–/–
Variable-speed	3/–/–/–/–/–	2/–/–/–/–/–	3/3/2/2/2/–
$TAC/10^6$ €	2.40	2.40	2.67
Gap ε/%	1.67	1.37	11.9
Time/s	5455	10^4	10^5

RiSES^4 shows very good convergence and yields designs with excellent quality for all instances of case study $\boldsymbol{PS}$, as in case study $\boldsymbol{ES}$. RiSES^4 outperforms the benchmark with CPLEX in all instances in terms of solution time and solution quality. Moreover, RiSES^4 always provides solutions satisfying the desired optimality gap ε_{RiSES} before the benchmark even found a feasible solution. In contrast to case study $\boldsymbol{ES}$, the lower bound $TAC^{\mathrm{LB}_{R\&A}}$ by time-series aggregation and relaxation provides the optimality proof. Before solving a specific problem, it is unknown whether the **B&C** branch or the **R&A** branch will provide tighter lower bounds. In general, the **B&C** branch will provide good lower bounds if the impact of binary decision variables is small. Thus, in case study $\boldsymbol{PS}$, the impact of binary decision variables is more important than for case study $\boldsymbol{ES}$. This finding shows the importance of the 2 competitive lower bound methods of RiSES^4.

4.4 Summary and conclusion

The synthesis of low-carbon energy systems typically includes multiple large time series. Consequently, the synthesis results in large-scale MILP optimization problems, which are computationally challenging and often not solvable within reasonable computational time or memory limits (Challenge 2, Chapter 2.4). Time-coupling constraints, e. g., due to storage systems, further increase the complexity of the synthesis problem.

We propose the rigorous synthesis method RiSES^4 to solve Challenge 2. RiSES^4 provides feasible solutions via time-series aggregation to typical periods to handle time-coupling constraints and variables. RiSES^4 includes a general under- and overestimation of input parameters to rigorously solve synthesis problems, including all time-dependent input parameters, such as energy demands, volatile prices, and renewable resources due to low-carbon technologies. To improve convergence of the optimality proof, RiSES^4 employs parallel computing and 2 competitive lower bounds, via time-series aggregation and linear-programming relaxation. The method is generally applicable to two-stage synthesis problems, including storage systems and low-carbon technologies, in which only the second-stage operational problem depends on large time series.

RiSES^4 is applied to 2 real-world industrial synthesis problems, including low-carbon technologies. The results are further validated in computational studies. The results show that few typical periods are sufficient to provide feasible solutions with excellent solution quality. Thereby, the problem size is reduced by up to 3 orders of magnitudes and the computation time is reduced by a factor of about 65. As shown by the 2 case studies, both types of lower bounds are required to evaluate the solution quality of the upper bounds. RiSES^4 provides fast convergence, outperforming a state-of-the-art solver. Thereby, RiSES^4 renders the rigorous synthesis of low-carbon energy systems applicable in reasonable computational time and memory limits.

In this chapter, the model scope is the rigorous synthesis optimizations of energy systems. By the method RiSES^4, we successfully extended the model scope to handle time-coupling constraints in the rigorous synthesis.

In the next chapter, we further extend the model scope to investigate multi-objective synthesis optimization problems.

Chapter 5

Multi-objective synthesis of energy systems for climate-friendly demand-side management

In this chapter, we extend the scope of multi-objective synthesis problems to consider the time-dependent nature of greenhouse gas (GHG) emissions from electricity generation. Thus, we solve Challenge 3, cf. Chapter 2.4. We economically and ecologically optimize designs of industrial energy systems. This model scope allows only for a medium model detail, Fig. 2.1. Thus, in contrast to Chapter 3, we assume constant part-load behavior and peak-power prices and, compared to Chapter 4, we employ time-series aggregation without optimality and feasibility proof.

To extend the scope, we investigate the impact of time-dependent grid emissions on the potential of low-carbon energy systems to reduce GHG emissions. For a holistic assessment of the potential of climate-friendly demand-side management, we consider a multi-objective synthesis problem. We investigate both time-dependent grid mix emissions and time-dependent marginal grid emissions to resolve the actual consequences of demand-side management. In Chapter 5.1, we present the multi-objective synthesis problem. To consider the impact of time-dependent grid emissions, we introduce methods to calculate time-dependent grid emissions, Chapter 5.2. In a real-world industrial case study, we employ time-dependent grid emissions in multi-objective optimization, obtaining design trade-offs between economic and ecological objectives, Chapter 5.3. In Chapter 5.4, we summarize and conclude this chapter.

Major parts of this chapter are reproduced by permission of Elsevier from:

Baumgärtner, N., Delorme, R., Hennen, M., and Bardow, A. (2019). Design of low-carbon utility systems: Exploiting time-dependent grid emissions for climate-friendly demand-side management. *Applied Energy*, 247, 755-765.

Contribution report: Writing the draft, principal author, development and implementation of the model, conceptual development of the method for time-dependent grid emissions, calculation and evaluation of results.

5.1 Design model for low-carbon industrial energy systems

In this section, we extend the scope of the single-objective synthesis problem stated in Chapter 4.1 to a multi-objective synthesis. For this purpose, we include a second ecological objective for optimization. Ecological friendliness can be evaluated based on life-cycle assessment, which employs many metrics such as toxicity and eutrophication (ILCD, 2010). However, in the field of energy systems, most commonly, only 1 ecological objective is taken into account in addition to an economic criterion (Alarcon-Rodriguez et al., 2010). Due to the current emphasis on climate change (IPCC, 2018), the global warming impact (GWI) is most often used as ecological performance metric. The GWI aggregates CO_2 and other greenhouse gas emissions according to their global warming potential (Alarcon-Rodriguez et al., 2010). Thus, as ecological objective, we choose to employ the global warming impact GWI as well,

$$\min_{\dot{V}_{n,t},\delta_{n,t},\dot{V}_n^{\mathrm{N}},\gamma_n,\dot{V}_{\mathrm{grid}}^{\max},\dot{V}_{\mathrm{grid},t},x,y} GWI. \tag{5.1}$$

We only consider the global warming impact GWI due to operation, since the impact of operation typically exceeds the impact of construction by multiple orders for fossil-fueled units (Guillén-Gosálbez, 2011). Obviously, for renewable-based units, the ratio between impacts for operation and construction is smaller. However, we focus on climate-friendly demand-side management (DSM). As DSM are changes in operation, focusing on the global warming impact GWI due to operation is reasonable.
The global warming impact GWI, Eq. (5.2), is defined as the sum of the input power $\dot{U}_{n,t}$ of every unit $n \in \mathcal{C}$ in every time step $t \in \mathcal{T}$ multiplied by the specific operational emission factor $\mathrm{e}^{\mathrm{o}}_{n,t}$ and the duration $\Delta \mathrm{t}_t$ of a time step,

$$GWI = \sum_{t \in \mathcal{T}} \left(\Delta \mathrm{t}_t \sum_{n \in \mathcal{C}} \mathrm{e}^{\mathrm{o}}_{n,t} \dot{U}_{n,t} \right). \tag{5.2}$$

The resulting bi-objective optimization problem is solved using the augmented ε-constraint method (Mavrotas, 2009), resulting in Pareto-optimal trade-off curves for total annualized costs TAC versus global warming impact GWI. The optimization problem is subject to the same constraints as in Chapter 4.1. The detailed MILP model of the low-carbon industrial energy system, including part-load behavior, minimal loads, and linearizations, is presented in Appendix A.
In the next section, we present methods to calculate the GHG emission factor $\mathrm{e}^{\mathrm{o}}_{n,t}$ of grid electricity.

5.2 Calculation of time-dependent GHG emission factors

When assessing emissions due to grid electricity consumption, average annual emissions are usually used. However, employing average annual emissions limits the validity of environmental conclusions drawn from these assessments. Actions resulting from such conclusions are often misleading and lead in general to suboptimal decisions regarding real emission reductions (Kono et al., 2017). Real grid emissions are time-dependent due to a changing generation mix, and thus should be described by a time-dependent emission factor. Time-dependent emission factors gained more interest recently as they enable emission reductions by optimized electricity consumption, e. g., climate-friendly demand-side management (Kopsakangas-Savolainen et al., 2017) and emission-minimizing production schedules (Kelley et al., 2019).

Despite this importance, no general accepted time-dependent grid emission factors are publicly available. In Chapter 5.2.1, we present a method to compute the time-dependent grid *mix* emission factor $\mathrm{XEF}_t^{\mathrm{el}}$ based on the current technology mix. The grid mix emission factor $\mathrm{XEF}_t^{\mathrm{el}}$ allows to assess the current emissions due to electricity exchange with the grid.
However, changes in electricity consumption also change the electricity grid mix, since not all electricity-generating units respond to a change in demand proportionally. Hence, emission reductions through changes in electricity consumption should be assessed by a time-dependent *marginal* emission factor, resulting from only the responding electricity-generating units (Voorspools and D'haeseleer, 2000). Ekvall and Weidema (2004) therefore, suggest to identify a marginal generator to evaluate emission reductions through changes in electricity consumption. Thus, in Chapter 5.2.2, we present a method to identify the marginal generator to compute the *marginal* emission factor $\mathrm{MEF}_t^{\mathrm{el}}$. This marginal emission factor $\mathrm{MEF}_t^{\mathrm{el}}$ allows to assess the actual emissions due to changes in electricity consumption.

It depends on the problem at hand whether to employ grid mix emissions $\mathrm{XEF}_t^{\mathrm{el}}$ or marginal emissions $\mathrm{MEF}_t^{\mathrm{el}}$ for the ecological assessment (Ryan et al., 2016): Life-cycle assessment distinguishes between attributional and consequential studies (Finnveden et al., 2009). Grid mix emissions $\mathrm{XEF}_t^{\mathrm{el}}$ should be used in attributional studies, e.g., for accounting (Plevin et al., 2014). In contrast, consequential studies focus on emissions changes due to system effects, and thus should use marginal emissions $\mathrm{MEF}_t^{\mathrm{el}}$ (Regett et al., 2018).

As a rule of thumb, the grid mix emission factor $\mathrm{XEF}_t^{\mathrm{el}}$ may still be used for existing systems and small changes in electricity consumption. For new systems or larger changes in electricity consumption, the marginal emission factor $\mathrm{MEF}_t^{\mathrm{el}}$ should be used, following the idea of 'consequential modeling for short-term marginal' effects in life-cycle assessment (ILCD, 2010).

In this thesis, we evaluate the impact of time-dependent grid emissions in general. Thus, we employ both emission factors for the same case study. However, for a specific case, the type of emission factor should be carefully chosen following the ILCD (2010) handbook ruels. As general accepted time-dependent grid emission factors are not publicly available, in the following, we present our methods to calculate time-dependent emission factors for electricity and apply them to a case study considering the German electricity mix in 2016.

5.2.1 Time-dependent GHG grid mix emissions based on the current technology mix

In this section, we present a method to compute time-dependent grid mix emission factors $\mathrm{XEF}_t^{\mathrm{el}}$ based on the current technology mix using the following equations,

$$\mathrm{XEF}_t^{\mathrm{el}} = \frac{\text{emissions due to electricity production}(t)}{\text{electricity consumption}(t)}. \tag{5.3}$$

The system boundaries typically correspond to a country. We assume that all electricity flows within the system boundaries are homogenous. Thus, exported and consumed electricity, as well as electricity lost due to transmission losses, are assessed by the same time-dependent grid mix emission factor $\mathrm{XEF}_t^{\mathrm{el}}$. We reformulate Eq. (5.3) to:

$$\mathrm{XEF}_t^{\mathrm{el}} = \frac{\sum_k \mathrm{EF}_k^{\mathrm{el}} P_{k,t}^{\mathrm{net}} + \sum_l \mathrm{EF}_l^{\mathrm{el}} P_{l,t}^{\mathrm{Imp}}}{\sum_k P_{k,t}^{\mathrm{net}} + \sum_l P_{l,t}^{\mathrm{Imp}} - P_t^{\mathrm{losses}}}. \tag{5.4}$$

Here, $P_{k,t}^{\mathrm{net}}$ is the net energy production per aggregated energy carrier k. Aggregated energy carriers k, e.g., lignite, are used instead of individual power plant production due to data availability. $\mathrm{EF}_k^{\mathrm{el}}$ are technology-specific emission factors. $P_{l,t}^{\mathrm{Imp}}$ is the imported electricity per country l, and P_t^{losses} are the transmission and distribution losses for each time step t.

To further simplify Eq. (5.4), we follow Kono et al. (2017) and apply a constant transmission efficiency η_T to consider all transmission and distribution losses.

$$\eta_T = \frac{\sum_k P_k^{\text{net}} + \sum_l P_{l,t}^{\text{Imp}} - P_t^{\text{losses}}}{\sum_k P_k^{\text{net}} + \sum_l P_{l,t}^{\text{Imp}}}. \tag{5.5}$$

With Eq. (5.5), Eq. (5.4) reduces to:

$$\text{XEF}_t^{\text{el}} = \frac{\sum_k \text{EF}_k^{\text{el}} P_{k,t}^{\text{net}} + \sum_l \text{EF}_l^{\text{el}} P_{l,t}^{\text{Imp}}}{\eta_T \left[\sum_k P_{k,t}^{\text{net}} + \sum_l P_{l,t}^{\text{Imp}} \right]}. \tag{5.6}$$

We employ the last published German value for a constant transmission efficiency $\eta_T = 96.1\,\%$ from 2014 (The World Bank, 2014). The value for the transmission efficiency η_T was also about 96.1 % for the years 2012 and 2013. Thus, we assume that this value still holds for the considered year 2016.
To illustrate the calculation of grid mix emission factors XEF_t^{el}, we use data for the German net electricity feed-in with a resolution of 15 minutes and as hourly resolved data for the German import and export from ENTSO-E (2017). We compute the time-dependent grid mix emission factors with hourly resolution. The published electricity feed-in data is aggregated for various energy carriers, see Table 5.2. For those aggregated energy carriers, we employ constant cradle-to-grave emission factors EF_k^{el} from eco-invent 3.3 (Wernet et al., 2016). The exact assignment is shown in the Appendix F.1. For electricity import, we employ average emission factors from IPCC (2013) for each neighboring country, see Table 5.1. Data for the electricity import $P_{l,t}^{\text{Imp}}$ is taken from the ENTSO-E (2017). Since some data points of the year 2016 are not available, we replaced the gaps by interpolation or substitution. The exact procedure is shown in Appendix F.2.

Table 5.1: Specific GWP_{100a} emission factors EF_l^{el} for the electricity import by countries l (IPCC, 2013)

import from	EF_1^{el} g_{CO_2eq}/kWh_{el}
Sweden	40.4
Switzerland	89.7
France	103.5
Denmark	345.9
Austria	362.4
Netherlands	542.2
Luxembourg[1]	590.4
Czech Republic	789.1
Poland	1009.0

[1] Listed for completeness. No import and export data for the year 2016.

Employing the electricity data $P_{k,t}^{\text{net}}$ and $P_{l,t}^{\text{Imp}}$ from ENTSO-E (2017) and the emission data EF_k^{el} and EF_l^{el} from eco-invent 3.3 and Table 5.1 in Eq. (5.6), we compute the hourly-resolved grid mix emission factor XEF_t^{el} for the year 2016, Fig. 5.1. The grid mix emission factor XEF_t^{el}, Fig. 5.1, is volatile and shows daily, as well as some weekly and seasonal patterns.

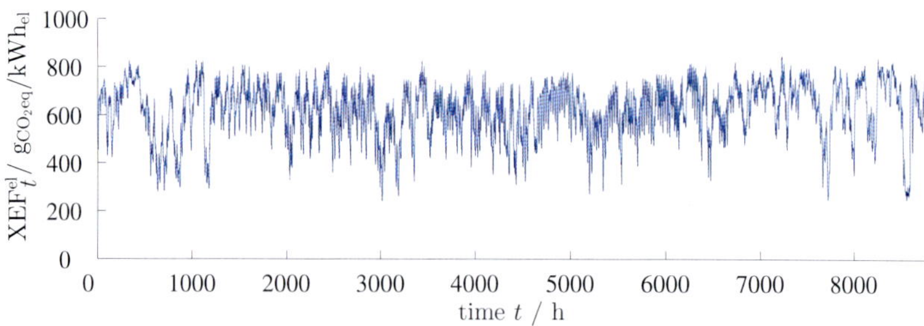

Figure 5.1: Grid mix emission factor XEF_t^{el}, Eq. (5.6), based on the employed technology mix in Germany for the year 2016.

The time-independent *average* emission factor AEF^{el} is the mean value of the grid mix emission factor XEF_t^{el} for the entire year:

$$\text{AEF}^{\text{el}} = \frac{\sum_{t=1}^{t=T} \text{XEF}^{\text{el}}}{T} = 620\ g_{CO_{2}eq}/kWh_{el}. \tag{5.7}$$

Our calculated value $\text{AEF}^{\text{el}} = 620\ g_{CO_{2}eq}/kWh_{el}$ is about 6 % higher than the reported value of 581 $g_{CO_{2}eq}/kWh_{el}$ from the German Environment Agency for the year 2016 (Icha and Kuhs, 2018). Probably, the reason for this difference lies in higher employed average emission factors EF_k^{el} for the aggregated energy carriers k from eco-invent 3.3. In the course of the energy transition, the average emission factor AEF^{el} will decrease, whereas the volatile character of the grid mix emission factor XEF_t^{el} will increase. The volatile nature of the emission factor provides an opportunity for ecological optimization by climate-friendly demand-side management.

However, to determine the impact of changes in electricity consumption a marginal emission factor MEF_t^{el} is needed, which is discussed in the next section.

5.2.2 Time-dependent marginal GHG emissions

Not all power plants react proportionally to changes in electricity consumption (Psomopoulos et al., 2010). Thus, changes in emissions due to changes in consumption are not accurately captured by the grid mix emission factor XEF_t^{el} based on the technology mix (Hawkes, 2010). Hence, to evaluate changes in consumption, it is necessary to identify the power plants that react to such changes, and their emission factor (Ekvall and Weidema, 2004). A change in electricity consumption is likely to induce changes in the mixture of power plants (Lund et al., 2010). The power plants that react to a change in power consumption are called marginal power plants. These reacting power plants determine, in general, the marginal emission factor. In several studies (Voorspools and D'haeseleer, 2000; Marnay et al., 2002) and (Bettle et al., 2006), emissions due to changes in power consumption are represented by constant marginal emission factors. These constant marginal emission factors are computed by regression of annual load duration curves. Thus, this factor is independent of time. However, studies investigating the impact of a change in electricity consumption have increasingly recognized the temporal variation of marginal emission factors (Kopsakangas-Savolainen et al., 2017) and (Roux et al., 2016). The international trade of electricity adds a spatial dimension to the determination of the reacting power plants (Graff Zivin et al.,

2014). Both spatial and temporal effects will increase variations in the marginal emission factors in the future (Olkkonen and Syri, 2016). Thus, it is difficult to identify the power plants reacting to a certain change in electricity consumption in space and time. Ryan et al. (2016) review methods to compute marginal emission factors and distinguish between 'Empirical Data & Relationship Models' and 'Power System Optimization Models.' 'Empirical Data & Relationship Models' correlate historical data of electricity demand, supply, and corresponding emissions. These models are mostly used for grid mix emission factors. In the 'Power System Optimization Model' class, economic dispatch, which is based on a merit order, is particularly promising to compute time-dependent marginal emission factors (Hawkes, 2010) and (Marnay et al., 2002). The approach simplifies the set of reacting power plants to one main marginal technology defined by its marginal costs (Hawkes, 2010) and (Siler-Evans et al., 2012). Long-term effects are neglected, such as the construction of new power plants and restrictions, e. g., due to shutdowns.

In this thesis, a merit-order-based approach is used to compute a marginal emission factor $\mathrm{MEF}_t^{\mathrm{el}}$. To compute the merit order, we sort the power plants by their marginal costs. Renewable generation technologies have zero marginal costs and are therefore not part of the merit order, but renewably generated electricity reduces the residual electricity demand. The power plants with the lowest marginal costs will generate the power to supply the residual electricity demand. Of those residual power plants, the technology with the highest marginal costs is identified as marginal technology. The identified marginal technology determines the marginal emission factor $\mathrm{MEF}_t^{\mathrm{el}}$. Here, the marginal costs $\mathrm{c}_{pp}^{\mathrm{mar}}$ of a power plant pp are defined as:

$$\mathrm{c}_{pp}^{\mathrm{mar}} = \underbrace{\frac{x_k}{\eta_p^{\mathrm{el}} p}}_{fuel\ costs} \underbrace{+ \mu_k^{\mathrm{CO_{2eq}}} \frac{\mathrm{c}^{\mathrm{CO_{2eq}}}}{H_{u,k} \eta_p^{\mathrm{el}} p}}_{GHG\ costs}. \tag{5.8}$$

The marginal costs $\mathrm{c}_{pp}^{\mathrm{mar}}$ consist of two parts, the fuel costs and the costs for GHG emissions. The fuel costs are the specific energy carrier costs x_k divided by the power plant efficiency $\eta_p^{\mathrm{el}} p$. The costs for GHG emissions are the specific GHG emissions per energy carrier $\mu_k^{\mathrm{CO_{2eq}}}$ multiplied by the CO_2-certificate price $\mathrm{c}^{\mathrm{CO_{2eq}}}$ and divided by the power plant efficiency $\eta_p^{\mathrm{el}} p$ and the calorific value $H_{u,k}$. For the CO_2-certificate price, we use the mean value of the third trade period for the year 2016 of $\mathrm{c}^{\mathrm{CO_{2eq}}} = 5.84€/t_{CO_2eq}$ (DEHSt, 2017). The efficiency $\eta_p^{\mathrm{el}} p$ of each power plant pp is calculated by a logarithmic size-dependent regression of existing power plants based on Heuck et al. (2013), Appendix F.3.

Based on the merit order, the marginal technology can be determined for a known electricity price or a known conventional net electricity production. Here, we determine the marginal technology based on the conventional net electricity production. We employ the conventional net electricity production to avoid market effects such as negative prices, which cannot be covered with a simple merit-order-based approach using the electricity price. However, if negative prices occur, the most expensive generator will still balance the system and, thus, be the marginal generator. The most expensive generator is always identified by the merit order based on the conventional net electricity production. Thus, even with negative prices, the approach yields marginal emission factors. If the conventional net electricity production is equal to zero, the marginal emission factor is determined by a balancing renewable generator — and thus zero. We use data for the conventional net electricity production for the year 2016 from ENTSO-E (2017). In the merit order, we distinguish conventional technologies based on their employed energy carriers k: nuclear, lignite, hard coal, gas (combined cycle), gas, and fuel oil. Further, we assume constant properties per energy carrier k, Table 5.2. The specific emissions of the identified marginal power plant define the marginal emissions. As for the marginal costs c_{pp}^{mar}, a size-dependent efficiency $\eta_p^{\text{el}}p$ for each power plant p is used based on Heuck et al. (2013), Appendix F.3, to calculate a power-plant specific emission factors $\text{EF}_p^{\text{el}}p$.

$$\text{EF}_{\text{el},pp} = \overline{\text{EF}_k^{\text{el}}} \frac{\overline{\eta_k^{\text{el}}}}{\eta_p^{\text{el}}p}. \tag{5.9}$$

For the energy carriers k, the average emission factors $\overline{\text{EF}_k^{\text{el}}}$ for operation of the conventional technologies are given in Table 5.2 and the related mean efficiencies $\overline{\eta}_k^{\text{el}}$ from eco-invent 3.3 (Wernet et al., 2016).

Table 5.2: Data for the merit order for the energy carriers k, nuclear, lignite, hard coal, gas (combined cycle), gas and fuel oil. Average emission factors $\overline{\mathrm{EF}_k^{\mathrm{el}}}$ and mean efficiencies $\overline{\eta_{\mathrm{el},k}}$ are from eco-invent 3.3 (Wernet et al., 2016).

energy carriers	specific costs	specific GHG emissions	calorific value	mean efficiency
k	x_k	$\mu_k^{\mathrm{CO_{2eq}}}$	$H_{u,k}$	$\overline{\eta_k^{\mathrm{el}}}$
	€$/MWh$	varying	varying	%
fuel oil[a]	38.63[b]	$2.71\,t_{CO_{2eq}}/m^3$[c]	$42.60\,MJ/m^3$[d]	37.3
$\mathrm{gas}_{\text{combined cycle}}$	26.71[e]	$1.93\,t_{CO_{2eq}}/m_k^3$[f]	$38.27\,MJ/m_k^3$[f]	50
gas	29.81[e]	$1.93\,t_{CO_{2eq}}/m_k^3$[f]	$38.27\,MJ/m_k^3$[f]	34.9[g]
hard coal	10.12[e]	$2.60\,t_{CO_{2eq}}/t_k$[f]	$26.47\,MJ/kg$[f]	37.7
lignite	6.30[e]	$1.39\,t_{CO_{2eq}}/t_k$[f]	$14.99\,MJ/kg$[f]	35.6
nuclear	4.18[e]	$0\,t_{CO_{2eq}}/t_k$	$41667\,MWd/t$[h]	32

[a]Light oil with a density of $\rho = 860\,kg/m^3$ is considered.
[b]Destatis (2019)
[c]DEHSt (2006)
[d]Based on industrial data
[e]Konstantin (2017)
[f]EPA (2016)
[g]Average efficiency based on logarithmic regression is used, Appendix F.3
[h]Just for completeness, as no specific emissions $\mu_{\mathrm{nuclear,CO_{2eq}}}$ are considered.

To determine the merit order, we use all power plants pp listed in the power-plant list of the Umweltbundesamt (2018). In this power plant list, all power plants with a minimum capacity of $100\,MW_{el}$ are listed for Germany in 2016. In Fig. 5.2, we show the merit order based on the installed power from the power plant list of Umweltbundesamt (2018), the marginal costs c_{pp}^{mar}, and the power-plant specific emissions $\mathrm{EF}_p^{\mathrm{el}}$. Because fewer power plants report to ENTSO-E (2017) as are listed in the power plant list of the Umweltbundesamt (2018), we scale the reported residual conventional electricity production from ENTSO-E (2017). With an installed capacity of $76.9\,GW_{el}$ in UBA 2017 and a maximal reported residual conventional electricity production of $59.0\,GW_{el}$, the scaling factor is 76.9/59.0.
The marginal emission factor $\mathrm{MEF}_t^{\mathrm{el}}$ is then calculated employing the merit order and the scaled conventional electricity production for the year 2016 based on data from ENTSO-E (2017), Fig. 5.3.

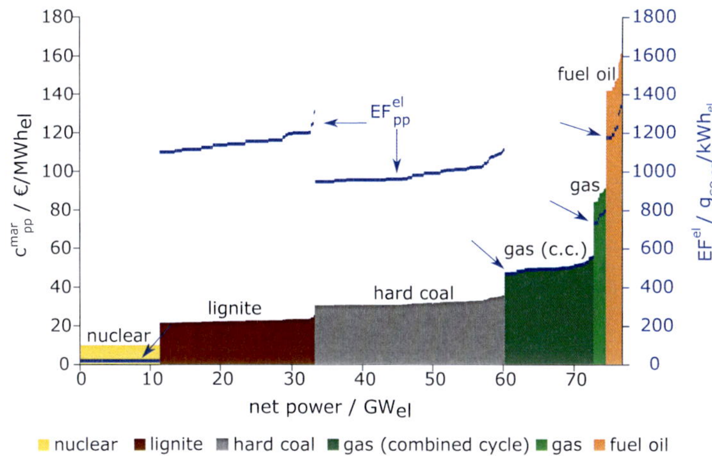

Figure 5.2: Calculated merit order for Germany (2016) using the power plants from Umweltbundesamt (2018) and the corresponding emission factors of the aggregated energy carriers nuclear, lignite, hard coal, gas, and fuel oil.

Figure 5.3: Flow chart to calculate the time-dependent marginal emission factor. The time-dependent residual load is taken from ENTSO-E. Using the residual load in the merit order, the marginal generator is identified. For this marginal generator, the power-plant-specific emission factor is defined by Eq. (5.9). This is the time-dependent marginal emission factor.

Fig 5.4 shows the marginal emission factors $\mathrm{MEF}_t^{\mathrm{el}}$ of 2016, based on the presented merit-order approach. The mean of the marginal emission factor, $1042\, g_{CO_2eq}/kWh_{el}$, exceeds the average emission factor $\mathrm{AEF}^{\mathrm{el}}$, $620\, g_{CO_2eq}/kWh_{el}$, which is often the case for marginal emission factors, because renewable energies drop out of the merit order (Finenko and Cheah, 2016). However, depending on the structure of the electricity system, the marginal emission factor might be lower than the average emission factor (Thind et al., 2017). We see a clear pattern by season: in summer, only lignite and hard coal power plants are the marginal technologies, due to the lower residual electricity demand. Besides this seasonal behavior, the marginal emission factor shows less daily or weekly patterns than the grid mix emission factor $\mathrm{XEF}_t^{\mathrm{el}}$, Fig. 5.1.

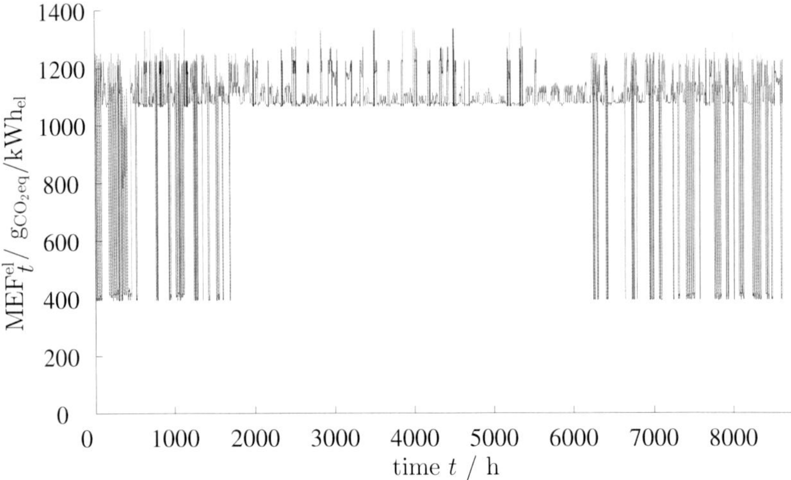

Figure 5.4: Time-dependent marginal emission factor MEF^{el}_t in Germany for the year 2016 based on the merit order, Fig. 5.2. The same color code, as in Fig. 5.2, is used for the aggregated energy carriers: nuclear (yellow), lignite (brown), hard coal (grey), gas (green), and fuel oil (orange). Gas includes both: simple gas turbines and combined-cycle plants, which are the marginal technology with about $400\, g_{CO_{2eq}}/kWh_{el}$.

In the next section, we evaluate the decarbonization potential of low-carbon energy systems and the impacts of the calculated time-dependent emission factors in an industrial case study.

5.3 Case study: Multi-objective design

In this section, we apply the time-dependent emission factors in the multi-objective synthesis problem of a low-carbon industrial energy system presented in Chapter 5.1 and Appendix A. We solve the multi-objective synthesis problem to identify trade-off curves between economic and ecological designs. In Chapter 5.3.1, we identify the trade-offs for the direct emissions without emissions due to electricity usage, the so-called Scope I emissions. Additionally, we investigate the impact of time-dependent grid emissions on the design of low-carbon industrial energy systems for direct and indirect emissions, including the emissions due to electricity usage, the so-called Scope II emissions.

Based on preliminary investigations, we use a superstructure consisting of 3 absorption chillers, 1 boiler, 3 CHP engines, 3 compression chillers, 1 electrode boiler, 2 heat pumps, 1 heat exchanger, 2 inverter stations, 1 organic Rankine cycle, and 1 wind converter. Additionally, we consider roof-top PV, roof-top solar thermal panels, a battery and one storage tank for each hot and cold water, Fig. 5.5. All energy-conversion technologies are modeled with constant part-load efficiency, as in Chapter 4. The model allows continuous sizing of all conversion technologies.

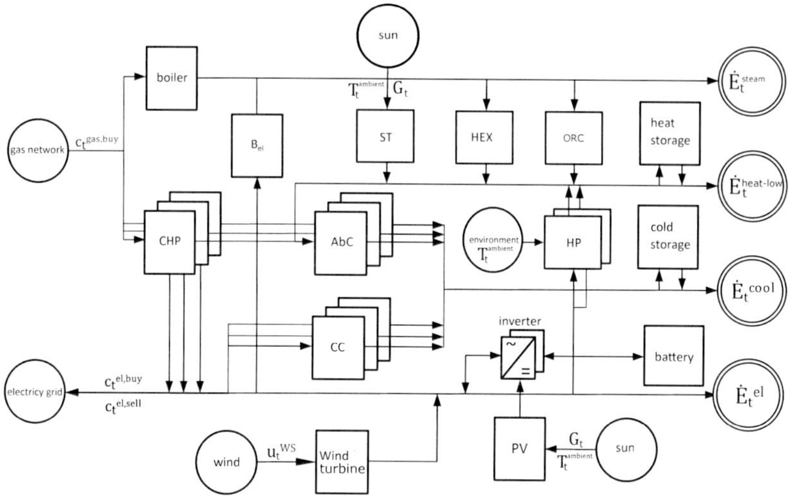

Figure 5.5: Superstructure of the case study: absorption chillers (AbC), battery (BAT), boilers (B), combined-heat-and-power engines (CHP), compression chillers (CC), electrode boilers (B_{el}), heat exchanger (HEX), heat pumps (HP), inverter stations, organic Rankine cycle (ORC), photovoltaic (PV), solar thermal (ST), thermal storage systems and wind turbines (WT). Redundant units of utilities are shown. The model considers time-varying energy demands ($\dot{E}_{cool,t}$, $\dot{E}_{heat-low,t}$, $\dot{E}_{steam,t}$, $\dot{E}_t^{\mathrm{el}}$), electricity prices ($c_t^{\mathrm{el,buy}}$, $c_t^{\mathrm{el,sell}}$), ambient temperature (T_t^{ambient}) and resources of renewable energies (G_t, u_t^{WS}).

Time series with hourly resolution are available for steam, low-temperature heating, cooling, and electricity demand, Fig. 3.5a. In this thesis, the demands of the industrial processes are fixed. As a result, climate-friendly demand-side management is achieved by the flexibility within the low-carbon industrial energy system only. Exploiting potential flexibilities within the industrial processes could further enhance the benefits of climate-friendly demand-side management, as shown for economic optimizations (Leenders et al., 2019). Moreover, we consider solar radiation and wind speed, time-varying electricity purchase and selling prices, and the ambient-temperature as time series, Fig. 3.5b-d. The gas price and resulting GHG emissions are assumed to be constant, $c^{o}_{\text{gas}} = 6\,cent/kWh$ and $e^{o}_{\text{gas}} = 244\,g_{CO_2eq}/kWh$.

The single-objective synthesis problem is already a large-scale MILP problem, which is computationally challenging, cf. Chapter 4. We extend the scope of the single-objective synthesis to a multi-objective synthesis. Hence, to render the computations tractable, we decrease the model detail, Fig. 2.1. Often, time-series aggregation is applied to reduce the model detail of synthesis problems, e. g., for the synthesis of chemical processes (Maravelias and Sung, 2009) and energy systems (Mancarella, 2014). Thus, to reduce model complexity, we employ time-series aggregation based on Bahl et al. (2018b): First, clustering is used to identify typical days of the original time series. Second, within these typical days, consecutive hours are further aggregated to segments, for details see Chapter 4.2.1. For all calculations, we use 5 typical days consisting of 12 segments each. As shown in the previous chapter and in previous work (Bahl et al., 2018a) and (Baumgärtner et al., 2018, 2019a,c), only few time steps are necessary to obtain solutions with excellent solution quality. Thus, the error introduced due to the lower model detail by time-series aggregation is assumed to be small.

To solve the multi-objective optimization problem, we employ the augmented ε-constraint method (Mavrotas, 2009). For the bi-objective optimization problem, the augmented ε-constraint method first computes both single objective optimization problems separately. The objective values of these single objective optimization problems are set as constraint and the 2 optimization problems are reoptimized regarding the other objective function to guarantee Pareto optimality. Between these 2 solutions, the objective space is split in equidistant sections by constraining 1 of the objective function values. For each section, a single objective optimization problem is solved regarding 1 objective with a constraint for the other objective function. To improve computational speed and for better-distributed Pareto curves, more advanced methods could be used. For example, Asrari et al. (2016) propose a combination of the fuzzy Pareto dominance with the shuffled frog-leaping algorithm to reduce the computational time significantly, and Chiu et al. (2015) propose a genetic, multi-objective artificial-immune-system algorithm to generate well-distributed Pareto curves. However, this work aims to show the potential of climate-friendly demand-side management. Thus, the augmented ε-constraint method is sufficient to compute the complete Pareto curve. The energy system model is modeled in GAMS (GAMS Development Corporation, 2019) and the synthesis problem is solved with CPLEX 12.6.3.0 (IBM, 2016) to an optimality gap of 2 %.

5.3.1 Direct emissions

In this section, we identify the trade-off curve between economical and ecological designs of the industrial energy system for the direct Scope I emissions, Fig. 5.6. The direct emissions do not include the grid emissions. Thus, the emission factor of grid electricity is $e^{o}_{n,t} = 0\, g_{CO_{2eq}}/kWh_{el}$ and does not influence the trade-off curve.

The most economical design costs $7.9\, M€/a$. The operation of this design induces direct emissions of $29.4\, kt_{CO_{2eq}}/a$. The most ecological design is 57 % more expensive with $12.6\, M€/a$. By employing this design, the direct emissions can be reduced to $0\, kt_{CO_{2eq}}/a$, by full electrification. The trend of the trade-off curve is slightly curved without any kinks. The costs for avoided CO_{2eq} rise from $87\, €/t_{CO_{2eq}}$ to $407\, €/t_{CO_{2eq}}$, with an average of $150\, €/t_{CO_{2eq}}$ for full decarbonization.

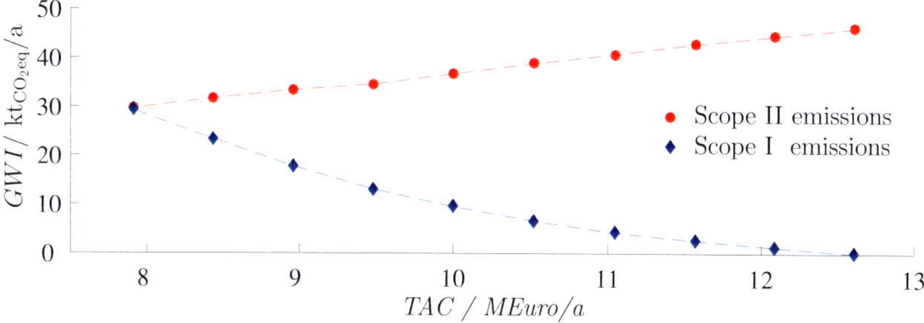

Figure 5.6: Trade-off curve between economical and ecological designs in terms of the global warming impact *GWI* and the total annualized costs *TAC*, assessed for Scope I emissions. The trade-off curve consists of 10 Pareto-optimal solutions. Additionally, the corresponding Scope II emissions are shown. Lines are included to guide reader's eyes.

For every solution on the trade-off curve, Fig. 5.7 shows the mix of technologies, which supplies the energy demands cooling, heating and steam. Additionally, the purchased electricity from the grid is shown. In the most economical design, cooling is supplied by a mixture of absorption and compression chillers. CHP engines supply the absorption chillers with heat and cover the low-temperature heat demand. The produced power satisfies the electricity demand and partly supplies the compression chillers. Only little additional electricity has to be bought from the grid. A conventional boiler supplies the steam demand. Grid electricity does not contribute to the Scope I emissions. Thus, from the most economical design (left) to the most ecological design (right), more and more units are electrified using electricity from the grid. The first measure to reduce Scope I emissions is to replace absorption chillers with compression chillers successively and reduce emissions by the CHP engines, which supply the absorption chillers with heat. After the cooling supply is electrified, the capacity of CHP engines is further reduced and successively replaced by heat pumps to supply the low-temperature heat demand. The last measure to reduce the Scope I emissions is to replace the conventional boilers with electrode boilers. In the most ecological design, all demands are fully supplied by electric units using grid electricity. No renewable-based energy converters, such as wind turbines, roof-top PV, or solar thermal panels, are built in any design minimizing the Scope I emissions. Considering only Scope I emissions renders using electricity from the grid the economical and ecological superior solution.

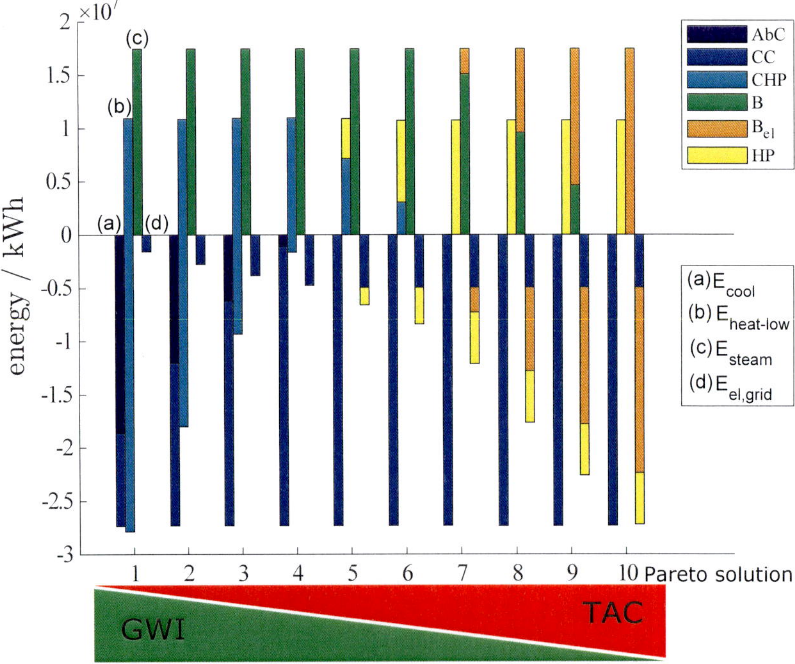

Figure 5.7: Supply structure for every Pareto-optimal solution on the trade-off curve for the demands (a) cooling, (b) low-temperature heating, and (c) steam for the built energy-conversion units (absorption chiller (AbC), compression chiller (CC), CHP engine (CHP), boiler (B), electrode boiler (B_{el}), heat pumps (HP)) . Additionally, (d) the purchased electricity from the grid is shown.

The full electrification of an industrial energy system is possible, but more expensive than conventional energy supply using, e. g., gas. Thus, the direct Scope I emissions could be reduced to zero, but with average cost of $150\,€/t_{CO_{2eq}}$. At the same time, as more electricity is bought from the grid, Scope II emissions rise significantly, Fig. 5.6. In the next section, we therefore investigate the ecological impact of low-carbon industrial energy systems, including emissions for grid electricity.

5.3.2 Indirect emissions: time-dependent GHG grid mix emissions $\mathrm{XEF}_t^{\mathrm{el}}$

In this section, we include Scope II emissions in the trade-off curve between economical and ecological designs. For this purpose, we consider time-dependent grid mix emissions $\mathrm{XEF}_t^{\mathrm{el}}$ based on the current technology mix, Eq. (5.6) and Fig. 5.1.
The emissions from the optimal designs considering the actual grid mix emissions $\mathrm{XEF}_t^{\mathrm{el}}$ provide the '*minimal emissions*,' Fig. 5.8. The emissions range for the most economical design from $29.2\,kt_{CO_{2eq}}/a$ to $17.6\,kt_{CO_{2eq}}/a$ for the most ecological design. The costs increase from $7.9\,M€/a$ for the economical design by 36 % to $10.7\,M€/a$ for the most ecological design. Noteworthy, the emissions can no longer be reduced to zero with the current superstructure. The cost for avoided $\mathrm{CO_{2eq}}$ range from $96\,€/t_{CO_{2eq}}$ to $415\,€/t_{CO_{2eq}}$ for the first 8 designs of the trade-off curve, Fig. 5.8 from left to right. For the last 3 designs, the cost for avoided $\mathrm{CO_{2eq}}$ rises drastically up to $3236\,€/t_{CO_{2eq}}$, as few emissions are reduced but at high costs. The most economical design considering the Scope II emissions is the same as considering the Scope I emissions. The corresponding emissions differ only slightly by $0.2\,kt_{CO_{2eq}}/a$ as electricity exchange with the grid is very small, Fig. 5.7 and Fig. 5.8.

As benchmark, the designs using grid mix emissions $\mathrm{XEF}_t^{\mathrm{el}}$ are compared to the designs using the average grid emissions $\mathrm{AEF}^{\mathrm{el}}$. However, the industrial energy systems optimized for average grid emissions $\mathrm{AEF}^{\mathrm{el}}$ will actually be operated on a grid with time-dependent grid mix emissions $\mathrm{XEF}_t^{\mathrm{el}}$. Thus, results using the average emissions $\mathrm{AEF}^{\mathrm{el}}$ provide only the '*assumed emissions*' whereas the '*actual emissions*' are obtained by evaluating the $\mathrm{AEF}^{\mathrm{el}}$-optimal system with the grid mix emissions $\mathrm{XEF}_t^{\mathrm{el}}$.
The difference between the '*assumed emissions*' and the '*actual emissions*' is small for the complete trade-off curve. However, 1 solution on the trade-off curve for the '*actual emissions*' is not Pareto-optimal, because the optimal solution employing the average emission factor $\mathrm{AEF}^{\mathrm{el}}$ is only assessed with the grid mix emissions $\mathrm{XEF}_t^{\mathrm{el}}$ and not optimized again. Thus, some operational strategies resulting from the average emission factor $\mathrm{AEF}^{\mathrm{el}}$ lead to higher emissions at higher costs.

Proper accounting for time-dependent emissions allows to reduce ecological impacts. The additional achievable emission reduction corresponds to the difference between the '*actual emissions*' and the '*minimal emissions*.' This difference increases with decreasing Scope II emissions up to $1.3\,kt_{CO_{2eq}}/a$, Fig. 5.8. This reduction in emissions corresponds to 6 % of the minimal emissions of this low-carbon industrial energy system at the same costs.

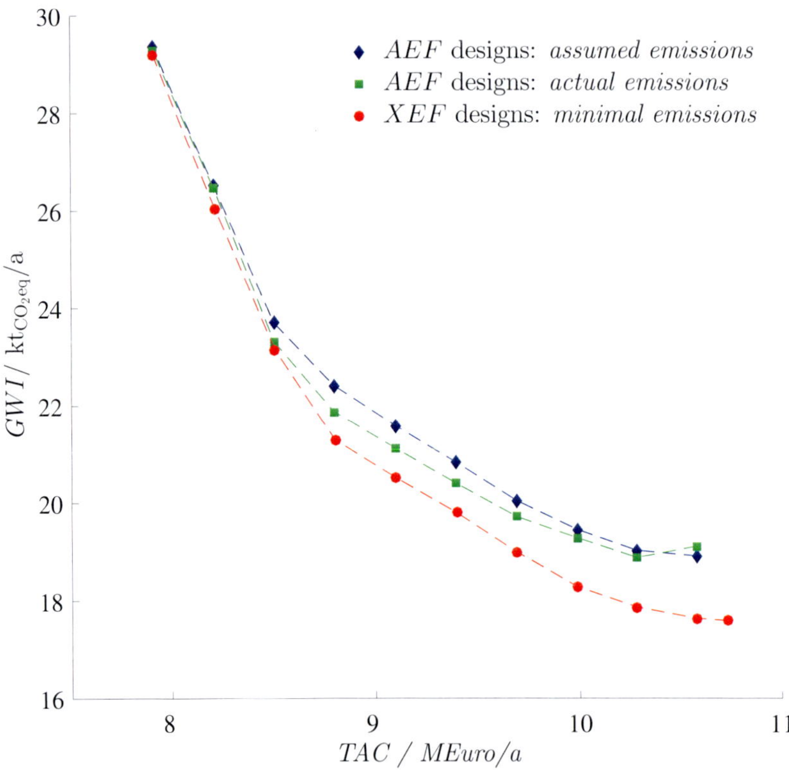

Figure 5.8: Trade-off curve between economical and ecological designs in terms of Scope II global warming impact GWI and total annualized costs TAC for time-dependent grid mix emissions $\mathrm{XEF}_t^{\mathrm{el}}$ ('*minimal emissions*'). As benchmark, solutions optimized for the annual average time-independent emissions $\mathrm{AEF}^{\mathrm{el}}$ ('*actual emissions*'), and based on the annual average time-independent emissions $\mathrm{AEF}^{\mathrm{el}}$ ('*assumed emissions*') are shown. Lines are included to guide reader's eyes.

Considering average emissions $\mathrm{AEF}^{\mathrm{el}}$ or grid mix emissions $\mathrm{XEF}_t^{\mathrm{el}}$ results in almost the same designs. However, climate-friendly DSM uses flexibilities in the operation to exploit times of high and low grid emissions. All designs include more units with decreasing emissions, independent from the employed emission factors $\mathrm{AEF}^{\mathrm{el}}$ or $\mathrm{XEF}_t^{\mathrm{el}}$: First, larger absorption chillers and storage systems are used and a small wind turbine is added to the design. Second, the CHP engines and the wind turbine are increased in size, up to the limits given by the superstructure. Then, heat pumps, roof-top PV, and inverter stations are included in the designs. Finally, battery systems are added to the designs with the least emissions. However, battery systems reduce emissions only slightly and have high costs. No design includes solar thermal panels or organic Rankine cycles. Solar thermal panels compete with roof-top PV for the limited roof area. Thus, the combination of roof-top PV and heat pumps is superior to solar thermal panels for this low-carbon industrial energy system. As no free waste heat source is considered in the industrial energy system, it is not optimal to build an organic Rankine cycle using steam to supply low-temperature heat and electricity.

As discussed in Chapter 5.2, grid mix emissions $\mathrm{XEF}_t^{\mathrm{el}}$ may only be used for small changes in consumption, e. g., for operation of an existing low-carbon industrial energy system. To evaluate large changes in consumption of a low-carbon industrial energy system, e.g., for a new plant, the marginal emissions should be used (Regett et al., 2018), as applied in the next section. Still, the actual choice of whether to employ time-dependent grid mix or marginal emissions is case-study specific and should be carefully conducted using the ILCD handbook guidelines (ILCD, 2010).

5.3.3 Indirect emissions: time-dependent marginal GHG emissions $\mathrm{MEF}_t^{\mathrm{el}}$

Analogous to Chapter 5.3.2, we identify the trade-off curve between economical and ecological designs of the low-carbon industrial energy system for Scope II emissions. However, here we consider time-dependent *marginal* emissions $\mathrm{MEF}_t^{\mathrm{el}}$, Fig. 5.4.

Considering the marginal emission factor $\mathrm{MEF}_t^{\mathrm{el}}$ for the optimal design leads to the '*minimal emissions*.' These '*minimal emissions*' range from $27.9\,kt_{CO_{2eq}}/a$ for the most economical design to $-0.5\,kt_{CO_{2eq}}/a$ for the most ecological design, Fig. 5.8. Negative overall emissions are possible due to on-site production and selling of electricity to the grid in times of high marginal emissions. The most economical design is practically the same as in Chapters 5.3.1 and 5.3.2, with costs of $7.9\,M€/a$. The costs increase from the most economical design to $11.3\,M€/a$ for the most ecological design. The cost for avoided CO_{2eq} is lower considering the marginal emission factor MEF_t^{el} than for the grid mix emission factor XEF_t^{el} in Chapter 5.3.2. The costs of avoided CO_{2eq} range from only $32\,€/t_{CO_{2eq}}$ to $242\,€/t_{CO_{2eq}}$ for the first 8 designs of the trade-off curve, Fig. 5.9 from left to right. However, again, for the last 3 designs, the cost for avoided CO_{2eq} rise even higher to $5098\,€/t_{CO_{2eq}}$. The costs are higher, since the marginal grid emissions show more individual peaks than the grid mix emissions that can be exploited at higher costs, cf. Fig. 5.1 and 5.4.

To resolve the impact of the marginal emissions, we compare the designs using the marginal emissions MEF_t^{el} to designs based on average grid emissions AEF^{el}. As in Chapter 5.3.2, the energy systems optimized for average grid emissions AEF^{el} will actually be operated on a grid with marginal grid emissions MEF_t^{el}. Hence, designs using the average emissions AEF^{el} results only in the '*assumed emissions*' whereas the '*actual emissions*' are obtained by evaluating the AEF^{el}-optimal system with the marginal emissions MEF_t^{el}.
The difference between these '*actual emissions*' and the '*minimal emissions*' varies along the trade-off curve. The emissions can be reduced by maximal $5.5\,kt_{CO_{2eq}}/a$, Fig. 5.9 ('*actual*' vs. '*minimal emissions*'). This reduction corresponds to 60 % of the emissions of this low-carbon industrial energy system at the same costs. On average, the emissions can be reduced by $3.0\,kt_{CO_{2eq}}/a$, corresponding to a 50 % reduction. Thus, the impact of marginal emissions is important when assessing Scope II emissions.

Assuming an average grid emission AEF^{el} factor instead of the marginal emission factor MEF_t^{el} leads to major differences in emissions by up to $17.5\,kt_{CO_{2eq}}/a$, Fig 5.9 ('*assumed*' vs. '*actual emissions*'). Counter-intuitively, although the average marginal grid emission factor MEF^{el} ($1042\,g_{CO_2eq}/kWh_{el}$) is larger than the average time-independent emission factor AEF^{el} ($620\,g_{CO_2eq}/kWh_{el}$), the emissions of the energy systems based on the marginal emissions are lower ('*assumed*' vs. '*actual emissions*'). The emissions are lower, since for the more ecological designs, more grid electricity is sold than purchased. For the more economical designs, the electricity purchase (selling) correlates slightly with low (high) marginal emission factors. Thus, in this case study, the average emission factor AEF^{el} overestimates the emissions of newly build low-carbon industrial energy systems.

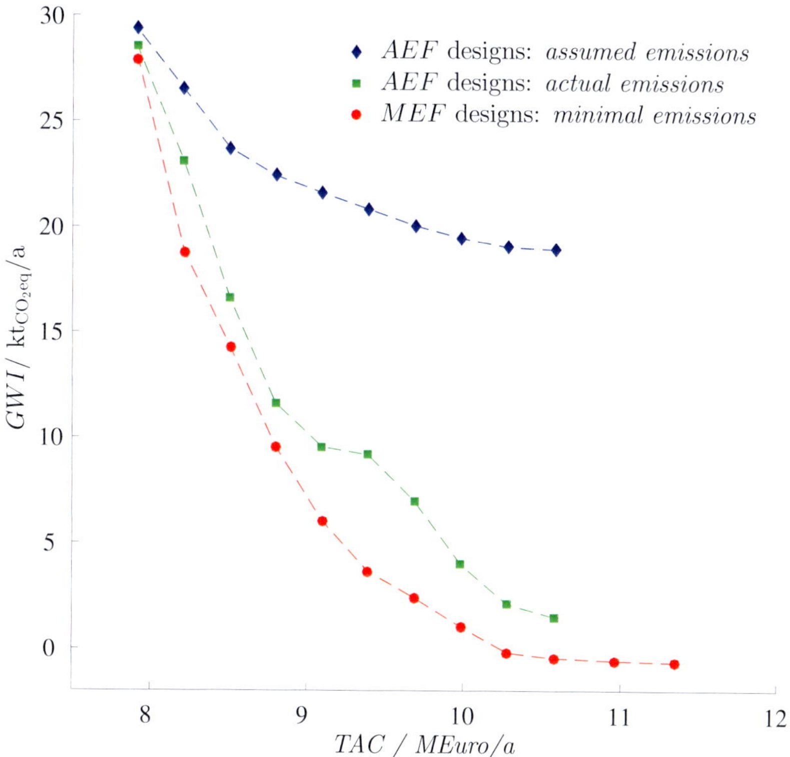

Figure 5.9: Trade-off curve between economical and ecological designs in terms of Scope II global warming impact GWI and total annualized costs TAC for time-dependent marginal grid emissions MEF_t^{el} ('*minimal emissions*'). As benchmark, solutions optimized for the annual average time-independent emissions AEF^{el} ('*actual emissions*'), and based on the annual average time-independent emissions AEF^{el} ('*assumed emissions*') are shown. Lines are included to guide reader's eyes.

The designs considering marginal emissions MEF_t^{el} are almost the same as the designs considering average grid emissions AEF^{el} and grid mix emissions XEF_t^{el}, Chapter 5.3.2. However, the operation of the low-carbon industrial energy systems differs significantly: Mainly, the interaction with the electricity grid increases when employing time-dependent grid emissions, both for buying and selling electricity to the grid, to exploit the volatile nature of the marginal emissions.

5.4 Summary and conclusion

Multi-objective design of low-carbon energy systems requires the computation of its greenhouse gas (GHG) emissions. For grid-connected industrial energy systems, electricity grid emissions are important. Today, grid emissions are commonly accounted for by annually-averaged emission factors, even though they are time-dependent. In this chapter, we investigate the impact of time-dependent grid emissions on the optimal design of low-carbon energy systems. We propose methods to obtain time-dependent grid emissions for the current mix of generation technologies and time-dependent marginal grid emissions based on a marginal technology.

We obtain trade-off curves between economical and ecological designs, using multi-objective optimization. We investigate both direct Scope I and indirect Scope II emissions of an industrial energy system. Considering only the Scope I emissions, the results show that 100 % decarbonization is possible. For this purpose, first, the cooling supply is electrified, followed by the low-temperature heat supply and the steam supply. However, the Scope I picture is incomplete and greatly increases grid emissions. Thus, Scope II emissions should also be considered. For our case study, the Scope II emissions can be significantly reduced at the same costs: the optimal eco-friendly design considering time-dependent GHG grid mix emissions reduces emissions by up to 6 %. The optimal design considering time-dependent marginal GHG grid emissions reduces emissions by even up to 60 %. In addition, using time-independent GHG grid emissions greatly overestimated emissions for the investigated energy system.

The resulting designs of the low-carbon energy system are almost the same. Thus, no technology change is induced by time-dependent grid emissions yet. However, flexibilities in operation are used differently to shift high electricity demand to times with low specific grid emissions resulting in climate-friendly demand-side management (DSM). In general, the reduction potential of climate-friendly DSM described in this chapter is likely to increase further in the future, as the share of variable renewable generation in the European energy system increases. Thus, the time-dependent emission factor should be frequently updated. The increase in renewables might also induce technology changes in energy systems, as more flexibility in operation might be necessary. This operational flexibility has to be implemented in daily schedules, while planning horizons remain long, which increases the planning effort for practical applications. However, as shown in Chapter 3, the increased planning effort can be handled by our fast solving method DeLoop.

In this chapter, the model scope is the multi-objective synthesis optimization of industrial energy systems. We successfully extended the model scope to assess GHG emissions correctly. Thus, we solve Challenge 3, cf. Chapter 2.4. In the next chapter, we extend the model scope to investigate national synthesis optimization problems.

Chapter 6

Synthesis and life-cycle assessment of national energy systems

In this chapter, we extend the scope of national energy system models by life-cycle assessment (LCA) to consider environmental burden shifting. Thus, we overcome Challenge 4, cf. Chapter 2.4. We propose a national energy system model optimizing the transition towards a low-carbon future. The model includes the sectors with most GHG emissions (Federal Ministry for the Environment, Nature Conservation and Nuclear Safety, 2018): electricity, heat, and private transportation. The sectors are modeled with high temporal and spatial resolution resulting in a large-scale model. We further extend the scope of this large-scale model by integrating LCA. The large scope allows only for a low degree of model detail, Fig. 2.1. Thus, we assume constant part-load behavior, no minimum load, and no minimal capacities, which converts the MILP to an LP problem. Further, we employ time-series aggregation with proven feasibility, but without optimality proof.

In Chapter 6.1, we present the proposed model and LCA method. In Chapter 6.2.1, we compute a cost-optimal pathway for the energy transition to achieve the German GHG reduction goal of 85 % until 2050 compared to 1990. In Chapter 6.2.2, we compute environmental impacts for electricity, heat, and private transportation for the optimal, sector-coupled energy transition. We show that the transition towards a low-carbon national energy system leads to many co-benefits in environmental impacts. However, some impacts such as metal and mineral resource depletion and freshwater ecotoxicity increase significantly. In Chapter 6.3, we summarize and conclude.

Major parts of this chapter are reproduced from:

Baumgärtner, N., Deutz, S., Reinert, C., Nolzen, N., Küpper, E., Hennen, M., Hollermann, D., and Bardow, A. (2020). Life-cycle assessment of sector-coupled national energy systems: Environmental impacts of future electricity, heat & transportation in Germany. *submitted.*

Contribution report: Writing the draft, 1 of 2 principal authors, development of the energy system model, co-development of the life-cycle assessment, calculation and evaluation of results.

6.1 Sector-coupled national energy system model

The Sector-coupled energy system MODel (SecMOD), established in this chapter, calculates the cost-optimal transition of a national energy system for a given GHG-emission target. For a holistic evaluation of the resulting energy systems, SecMOD incorporates life-cycle assessment (LCA) using the categories of the Environmental Footprint 2.0 life-cycle impact assessment (LCIA) (JRC, 2018). In this thesis, we parametrize SecMOD for the German energy transition from the year 2016 until 2050.

In Chapter 6.1.1, we give a concise overview of SecMOD. In Chapter 6.1.2, we present the life-cycle assessment (LCA) integrated into SecMOD. The detailed model formulation of SecMOD is given in Appendix C.

6.1.1 Model overview

SecMOD computes a transition path towards a sustainable low-carbon energy system by an economic optimization satisfying GHG-reduction targets. SecMOD is formulated as linear programming (LP) problem in GAMS (GAMS Development Corporation, 2019) and solved with CPLEX 12.9.0.0 (IBM, 2016).
In SecMOD, the energy system is locally resolved by nodes. At each node and time step, demands need to be fulfilled. In the current version of SecMOD, the demands are electricity, centralized and decentralized heating on 4 temperature levels (domestic heating, and industrial heat below 100 °C, between 100-400 °C, and above 400 °C), and private short- and long-distant transportation. The electricity demand is based on Egerer (2016); the heating demand is based on household (Federal Statistical Office, 2015) and industrial data (Öko-Institut and ISI, 2015); and the demand for private transportation is taken from Federal Motor Transport Authority (2017a). We assume constant electricity demand, besides sector-coupling technologies and a constant heating demand. However, the heating demand can be reduced within the model by thermal insulation. For the transportation demand, we assume a slight increase till 2030 and then a constant demand. The detailed demand structure is given in Appendix C.

SecMod can choose between various technologies (sector-coupling technologies are highlighted in ***bold-italic***) to fulfill the demands. For private transportation, SecMOD includes the following vehicle technologies:

- diesel (and ***synthetic diesel***), gasoline, natural gas (and ***synthetic methane***), ***batteries***, ***plug-in hybrids***, and ***fuel cells***.

For centralized and decentralized heating, SecMOD includes the following technologies:

- domestic thermal insulation, ***electrode boilers***, gas boilers, oil boilers, ***heat pumps***, and ***heat cogeneration from power plants***

For electricity demand, SecMOD includes the following technologies:

- biomass plants, combined-cycle power plants, coal plants, gas plants, geothermal generation, ***H_2-electrolysis*** with H_2-fuel cell, lignite plants, lithium-ion batteries, nuclear plants, oil plants, ***power-to-methane***, ***power-to-diesel***, pumped-hydro-storage plants, run-of-river power plants, photovoltaics, transmission grids, waste incineration plants, wind (onshore and offshore).

The primary data source for technology parameters, such as costs, is the study from Deutsche Energie-Agentur (2018). All sources are given in Appendix C. Climate change itself can affect the performance of technologies positively or negatively (Cronin et al., 2018). This effect is not reflected in our model. For our model scope in central Europe, the review of Cronin et al. (2018) suggests that the impact of climate change should be strongest for bio-energy, which, however, is not expandable in our model. For the investment costs and efficiencies, projects of future development are implemented to model cost degression for specific technologies with ongoing improvements, see Appendix C. Time series for renewable generation and conventional availability are taken from Egerer (2016). The potential for renewable generation, capacities for onshore wind, offshore wind, and roof-top photovoltaics is regionally limited. The regional limits are included for onshore wind energy with values based on McKenna et al. (2015), for offshore wind on Gerhardt et al. (2015), and photovoltaics on Mainzer et al. (2014). The limits are based on available area for each municipality considering political restrictions. The resulting overall limits are 698 GW^{on} onshore wind, 36 GW^{of} offshore wind, and 235 GW^{pv} PV. The overall limits are stated as the maximal installed nominal capacity.

The power sector includes the high-voltage transmission grid based on ELMOD-DE from 2012 (Egerer, 2016). We updated the network data for the year 2016. The German transmission grid is modeled by 416 nodes and 661 power lines. Following ELMOD-DE, SecMOD represents network flows with the DC-load-flow approximation by Overbye et al. (2004), which linearizes the AC-power flow. For this study, we spatially aggregated the nodes to 18 nodes representing the 18 regions proposed by the German energy agency Dena (2010). For grid expansion, power lines can be upgraded to a higher voltage level or new power lines can be installed in parallel.

As objective, we minimize the total annualized costs assuming a discount rate of 5 % as in the JRC-EU-TIMES model (Simoes et al., 2013). Investment costs in technologies are annualized over their economic lifetime also with an interest rate of 5 %, see Appendix C. The economic objective is minimized subject to the given GHG reduction targets taken from the German federal environmental agency (Salb et al., 2018). Besides the climate impact, 15 further environmental impacts are assessed, but not constrained in the optimization.
The system boundaries include the German sectors for power, heating, and private vehicles accounting for 77 % (690 Mt CO_{2eq}) of the total GHG emissions in 2016. Therefore, we scale the GHG-reduction targets by 77 %. Thus, a relative reduction of the GHG emissions by 85 % compared to 1990 (966 Mt CO_{2eq}) until the year 2050 corresponds to absolute emissions of 144 Mt CO_{2eq} in SecMOD.

For the GHG-reduction targets, we consider the operational climate impacts, see Chapter 6.1.2. The operational climate impacts comprise not only direct CO_{2eq} emissions from flue-gas stacks and exhausts in the energy system. In addition, we include indirect upstream emissions from fuel supply, e. g., mining of energy carriers. Current policies for emission reductions target the direct emissions only. However, both direct and indirect emissions need to be constrained to cover the entire climate impact of operation (Lopion et al., 2018). Thus, our emission targets are more ambitious as current policies as we incorporate the entire operational impact by life-cycle thinking.

To calculate the transition path till 2050, SecMOD is modeled as a multi-period problem considering 8 consecutive investment periods with a horizon of 5 years each, in line with state-of-the-art energy system models (Lopion et al., 2018).
In general, transition paths can be optimized either with perfect or myopic foresight (Lopion et al., 2018) and (Heuberger et al., 2018). In the following, we discuss our approach to optimize the transition path. Perfect foresight solves the whole optimization problem at once with complete information on all future parameters, whereas myopic foresight solves the investment periods sequentially with only current information on parameters (Ringkjøb et al., 2018). Perfect foresight leads to an optimal transition path; however, it is computationally demanding and does not capture reality, where future parameters are uncertain. While myopic foresight reduces the computation time,

it leads only to a near-optimal transition path (Heuberger et al., 2018). In SecMOD, we employ a rolling-horizon strategy with myopic foresight as decomposition method of the multi-period problem (Grossmann, 2012). In the rolling-horizon strategy, the optimization is carried out over a limited horizon which covers only a few investment periods. After each optimization, the variables of the first investment period within the horizon are fixed, and the next investment periods are recalculated with updated myopic foresight. For this thesis, we set the myopic foresight to 15 years. Preliminary studies with SecMOD have shown that the transition paths are almost the same with perfect foresight and 15 years of myopic foresight within the rolling horizon strategy. The rolling horizon strategy is described in detail in Appendix C.

Another challenge for energy system models is sufficient time resolution to capture the volatile nature of renewable energies (Mallapragada et al., 2018). SecMOD therefore employs an hourly time resolution. To reduce the computational effort, the two-step method from Bahl et al. (2018a) is employed as so-called primal heuristic to calculate near-optimal feasible solutions. In the first step, SecMOD solves a time-aggregated synthesis optimization in each investment period of the rolling horizon to expand the existing capacity. As aggregated time series, SecMOD considers an adjustable number of typical periods with typical time-slices. Here, we choose to represent every investment period by 10 typical periods with 12 typical time steps each. In the second step, after each synthesis optimization, the operation is optimized. The operational optimization is conducted with the full hourly resolution and capacities fixed of the current investment period. The operational optimization ensures that the energy system is feasible, thus, can fulfill the energy demand of the original time series. If the operational optimization reveals that the energy system is not feasible, additional power, heat or transportation capacities are added and the operational optimization is repeated. For details see Appendix C. Importantly, the operational optimization is more accurate in terms of costs and GHG emissions as the calculated operational costs and GHG emissions from the time-aggregated synthesis optimization in the first step. In summary, the two-step optimization allows SecMOD to provide a feasible near-optimal solution for the hourly resolved problem.
The model formulation in SecMOD corresponds to the current state-of-the-art in energy system models with its formulation as linear programming, time- and spatial-resolution, optimization horizon, and technology richness (Ringkjøb et al., 2018). To holistically evaluate environmental impacts of the transition path, SecMOD expands energy system modeling by LCA as presented in the following section.

6.1.2 Life-cycle assessment

LCA is a holistic methodology for the environmental assessment of products and services, taking into account their entire life-cycle (JRC, 2010). The life-cycle includes all activities from cradle-to-grave: the extraction of raw materials, transportation, production and product use, recycling, treatment, and final disposal of waste. Through the life-cycle, the activities result in material and energy flows that are exchanged with the environment. In life-cycle assessment, the material and energy flows are collected and interpreted regarding their environmental impacts.

The International Organization for Standardization (2006a,b) defined and standardized the LCA methodology in the ISO Standards 14040 and 14044. In the following, we discuss the general goal and scope definition for our study of low-carbon transition pathways, including the definition of system boundaries, functional unit, life-cycle inventory (LCI) and methods for life-cycle impact assessment (LCIA).

Goal, system boundaries, functional unit and multi-functionality

Our LCA pursues two main goals: 1) to determine environmental burden-shifting caused by the German low-carbon transition path and 2) provide specific environmental impacts for the services power, heat, and private transportation in 5 year steps according to the German low-carbon transition path till 2050.

Based on our study's two goals, the system boundary includes all processes associated with the low-carbon transition path of the German energy system. The considered German energy system contains the sectors power, heating and private transportation, while international electricity trade is not considered.

In LCA, services or products are compared based on a so-called functional unit that quantifies the function of the investigated product systems and serves as a basis for comparison (International Organization for Standardization, 2006a,b). The function of the considered energy system is to provide the services power, heat, and private transportation. To determine the environmental impacts along the low-carbon transition path, we use the functional units: power: 1 $\text{MWh}_{\text{power}}$, heat: 1 MWh_{heat} and private transportation: 1 vkm (vehicle kilometer), for the German energy system.

The environmental impacts of the electricity used in the heating and private transportation sector are accounted for within the heating or private transportation sector. Synthetic natural gas can be used in the power, heating and private transport sectors. We assume that the produced synthetic natural gas is feed into the natural gas grid. Therefore, we allocate the environmental impact of synthetic natural gas to the individual sectors in proportion to the amount of the used natural gas.

In LCA, processes are often multi-functional, producing more than one product. Multi-functionality is a well-known problem that can be solved using established methods such as system expansion, avoided burden and allocation physical relationships or other criteria such as mass, energy or economic value (JRC, 2011) and (International Organization for Standardization, 2006a). Some of the existing conventional power plants in SecMOD generate both electricity and heat (combined heat and power generation). We assume heat as waste and assign all environmental impacts to electricity generation. This choice does not affect the transition pathway, and has a minor impact on the distribution of the environmental impacts on the heating and electricity sector. In general, the contribution of heat from co-generation is low for the considered low-carbon transition path: 4 % and 0.5 % of the heating demand in 2016 and 2050, respectively.

Life-cycle inventory

The life-cycle inventory (LCI) contains all mass and energy balances within the system boundaries required for the LCA. SecMOD computes the flows between the units and to fulfill the final demand. For the considered technologies themselves, see Chapter 6.1.1. The LCI datasets are based on the LCI database ecoinvent version 3.5 APOS (Wernet et al., 2016). We chose the ecoinvent system model APOS to consistently consider recyclable materials. Recycling of materials is an important issue in energy system design due to the extensive infrastructure. Recycling rates for all metals are given in the supplementary material. For all other materials, we assume treatment and final waste disposal documented in the supplementary material with the LCI data. The LCI data of novel technologies missing in ecoinvent are modeled based on literature data with ecoinvent as background system. The modeled LCI datasets are described in detail in Appendix C.

In general, we distinguish between infrastructural and operational LCI datasets. The operational LCI datasets describe impacts resulting from the operation of a technology, e. g., due to the consumption and combustion of fuels, and auxiliaries for maintenance. The operational LCI datasets quantify the environmental impact per functional unit: $\mathrm{MWh_{power}}$, $\mathrm{MWh_{heat}}$, and vkm. In contrast, the infrastructural LCI datasets of a technology cover the construction and subsequent dismantling, potential recycling, treatment and final disposal of waste. The infrastructure of the technology is assessed based on the installed capacity in MW or number of vehicles. This distinction between infrastructure and operation enables to differentiate between environmental impacts that arise on-site and those that occur (partly) outside of Germany. Furthermore, the operational impacts are used to constrain the German GHG emissions in SecMOD. The current reduction targets are taken from Salb et al. (2018).

From the synthesis optimization, we determine the infrastructural LCI datasets of existing and added capacity of each technology, for each investment period. To distribute the environmental burdens over the investment periods, the infrastructural LCI datasets are annualized over the minimum of technology lifetime and 30 years, see Chapter 6.1.1. As in Rauner and Budzinski (2017), we assume a discount rate of 0 % for the LCI datasets. A discount rate of 0 % implies that future environmental damage is equivalent to current damage.
The operational optimization calculates the operational LCI datasets of all technologies for each hour of operation within the investment period. The infrastructural and operational LCI datasets are subsequently classified into the sectors power, heat and private transportation based on the functional units.
An important issue for LCA in sector-coupled energy system model is double counting (Lenzen, 2008). Double counting occurs since LCI datasets contain energy-related environmental impacts from production steps in Germany. However, energy-related environmental impacts of the German industry are already included within the national energy model. As a result, such energy-related environmental impacts would be double-counted. For example, electricity needed to manufacture German vehicles is considered as part of the industrial demands. However, the same electricity is also part of the LCI datasets of vehicles. In SecMOD, we solve the problem of double-counting by deleting impacts in the LCI, which we already account for in the German energy system following the approach of Volkart et al. (2018) and Reinert et al. (2020b).

Life-cycle impact assessment

The LCI data provides the basis for the life-cycle impact assessment (LCIA). In the LCIA phase, all mass and energy flows are analyzed concerning their environmental impacts. For the low-carbon transition path, the main goal is to reduce GHG emissions, i. e., the climate impact. However, besides climate impacts, the energy system leads also to other environmental impacts. To identify potential environmental trade-offs and to determine burden-shifting according to the first goal of our study, we consider 15 additional environmental impacts, Table 6.1. We assess the environmental impacts according to the LCIA-method *Environmental Footprint 2.0* recommended by the European Commission's Joint Research Center (JRC, 2018). Environmental Footprint 2.0 is the latest recommended method by the European commission implemented in the LCI database ecoinvent. The recommended environmental impacts are classified into 3 quality levels (JRC, 2011) and updated in JRC (2018) with regard to completeness, relevance, robustness, transparency, applicability, acceptance, and suitability. The quality classification of each environmental impact (Table 6.1) should be considered when interpreting LCA results.

Table 6.1: Considered environmental impacts in SecMOD following the recommendation of the European Commission's Joint Research Centre in the European context, sorted by their quality level (JRC, 2011, 2018)

Quality level	Environmental impact
I: recommended and satisfactory	Climate impact (CI) Ozone depletion (OD) Particulate matter (PM)
II: recommended, some improvements needed	Acidification, terrestrial and freshwater (A_{tfw}) Eutrophication, freshwater (E_{fw}), terrestrial (E_t) and marine (E_m) Ionizing radiation (IR) Photochemical ozone formation (PCOF)
II/III	Human toxicity, cancer (HT_c) and non-cancer (HT_{nc}) Ecotoxicity, freshwater (ET_{fw})
III: recommended, but to apply with caution	Land use (LU) Water scarcity (WS) Resource depletion, energy (RD_e), and mineral and metal (RD_{mm})

According to the second goal of our study, we determine the specific environmental impacts of the energy services in the sectors: power, heating and private transportation. For this purpose, we also calculate the annualized infrastructural environmental impacts due to installed capacities based on the LCIs from the synthesis optimization. The operational optimization calculates the operational environmental impacts on an hourly basis. The infrastructural and operational environmental impacts are subsequently assigned to the sectors power, heat and private transportation.
Specific environmental impacts are achieved by dividing the total impacts by the energy output for each sector in MWh_{power}, MWh_{heat} and vkm.

6.2 Transition pathway of the German energy system

In SecMOD, we optimize the transition pathway to a low-carbon energy system in Germany in 2050. In Chapter 6.2.1, we first present a complete low-carbon transition pathway of the German energy system until 2050. The transition pathway results from an economic optimization with constrained GHG emissions. In Chapter 6.2.2, we show the environmental impacts of the provided energy services: electricity, heat, and transportation. The transition pathway is optimal under the given model assumptions that, however, represent only one possible transition scenario among many possibilities. Key assumptions are discussed in Section 6.2.3.

6.2.1 Energy transition pathway

Emission reductions of 85 % by 2050 are possible in the considered sector-coupled German energy system without international electricity trade, Fig. 6.1. In the first investment periods, emissions are reduced in the power and especially in the heating sector. The decarbonization of the transportation sector starts in later stages of the transition pathway. In 2050, to achieve the 85 % reduction target, the transportation sector is almost carbon-neutral and GHG emissions from both the heating and electricity sector are reduced to approximately a third of the emissions in 2016.

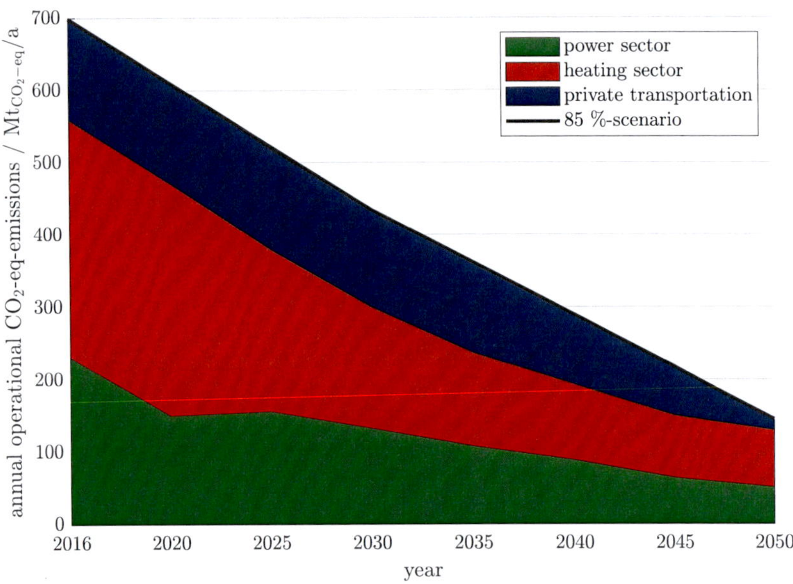

Figure 6.1: Operational GHG emissions for the transition path considering the 85 % reduction target compared to 1990 (966 Mt CO_{2eq}). The 85 %-Scenario is based on the targets of the German federal government of a 55 % reduction by 2030, 70 % by 2040, and a 80-95 % reduction by 2050. We assume a linear reduction of the emission limits. The emissions are assigned to the power, heating, and private transportation sector.

The resulting energy system is proven to be feasible by operational optimization with a time-resolution of 1 hour. The total annualized costs increase by a factor of 3 until 2050 compared to current costs, Fig. 6.2. The costs increase mainly due to the demand for storage capacity. The storage is necessary to cope with the volatile nature of renewable generation. The required amount of storage needed is high due to the considered high temporal and spatial resolution. Storage demand is often underestimated in energy system models which are not spatially resolved (Heuberger et al., 2020). In addition, Victoria et al. (2019) show that storage demand increases strongly for reduction targets beyond 80 %. For a decarbonized European energy system in 2050, Child et al. (2018) predict a cost-share of storage of about 30 %.

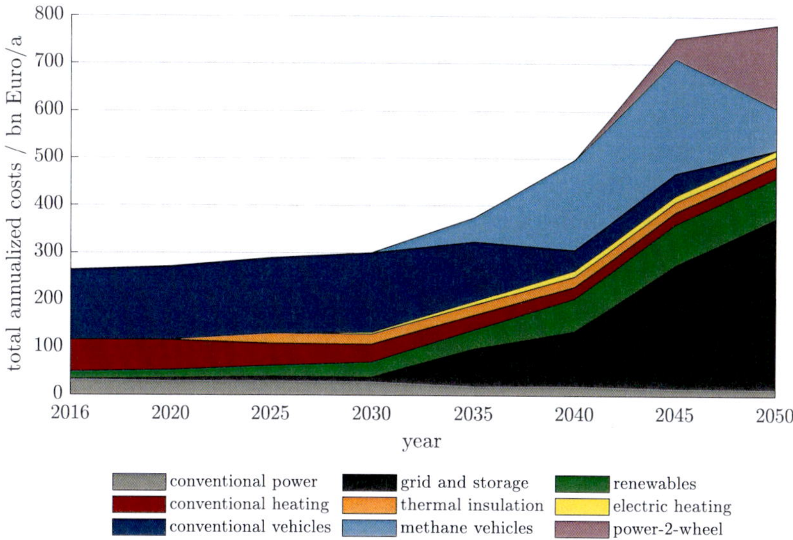

Figure 6.2: Total annualized costs for the energy-transition path reaching the 85 % reduction target. Technologies are lumped for clarity of presentation. Power-2-wheel includes battery electric vehicles, fuel cell vehicles, and synthetic diesel.

Energy services are provided by different technologies in the energy transition, since increasingly tighter emission constraints from 2016 to 2050 require the change to low-carbon technologies. First, we discuss the heating sector, Fig. 6.3. We then regard private transportation, Fig. 6.4. Finally, we describe the transition of the electricity sector that becomes increasingly important in the low-carbon energy system, Fig. 6.5.

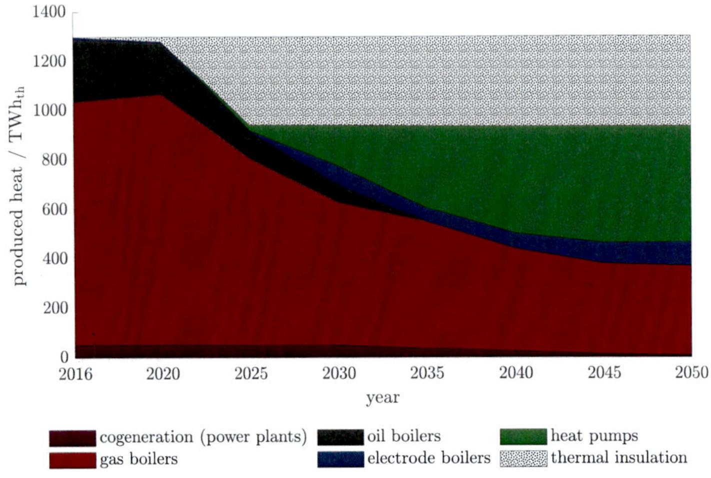

Figure 6.3: Provided heating demand in Germany during the energy transition from 2016 to 2050. Thermal insulation reduces the overall domestic heating demand and is included as provided heating demand. Temperature levels of the heating demand, as well as centralized and decentralized heating demands, are lumped for clarity of presentation. Gas boilers can be fueled by natural gas or synthetic natural gas.

The first measure to reduce GHG emissions in the heating sector is to reduce the heating demand by thermal insulation, Fig. 6.3. Thermal insulation is already known as one main driver of the energy transition in the heating sector (ifeu, Fraunhofer IEE, Consentec, 2019). In SecMOD, investment rates for all technologies are unlimited, i. e., no budget limit for investments is given. Thus, a rapid technology shift, as observed for thermal insulation, is possible. Starting from the year 2016, oil boilers are replaced by more efficient gas boilers. Beginning in the year 2025, a large amount of heating is electrified. First, the supply of domestic and industrial low-temperature heating is electrified by heat pumps. Beginning in the year 2030, electrification of industrial medium-temperature heating starts with electrode boilers, but with slower deployment rates. In the year 2040, all oil boilers are completely dismantled. In the year 2050, 75 % of the heating demand is directly electrified or reduced by thermal insulation. In SecMOD, we assume that industrial, high-temperature heat demand above 400 °C cannot be directly electrified. Still, starting in year 2035, the heating sector is further indirectly electrified by synthetic natural gas, see Fig. 6.5. The measures to decarbonize the heating sector can be categorized into 3 groups: First, low-temperature domestic heating demand is reduced by thermal insulation. Second, the heat supply is directly electrified via heat-pumps and electric boilers. Third, high-temperature heating supply is electrified indirectly by synthetic natural gas.

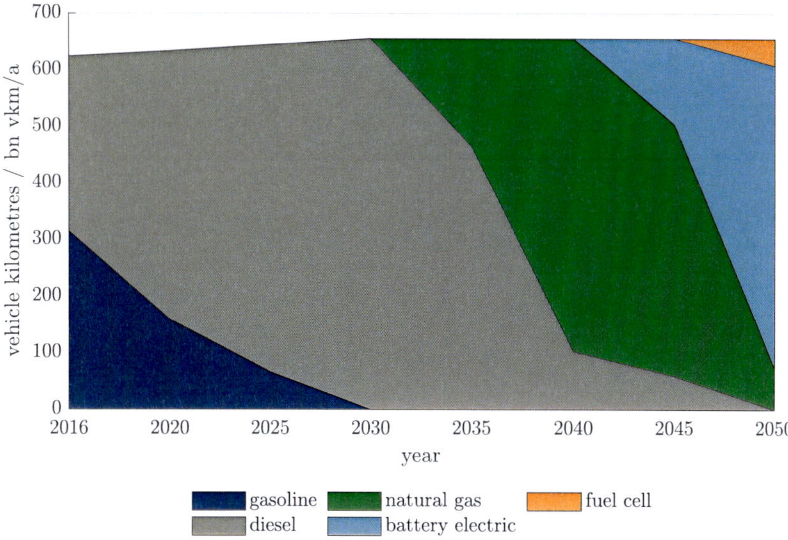

Figure 6.4: Provided transportation service per vehicle type. Natural gas vehicles can be fueled by natural gas or synthetic natural gas.

Significant decarbonization of the private transportation sector occurs late compared to the other sectors, Fig. 6.4. The reason is higher abatement costs compared to other sectors. During the first investment periods, the ratio between gasoline and diesel vehicles shifts towards more diesel vehicles, due to lower specific GHG emissions in operation of diesel than gasoline vehicles. Until 2030, all gasoline vehicles are replaced by diesel vehicles. Natural gas vehicles act as transition technology, due to even lower specific GHG emissions in operation compared to diesel vehicles at slightly higher costs. Starting in 2030, natural gas vehicles provide increasing parts of the transportation demand. Until 2040, about 85 % of the demand is provided by natural gas vehicles. Beginning in 2040, a large share of battery electric vehicles (BEV) replaces the remaining diesel vehicles. Battery electric vehicles also already start to replace old natural gas vehicles, integrating more renewable electricity directly in the transportation sector. In addition beginning from 2040, synthetic diesel is partly used in remaining diesel vehicles. In 2050, about 80 % of the transportation demand is provided by battery electric vehicles, while about 15 % of the demand is still provided by natural gas vehicles. In our model, Battery electric vehicles can only replace a certain share of conventional technologies (45 % in the year 2016 and 82 % in 2050), as long-distant transportation is limited due to battery capacities, see Appendix C. In 2045, fuel cell vehicles begin to replace the remaining natural gas vehicles, as low-carbon, long-distant transportation technology. To account for the costs of additional charging infrastructure, the employment of fuel cell and battery electric vehicles leads to a parallel investment in this infrastructure, Appendix C. For conventional vehicles, the cost for charging infrastructure is included within the fuel costs. In summary, decarbonization of the private transportation sector starts later than the decarbonization of the heating sector due to higher costs and employs a mix of technologies to cover short- and long-distance transportation.

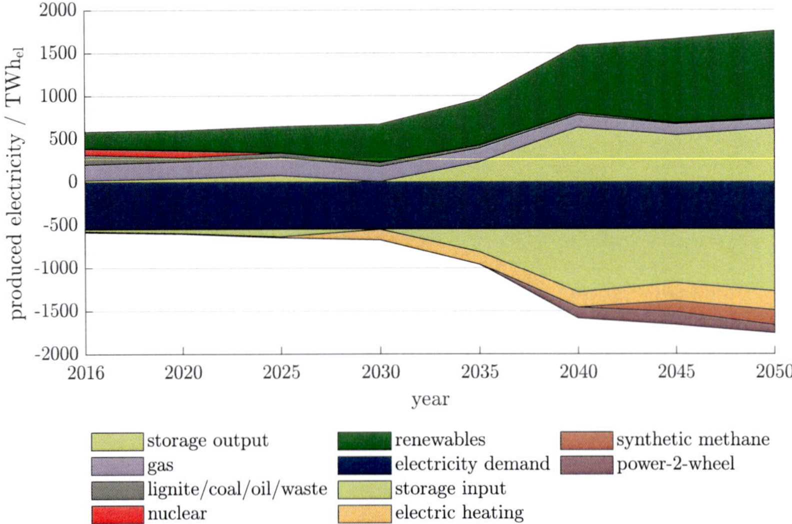

Figure 6.5: Produced electricity (positive) and consumed electricity in cross-sectoral technologies (negative) in the German energy transition. Gas includes gas turbines as well as combined cycle gas turbines. Power-2-wheel includes battery electric vehicles, fuel cell vehicles, and synthetic diesel.

The decarbonization of both transportation and, in particular, heating heavily depends on low-carbon electricity, Fig. 6.5. SecMOD computes that about one-third of the electricity is provided by renewable generation in Germany in 2016, which is very close to the actual value of 32 %, according to Fraunhofer ISE (Burger, 2017). To decarbonize the electricity sector, almost all conventional power plants continuously phase-out, except for gas-power plants. At the same time, these conventional power plants are substituted by renewable electricity production. Between 2025 and 2050, the electricity demand almost doubles to 1100 TWh (excluding storage input and output).
Due to the high electricity demand caused by sector coupling and distributed renewable generation, the grid capacity is almost doubled. To cope with the volatile nature of renewable electricity in 2050, about 720 TWh_{el} of electricity are stored and 620 TWh_{el} discharged during periods of low renewable electricity generation. The increase in storage and grid capacity keeps electricity curtailed low with 4 % (47.6 TWh_{el}) in 2050.
The storage periods are predominantly short and thus used to shift wind and solar electricity within a day. For longer storage periods and further decarbonization, synthetic natural gas is produced starting in year 2035. An overall amount of 180 TWh_{el} is used to produce synthetic natural gas in year 2050. Thereby, in 2050, about 16 % of the natural gas demand is satisfied by synthetic natural gas produced using low-carbon electricity. In summary, the decarbonization of the electricity sector results in a significant increase in renewable energy generation. The volatile renewable generation leads to an increase in transmission and storage capacity. Sector coupling increases the electricity demand, in particular for heating and synthetic natural gas. The findings are in line with previous studies, e. g., Breyer et al. (2018). In the following, we regard the expansion of renewable electricity generation in detail.

To decarbonize the electricity sector, a rapid expansion of solar power, onshore wind, and offshore wind is required, Fig. 6.6. First, regions with high full load hours per year for wind and solar are exploited. The capacities in these regions are, however, limited, Appendix C. Thus, with increasing need for renewable electricity, regions with lower full load hours are also employed. Therefore, the overall full load hours decline by about 35 %, Fig. 6.6. In 2045 and 2050, the same amount of low-carbon electricity is produced. In 2050, the low-carbon electricity is used more efficiently by direct electrification; however, with higher investment costs. The decline in full load hours and the high demand for short-term storage capacities are major contributors to burden-shifting, as discussed next.

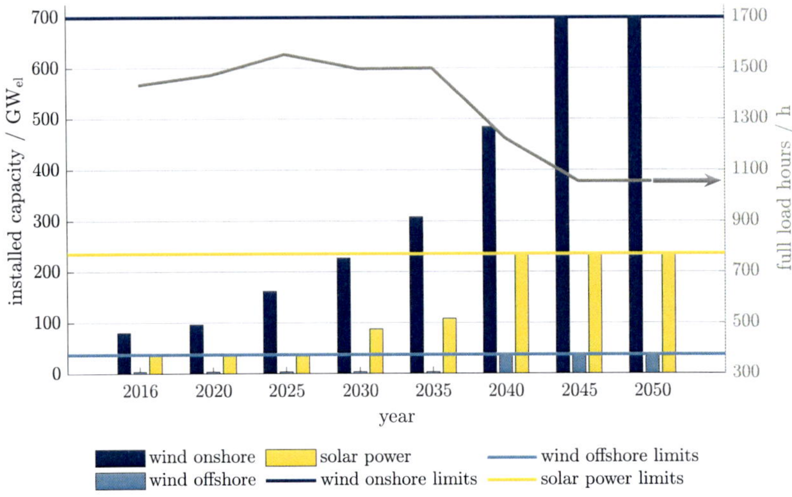

Figure 6.6: Installed capacity of onshore wind, offshore wind, solar, and full load hours per year for all investment periods.

6.2.2 LCA of the energy transition pathway

In the following, we discuss the environmental impacts per functional unit: MWh_{power}, MWh_{heat}, and vehicle kilometer (vkm). The environmental impacts of the future years are shown relative, using 2016 as reference year. The total environmental impacts include operation of technologies and infrastructure, cf. Chapter 6.1.2. In general, we observe a shift in the origin of the environmental impacts: In 2050, the infrastructure contributes to the impacts by 41-98 % for the electricity sector, between 10-58 % for the heating sector, and between 38-98 % for the transportation sector, Appendix G. Importantly, as a result, the GHG reduction target of 85 % in operational emissions corresponds only to a total reduction of 70 % if both operational and infrastructural emissions are accounted for. This drastic trend from impacts due to operation to infrastructure shows that both politics and energy system models must include LCA to assess environmental impacts holistically over the full life cycle.

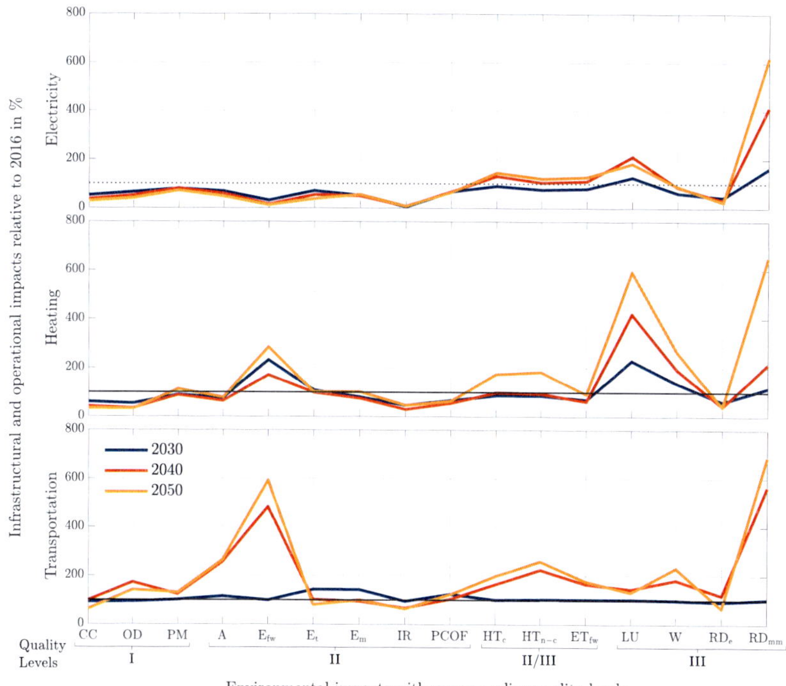

Figure 6.7: Operational and infrastructural environmental impacts for 1 MWh_{power} electricity, MWh_{heat} heating, and 1 vkm of private transportation. The impacts of year 2016 are used as reference (dotted black line: 2016 = 100 %). For Abbreviations and quality levels, see Table 6.1. Electricity used in heating and transportation sector is accounted for in these sectors.

In the following, we first discuss environmental impacts of each sector. Tables with the most important LCA results for 2016-2050 are given in the in Appendix G. In detail, we provide tables with:

- specific environmental impacts per functional unit,
- technology-specific contribution to the environmental impacts of the electricity sector in 2050,
- share of the infrastructural environmental impacts in each sector,
- share of the environmental impacts due to used electricity in the heating and transportation sector.

Electricity sector

In the electricity sector, the reduction of GHG emissions leads to co-benefits; thus, other environmental impacts are reduced. In 2050, environmental impacts are reduced most for the depletion of energy resources (RD_e, 78 %), eutrophication freshwater (E_{fw}, 90 %), and ionizing radiation (IR, 93 %). In total, 11 of 16 environmental impacts are reduced by 10-92 % within the power sector in 2050, Fig. 6.7. The fact that reductions of GHG emissions lead to co-benefits is in line with other studies and has recently been discussed in detail by Luderer et al. (2019).

However, the low-carbon transition of the electricity sector results in significant increases in other environmental impacts in 2050: metal and minerals resource depletion RD_{mm} increases by 520 %, ecotoxicity freshwater ET_{fw} by 30 %, human toxicities HT_c and $HT_{n\text{-}c}$ by 45 and 20 %, and land use LU by 80 %. Environmental impacts increase due to the expansion of stationary batteries and low-carbon electricity generation technologies, see Chapter 6.2.1, with a contribution between 7-51 % for batteries, 8-48 % for wind, and 12-66 % for photovoltaics, see Appendix G. All environmental impacts that significantly increased are at quality level II/III-III indicating high uncertainty. Thus, there is a need to improve LCA and LCIA methods to clarify the extent of increasing environmental impacts.

Heating sector

In the heating sector, 6 out of 16 environmental impacts are lower than in 2016, whereas 4 stay nearly the same in 2050 ($\leq$ 15 %). With rising shares of heat pumps and electrode boilers, the environmental impacts of heating are strongly determined by the environmental impacts of the electricity system. In 2050, environmental impacts in the heating sector are dominated by electricity used with values of up to 90 % in metal and minerals resource depletion RD_{mm}. Metal and minerals resource depletion RD_{mm} increases by 550 % and land use LU by 490 %. Further, freshwater eutrophication E_{fw} increases by 185 %.

Transportation sector

Private transportation shows a slower transition than the other sectors. The initial shift from gasoline to diesel vehicles changes environmental impacts for 1 vkm of private transportation by less than 50 % by 2030, due to the fuel types' similarity. The shift from diesel to natural gas vehicles after 2035 improves all environmental impacts. However, starting in year 2040, the transportation sector shifts to battery electric vehicles, small shares of synthetic diesel, and fuel cell cars. Therefore, by 2050, we observe a decrease in climate impact by about 45 %; however, also a significant increase in freshwater eutrophication by 490 % and in metal and minerals resource depletion RD_{mm} by 590 %. Again, it is important to consider the quality levels of increased environmental impacts. The increase of metals and minerals resource depletion and freshwater eutrophication have low quality levels. Electricity used in the transportation sector in the year 2050 has only minor effects on specific environmental impacts ($\leq$ 15 %), as impacts are dominated by infrastructure.

The trends of our results are in line with other studies which focused only on the electricity sector. For the electricity sector in Denmark, Turconi et al. (2014) show a reduction of all environmental impacts besides metal depletion ('Green Scenario 2030'). Metal depletion worsens by about 50 %. In our multi-sector investigation of Germany, we also observe a reduction in all environmental impacts in 2030 except metal and minerals depletion and land use. Metal and minerals depletion increases by about 65 % and land use by 25 %.
Until 2030, the sectors are not yet highly coupled, storage capacities are still small, and renewable sites with low full load hours have not yet been employed. Therefore, until 2030, studies only investigating the electricity sector with low spatial and temporal resolution lead to similar results as our study considering multiple sectors and high resolution.
Also, Rauner and Budzinski (2017) show in their most environmentally friendly scenario for 2050, increases in the same environmental impacts as in our study: metal depletion, land occupation, and ecotoxicity. However, the authors find an increase in metal depletion by merely 200 %, whereas the increase is about 520 % in our study. The comparison shows that only considering the electricity sector underestimates environmental impacts due to an increasing electricity demand caused by sector-coupling technologies, particularly after 2030 for the German case study.

6.2.3 Discussion of the results

All energy models attempt to represent reality. The larger and more complex the energy systems under consideration are, the more assumptions are needed. Besides, models that predict the future require more far-reaching assumptions, e. g., the development of efficiencies and costs. Here, we not only investigate the complex German energy system, but also look 30 years into an uncertain future. Accordingly, our model is based on many assumptions, which we carefully reviewed and documented.
Here, we discuss the influence of the 3 most important assumptions.

System boundaries

SecMOD combines many features to model the German energy system as accurately as possible, such as high spatial and temporal resolution and the most important sectors. This accurate representation of time and space leads to a great demand for storage capacity, which is the main cost driver of the low-carbon energy system, cf. Chapter 6.2.1. For the considered sectors, we assume constant or rising demands. Behavioral change might reduce these demands, which directly reduces system costs and environmental impacts. Besides, further sectors can influence the system costs and impacts, such as the chemical industry as a potentially large consumer of renewable energy and/or producer of fuel and storage provider.

In SecMOD, we consider the German energy system without international electricity trade. However, we expect that international electricity trade would reduce the storage demand and total annualized costs as suggested by Weitemeyer et al. (2016), (Child et al., 2018), (Louis et al., 2020), and Breyer et al. (2020). For example, Child et al. (2018) find that costs can be reduced by about 10 % for a highly integrated European electricity system. Storage demand and total annualized costs can further be slightly reduced by exploiting storage capacities of battery electric vehicles (Weitemeyer et al., 2016) and (Victoria et al., 2019).

Our modeled German energy system can only import conventional fuels, i. e., natural gas, gasoline, and diesel. However, in a low-carbon world, synthetic fuels are likely to be produced abroad and transported to energy-importing countries like Germany (Agora Verkehrswende et al., 2018) and (Hank et al., 2020). Synthetic fuels could enable the transition towards low-carbon energy systems, still based on current infrastructure, without large changes (The Royal Society, 2019). Thereby, synthetic fuels might reduce costs. Furthermore, synthetic fuels could also reduce environmental impacts such as NOx and soot (Deutz et al., 2018).

In summary, we expect that international trade of electricity reduces costs. Further cost reductions could arise by reducing demand by behavioral change, storage in BEVs, and importing synthetic fuels. The consideration of further sectors might reduce costs due to synergies or impose further costs and burdens as added consumers.

Annualization

Technologies are usually annualized over their lifetime. We use an interest rate of 5 % for annualization. Both parameters, time of annualization, and interest rate, are fixed over the entire optimization horizon. These parameters have a high impact on the total annualized costs and also on technology preference since they define the balance between operation and investment cost. However, the importance of operational costs declines in low-carbon energy systems. Thus, the right balance between operation

and investment costs becomes less important.
We also use annualized infrastructural impacts in the LCA. Without annualized impacts, the impacts would show peaks in the year of construction. However, these annualized impacts do not affect on the transition pathway, but only on the representation of results. Still, when interpreting results, we must consider that environmental impacts actually occur in the year of construction.

Technology parameters
80 % of the total annualized costs in 2050 are due to the investment costs of just 5 technologies: battery storage, battery-electric vehicles, natural gas vehicles, onshore wind, and photovoltaics. Thus, cost predictions for these technologies have a significant impact on the total annualized costs. To evaluate a potential cost range, we collected more optimistic cost projections, Table 6.2. These cost projections reduce the total annualized costs in 2050 by over 40 % to about 460 bn €/a (Table 6.2 and cf. Fig. 6.2). Thereby the total annualized costs increase only by a factor of 1.7, instead of 3, compared to current costs. Thus, the economic development of technologies has a large influence on future costs.
The ecological development of technologies can have a similar positive effect on our environmental results. For example, Cox et al. (2018) show that climate impact of future battery electric vehicles might decrease by 75 %, mainly due to cleaner resources. Our results show the strong need for low-carbon technologies with reduced needs for metal and minerals and less freshwater eutrophication. Hopefully, technology development will reduce both, costs and environmental impacts.

Table 6.2: Comparison of investment costs of specific technologies in SecMOD and more optimistic cost projections in 2050.

technology	**SecMOD**	**lower predictions**
battery systems (€/kW$_{el}$ and /kWh$_{el}$)	350	105 (Fasihi and Breyer, 2020)
battery electric vehicles (€ per vehicle)	29437	20843 (Khalili et al., 2019)
natural gas vehicles (€ per vehicle)	24289	20260 (Khalili et al., 2019)
onshore wind (€/kW$_{el}$)	938	524 (He et al., 2020)
photovoltaics (€/kW$_{el}$)	547	246 (Fasihi and Breyer, 2020)

All assumptions, and these 3 assumptions in particular, have a significant influence on the quantitative results. We believe that all our assumptions are well in line with the state-of-the-art and, thus, reasonable. Still, we recognize the uncertainty and focus our conclusion on underlying trends.

6.3 Summary and conclusion

In this chapter, we propose the national energy system model SecMOD. SecMOD considers the electricity sector, the domestic and industrial heating sector as well as the private transportation sector in Germany. SecMOD optimizes the low-carbon transition of a national energy system with a high spatial and temporal resolution, and expands the scope of the state-of-the-art by life-cycle assessment (LCA). Thus, we solve Challenge 4, cf. Chapter 2.4. SecMOD is parametrized for the German energy system; however, we believe that key findings and trends are valid for all energy systems in industrialized countries under transition.

We show a pathway optimized in a rolling-horizon approach to achieve the German GHG reduction goal of 85 % until 2050 compared to 1990. To achieve ambitious GHG reduction goals, 3 elements are key: deployment of low-carbon technologies, storage technologies, and the electrification of all sectors. The electrification of all sectors results in a doubling of electricity demand to 1100 TWh/a. To cope with the volatile and distributed nature of renewable energies, doubled electricity grid capacity and large storage capacity are necessary. We show that the energy transition is feasible, but, total cost increases by a factor of 3. However, these high costs might be reduced significantly by international trade of electricity, import of synthetic fuels, and technology development. Using alternative learning curves, reduces the increase already to a factor of 1.7.

In future, operation contributes less to overall environmental impacts, while infrastructural environmental impacts are gaining more importance. This shift to infrastructural impacts shows that energy system models have to include LCA to assess environmental impacts holistically and detect burden-shifting. We derive specific impacts for 1 MWh_{power} of power and MWh_{heat} heat and 1 vkm of transportation, for the computed energy transition satisfying the GHG-emission targets.
For the power sector, we observe many co-benefits of the energy transition: 11 of 16 environmental impacts decrease. However, some environmental impacts also increase: For instance, the transition increases the depletion of metal and minerals by a factor of 6, compared to 2016. The comparison with other studies shows that considering only the electricity sector underestimates the specific environmental impacts, also for the electricity sector. Thus, all sectors need to be considered, even if only electricity supply is of interest.
In the heating and transportation sector, also only few impacts increase with the largest increase for metals and minerals resource depletion, freshwater eutrophication and land use. These impacts are critical for future technology development.

To avoid burden-shifting, national GHG-reduction strategies should consider the mentioned environmental impacts rather than focusing on climate change only. The observed increases concern mainly impacts with high uncertainty. Further research in LCA should improve LCIA methods to reduce uncertainty for these important environmental impacts.

Chapter 7

Summary and conclusions

In this final chapter, we summarize the scope expansion realized in this thesis and we draw main conclusions. In Chapter 7.1, we then discuss potential future research directions to enhance the presented methods and models.

Mitigation of climate change requires significant reduction of greenhouse gas (GHG) emissions. The main contributor to anthropogenic GHG emissions is the energy sector, due to its dependence on fossil fuels. Thus, we need to transform the energy sector from carbon-based towards low-carbon technologies. The transformation relies on affordable and innovative energy system designs and well-suited operation strategies. Mathematical optimization is suited to find these designs and operation strategies.

However, mathematical optimization of energy systems with low-carbon technologies is challenging. Low-carbon technologies introduce more and increasingly volatile inputs to energy system optimization. Further, to properly manage the transformation process, GHG emissions and potential burden-shifting have to be correctly assessed. Both, the time-dependent inputs and the correct assessment, complicate optimal synthesis and operation. To cope with these challenges, we extended the scope of energy system optimization in this thesis.

The operation of energy systems, including low-carbon technologies, requires short-term flexibility to balance volatile renewable generation. Simultaneously, planning horizons are becoming longer due to the increased importance of emission limits, network constraints, and seasonal storage. Thus, optimal operation of low-carbon energy systems leads to long-term optimization problems with time-coupling constraints and high temporal resolution. In Chapter 3, we propose the solution method DeLoop, which extends the scope of solution methods to handle these extended planning horizons, while exploiting short-term flexibility and guaranteeing the desired solution quality. In an industrial case study, we first show the importance of long-term operational planning as significant cost reductions of about 10 % can be achieved. Second, we show that DeLoop outperforms a commercial solver, in terms of solution time, in a large computational study on average by a factor of 32. Due to the increased computing speed, DeLoop enables long-term planning and associated cost reductions to be realized in practice.

Not only the operation becomes more complex due to low-carbon technologies, but also the optimal synthesis of energy systems. The optimal synthesis needs to consider additional time-dependent input data that is at the same time increasingly volatile, e. g., renewable generation and time-variant prices. Further, the synthesis has to include the optimal design of seasonal storage systems. For both, time-dependent input data and optimal seasonal storage design, previous solution methods are not suited. Therefore, in Chapter 4, we propose the solution method $RiSES^4$, which extends the scope of synthesis optimization to handle time-dependent input data and seasonal storage design. Although, $RiSES^4$ employs time-series aggregation, the method maintains known optimal solution quality. In 2 industrial case studies, we show that the time-dependent input data can be aggregated to few typical periods, which reduces the problem size significantly, but still leads to near-optimal solutions. Due to the smaller problem size, solution time reduces by a factor of 65 compared to a commercial solver. $RiSES^4$ thus extends the scope of current solution methods and enables optimal synthesis of low-carbon energy systems, including seasonal storage systems.

To identify trade-offs between economic and ecological designs, multi-objective optimization has to be employed. For the multi-objective design of low-carbon energy systems, the GHG emissions are often considered as a second objective. State-of-the-art energy system optimization considers GHG emissions from grid electricity with constant emission factors. Whereas, in reality, these grid emissions are time dependent. In Chapter 5, we therefore propose methods to compute time-dependent electricity grid emission factors. In an industrial case study, we compute trade-off curves between economic and ecological designs of low-carbon energy systems. We show that using constant grid emissions misquantifies emissions and that considering time-dependent grid emissions reduces emissions by 6-60 %. In our vision, these emission factors are incorporated not only in future environmental assessments in academia, but also as part of industrial accounting, for example, within the European 'Emission Trading System.' Policies could enforce the use of time-dependent GHG grid emissions by a binding 'day-ahead emissions' forecast similar to the day-ahead market for electricity.

In the previously discussed chapters, we investigate the integration of low-carbon technologies into industrial energy systems. However, the transformation will also occur on the national scale and thereby impact environment, economy, and society, not only by GHG but also via other emissions. In Chapter 6, we therefore develop the national energy system model SecMOD to optimize the transition towards a low-carbon future. We extend the scope of the model by life-cycle assessment to consider further environmental impacts and potential burden-shifting. We show that a low-carbon energy system is possible in Germany and leads to many co-benefits. However, some environmental categories worsen significantly, such as metal and mineral resource depletion. Thus, we suggest to consider these categories when developing new low-carbon technologies and designing energy systems.

This thesis aims to facilitate the integration of low-carbon technologies in energy systems. For this purpose, we improve solution methods for complex optimization problems and propose methods to include holistic environmental assessment in the synthesis of energy systems. We thus extend the scope of energy system optimization for low-carbon technologies. However, there is still a need for further scope extensions. Thus, in the next section, we suggest possible expansions.

7.1 Future scope of energy system optimization

In this section, we suggest further scope expansions for energy system optimization. We rate the suggestions according to their potential benefit and effort to implement on a scale of 1-5. The suggestions are ordered by increasing benefits.

Extreme events in time-series aggregation: benefit 1, effort 1

In this thesis, synthesis of energy systems is always based on time-series aggregation. However, time-series aggregation removes peaks and other extreme events, which are crucial to achieve feasible designs. We use heuristics to include peaks and other extreme events in time-series aggregation based on Bahl et al. (2017a), which is sufficient for our case studies. The heuristic, however, does not account for storage systems, which might enhance feasibility. Thus, methods to include extreme events to ensure feasibility for the design of all kinds of energy systems are currently missing. Such methods would further improve the applicability of time-series aggregation to energy system optimization. Currently, Teichgraeber et al. (2020) is working on an overview and new methods to include extreme events in time-series aggregation.

Combination of RiSES^4 and DeLoop: benefit 2, effort 2
In Chapter 4, we develop RiSES^4, a two-stage algorithm for the rigorous synthesis of energy systems with coupling constraints due to seasonal storage systems. However, in the second-stage operational optimization of RiSES^4, we neglect additional coupling constraints by emission limits and network-connection fees. In Chapter 3, we deal with these coupling constraints in operational optimization using the decomposition method DeLoop. RiSES^4 could be extended to deal with these coupling constraints as well. For this purpose, DeLoop has to be integrated into the second stage of RiSES^4. With the combination of RiSES^4 and DeLoop, we could rigorously solve design problems with even more model detail, e. g., network-connection fees.

Multi-objective rigorous synthesis: benefit 3, effort 2
In Chapter 4, we develop the method RiSES^4 for the rigorous synthesis of energy systems regarding a single objective. For the synthesis of energy systems, however, often, many conflicting objectives exist. Thus, the scope of RiSES^4 could be extended to multi-objective rigorous synthesis. The extended method needs to compare Pareto frontiers of lower bounds and upper bounds. The frontiers could be iteratively improved using the same concepts as in RiSES^4, until a distance criterion between upper and lower bound frontiers is satisfied.
In Chapter 5, we solve a multi-objective synthesis problem with time-series aggregation, but lack a rigorous method to quantify the quality of the resulting energy systems. The extended multi-objective RiSES^4 method would enable us to solve the problem presented in Chapter 5 with more model detail, i. e., with optimality and feasibility proof.

Multi-dimensional rigorous synthesis (time/spatial): benefit 3, effort 3
In Chapter 6, we solve a multi-period problem for the optimal transformation of a national energy system, without optimality proof. RiSES^4 could be adapted to consider the multiple periods as additional time dimension to aggregate and relax. The solution of the aggregated and relaxed problem could serve as lower bound, whereas the current rolling-horizon approach can serve as feasible solution. This scope expansion would enable us to rigorously solve the multi-dimensional national energy system model presented in Chapter 6.
In the context of national energy system models, besides time resolution, spatial resolution is crucial. In current national energy system models, the spatial dimension is often aggregated without consideration of optimality or feasibility. RiSES^4 currently ensures the optimality and feasibility of designs obtained via time-series aggregation. We suggest to extend RiSES^4 to also ensure the optimality and feasibility of designs

obtained using spatial aggregation. This scope expansion would enable us to consider the national energy system model presented in Chapter 6, even with a higher spatial resolution. Very recently, Reinert et al. (2020a) proposed a heuristic approach to aggregate the spatial resolution with feasibility proof, but without optimality proof.

Uncertainty in energy system models: benefit 4, effort 4

In this thesis, all inputs are assumed to be certain. However, in general, optimization of energy systems should consider the uncertainty of input data, e. g., forecast uncertainty. The integration of uncertainty would be an important scope expansion, enhancing the optimal solutions's reliability and robustness.

The scope of operational optimization can be extended to handle uncertainty via a receding-horizon approach, where the operation schedule is frequently optimized based on updated inputs. As DeLoop enables fast computation, it could be used in a frequent receding-horizon optimization to account for uncertainty.

To account for uncertainty in the optimal design, our methods and models could be extended by methods from robust or stochastic optimization. However, robust and stochastic optimization also increase the complexity of the optimization. To reduce complexity, the time-series aggregation concept of RiSES[4] could be used to identify necessary inputs without losing optimality or feasibility.

Going open-source: benefit 5, effort 3

In this thesis, we publish all our assumptions, equations and a lot of the data used in our methods and models. Still, the software-implementation and data have not been published open-source and thus are time-consuming to reproduce and reuse for external stakeholders. Open-source means that data sets, results, and methods can be freely downloaded, applied, improved, and redistributed. Open-source energy models will lead to increased sharing and cooperation within our community of modelers, which reduces current massive duplication of modeling and method developing effort. Furthermore, results would be more credible and easier to compare, as all assumptions are public. Therefore, we suggest to develop methods and models already with the intention to publish open-source in the end.

Appendices

Appendix A

MILP model for the industrial energy system

Here, we first provide the literature overview and second present a detailed model description for the industrial energy system model used in this thesis. Here, we present the multi-objective synthesis model as the most comprehensive model used in Chapter 5. In Section A.14, we describe the model adaptations necessary for Chapter 3 and Chapter 4.

A.1 Literature overview

We conducted a literature review to determine: the type of model formulation; the necessary energy-conversion components; and the demands which are usually supplied by such energy systems. The literature review is summarized in Table A.1. The references are ordered by the model formulation type and within one type chronological.

Table A.1: Literature review on industrial energy system models. Used abbreviations are for the model type: mixed-integer (non) linear programming problem MI(N)LP, combination of successively solving a MILP and a non-linear simulation (S-MILP), operational optimizations are marked by *; abbreviations for the energy-conversion units: absorption chillers (AbC), boilers (B), batteries (BAT), electrode boilers (BEL), compression chillers (CC), combined heat and power engines (CHP), concentrated solar power (CSP), cooling towers (CT), fuel cells (FC), heat exchanger (HEX), heat pumps (HP), heat recovery (HR), hydrogen storage (HS), DC/AC inverter (INV), organic Rankine cycle (ORC), power engine (PE), photovoltaic (PV), steam compressor (SC), solar thermal (ST), thermal storage (TS), wind turbines (WIND), and turbines (T); abbreviations for demand types: cooling (c), electricity (el), heating (h), and steam (s). References are sorted according to model type and chronology.

type	units	demands	reference
S-MILP*	B, T	el, s	Varbanov et al. (2004)

S-MILP	B, HR, T	el, s	Varbanov et al. (2005)
S-MILP*	B, HR, T	el, s	Velasco-Garcia et al. (2011)
S-MILP	B, CHP, ST, TS, T	c, el, s	Varbanov and Klemeš (2011)
S-MILP	B, HEX, ST, PV, TS,	el, h	Wallerand et al. (2016)
MINLP*	B, T	el, s	Shamsi and Omidkhah (2012)
MI(N)LP	AbC, B, CHP, TS	c, el, h	Zhou et al. (2013)
MINLP	B, CHP, HP, PV, ST, TS	el, h	Falke et al. (2016)
MINLP	AbC, B, HP, T	c, el, h	Zhu et al. (2017)
MI(N)LP	AbC, B, CC, CHP, HEX, PV, TS, WIND	c, el, h	Timmerman et al. (2017)
MINLP	CT, HEX, PE, SC, T	c, el, h, s	Onishi et al. (2017)
MILP	B, T	el, h, s	Kim et al. (2010)
MILP*	AbC, B, CHP, FC, PV	c, el, h	Ren et al. (2010)
MILP	AbC, B, CC, CHP, TS	c, el, h	Ortiga et al. (2011)
MILP*	CHP, HP, HR, TS	el, h	Blarke and Dotzauer (2011)
MILP*	CHP, TS	el, h	Christidis et al. (2012)
MILP*	B, T	el, s	Luo et al. (2012)
MILP	B, CHP, FC, HP, PV	el, h	Fazlollahi et al. (2012)
MILP*	CHP	el, h	Mitra et al. (2013)
MILP	B, CHP, ST, TS	el, h	Buoro et al. (2013)
MILP	B, CHP	el, h	Bracco et al. (2013)
MILP	AbC, B, CC, CHP	c, el, h	Voll et al. (2013) Kirschbaum et al. (2013) Yang et al. (2015a)
MILP	AbC, B, CC, HP, TS	el, h	Haikarainen et al. (2014)
MILP*	AbC, B, CC, CHP, HP, TS	c, el, h	Bischi et al. (2014)
MILP	B, CHP, ST	el, h	Fazlollahi et al. (2014)
MILP	AbC, B, CC, CHP, PV, TS, WIND	c, el, h	Yang et al. (2015b)

MILP	B, BAT, BEL, CHP, HP, INV, PV, ST, TS	el, h	Harb et al. (2016)
MILP	B, CHP, PV, TS	el, h	Morvaj et al. (2016)
MILP	B, CC, CHP, HP, PV, ST, TS	c, el, h	Bracco et al. (2016)
MILP	B, CHP, ORC, PE	el, h	Cedillos Alvarado et al. (2016)
MILP	B, BAT, CHP, PV, ST, TS	el, h	Maroufmashat et al. (2016)
MILP	AbC, B, BAT, CHP, CC, PV, TS,	c, el, h	Li et al. (2016)
MILP	B, CC, CHP, ST , TS	c, el, h	Wu et al. (2016)
MILP	B, BAT, CHP, BEL, PV, TS	el, h	Atabay (2017)
MILP	B, CHP, HP, PV, TS	el, h	Morvaj et al. (2017)
MILP	AbC, B, CC, CHP, PV, WIND	c, el, h	Bahl et al. (2017a)
MILP	B, CC, HP, HS, PV, ST, TS	c, el, h	Wallerand et al. (2018)
MILP	B, BAT, BEL, CHP, HP, INV, PV, ST, TS	el, h	Schütz et al. (2018)
MILP	AbC, B, CC, CHP, TS	c, el, h	Bahl et al. (2018b)
MILP	B, BAT, CHP, HP, HS, PV, ST, TS	el, h	Gabrielli et al. (2018)
MILP	AbC, B, CC, CHP	c, el, h, s	Yokoyama et al. (2018a)
MILP	B, CHP	el, h	Yokoyama et al. (2018c)
MILP*	B, T, TS	el, s	Panuschka and Hofmann (2019)
MILP	B, BAT, CC, CHP, HP, PV ST, TS	c, el, h	Zatti et al. (2019)
MILP	CHP, FC,	el, h	Yokoyama et al. (2019)
MILP	B, BAT, CHP, PV, WIND	el, h	Roustaei et al. (2020)
MILP	AbC, B, BAT, CC, CHP, HP, ORC, PV, ST, TS	c, el, h	Martelli et al. (2020)

A.2 Model formulation

In this work, we follow most authors and formulate the synthesis of the low-carbon industrial energy system as piecewise linearized MILP problem. For the types of energy demand, the energy systems investigated in literature supply some combination of electricity, cooling, heating and steam, Table A.1. In our model, we consider all these four types of energy demand. We include most energy-conversion units considered in the literature in the proposed model.
Parts of the model are based on previous work of our group (Voll et al., 2013; Bahl et al., 2017a, 2018b). For readability, here, we present the complete model description, including all novelties. The section is structured as follows: First, the objectives of the low-carbon industrial energy system are presented. Second, the energy balances are presented. Third, all considered energy-conversion units are described in detail. To avoid repetition, we start with a general unit subsection and refer to this subsection for further units whenever possible.

A.3 Objectives

As objective functions, we employ the total annualized costs TAC, Eq. (A.1) and the global warming impact GWI due to operation, Eq. (A.2),

$$\min \ TAC$$

$$= \overbrace{\frac{1}{APVF} \cdot \sum_{n \in \mathcal{C}} I_n^{\mathrm{N}}}^{CAPEX} + \overbrace{\sum_{t \in \mathcal{T}} \left(\Delta \mathrm{t}_t \sum_{n \in \mathcal{C}} \mathrm{c}_{n,t}^{\mathrm{o}} \cdot \dot{U}_{n,t} \right) + \sum_{n \in \mathcal{C}} \mathrm{c}_n^{\mathrm{m}} \cdot I_n^{\mathrm{N}} + \mathrm{c}^{\mathrm{p}} \cdot \dot{V}_{\mathrm{grid}}^{\max}}^{OPEX}, \tag{A.1}$$

$$\min \ GWI = \sum_{t \in \mathcal{T}} \left(\Delta \mathrm{t}_t \sum_{n \in \mathcal{C}} \mathrm{e}_{n,t}^{\mathrm{o}} \cdot \dot{U}_{n,t} \right). \tag{A.2}$$

The total annualized costs TAC consist of the capital and operational expenditures, $CAPEX$ and $OPEX$. The annualized investment costs CAPEX, Eq. (A.1), are the sum of the units investment costs I_n^{N} for all units $n \in \mathcal{C}$ divided by the annualization factor $APVF = \frac{(i+1)^{T_{\mathrm{a}}}-1}{i(i+1)^{T_{\mathrm{a}}}}$. The interest rate is $i = 8\,\%$ and the time horizon $T_{\mathrm{a}} = 15a$. The unit investment costs I_n^{N} are calculated based on the nominal capacity of each unit $\dot{V}_n^{\mathrm{N}}$, Section A.5. The operational costs $OPEX$, Eq. (A.1), are defined as the

sum of energy costs, maintenance, and peak power costs. The energy costs are the input power $\dot{U}_{n,t}$ of all units $n \in \mathcal{C}$ in every time step $t \in \mathcal{T}$ multiplied by the specific operation cost $c^{\mathrm{o}}_{n,t}$ and the duration $\Delta \mathrm{t}_t$ of a time step. The specific operation cost $\mathrm{c}^{\mathrm{o}}_{n,t}$ are the gas costs for all gas consuming units, the electricity price for all electricity consuming units, and $\mathrm{c}^{\mathrm{o}}_{n,t} = 0$ for all other units. The maintenance costs depend on the investment costs and are calculated by multiplying the investment costs of each unit I^{N}_n by the corresponding maintenance factor $\mathrm{c}^{\mathrm{m}}_n$. Additionally, the peak power price is considered in the operational costs $OPEX$ by multiplying the electrical peak power $\dot{V}^{\mathrm{max}}_{\mathrm{grid}}$ by a specific grid fee $\mathrm{c}^{\mathrm{p}} = 70€/kW \cdot a$.
The global warming impact GWI, Eq. (A.2), is defined as the sum of the input power $\dot{U}_{n,t}$ of every unit $n \in \mathcal{C}$ in every time step $t \in \mathcal{T}$ multiplied by the specific emission factor $\mathrm{e}^{\mathrm{o}}_{n,t}$ and the duration $\Delta \mathrm{t}_t$ of a time step. The specific emission factor $\mathrm{e}^{\mathrm{o}}_{n,t}$ is the one of natural gas for all gas consuming units, grid emissions for all electricity consuming units, and $\mathrm{e}^{\mathrm{o}}_{n,t} = 0$ for all other units.

A.4 Energy balances

The energy system has to supply the energy demands $\dot{E}_{x,t}$ in each time step ($\forall t \in \mathcal{T}$):

$$\begin{aligned}
\dot{E}_{steam,t} &= \dot{V}_{B,t} + \dot{V}_{B_{el},t} - \dot{U}_{ORC,t} - \dot{U}_{HEX,t}, \\
\dot{E}_{heat-low,t} &= \dot{V}_{ORC_{th},t} + \dot{V}_{HEX,t} + \dot{V}_{CHP_{th},t} + \Delta\dot{V}_{TS_{heat},t} - \dot{U}_{AbC,t}, \\
\dot{E}_{cool,t} &= \dot{V}_{AbC,t} + \dot{V}_{CC,t} + \Delta\dot{V}_{TS_{cool},t}, \\
\dot{E}_{el_{AC},t} &= \dot{V}_{CHP_{el},t} + \dot{V}_{ORC_{el},t} + \Delta\dot{V}_{INV,t} + \dot{V}_{WIND,t} + \Delta\dot{V}_{grid,t}, \\
\dot{E}_{el_{DC},t} &= \dot{V}_{PV,t} - \Delta\dot{V}_{Bat,t} - \Delta\dot{U}_{INV,t}.
\end{aligned} \tag{A.3}$$

The steam demand $\dot{E}_{steam,t}$ can be supplied by the gas boiler (B) and electrode boiler (B_{el}). The heat exchanger (HEX) and the organic Rankine cycle (ORC) consume steam as input power $\dot{U}_t$. The low-temperature heat demand $\dot{E}_{heat-low,t}$ is supplied by the output power of the heat exchanger (HEX), the thermal output power of the organic Rankine cycle (ORC), and the combined heat and power engine (CHP). The absorption chiller (AbC) consumes low-temperature heat as input power $\dot{U}_{AbC,t}$. The net energy output $\Delta\dot{V}_{TS_{heat},t}$ of the low-temperature heat storage (TS_{heat}) consumes or supplies additional low-temperature heat in a specific time step t. The cooling demand $\dot{E}_{cool,t}$ is supplied by the output power of the absorption chiller (AbC) $\dot{V}_{AbC,t}$ and the compression chiller (CC) $\dot{V}_{CC,t}$. The net energy output $\Delta\dot{V}_{TS_{cool},t}$ of the cooling storage (TS_{cool}) consumes or supplies additional cooling in a specific time step t. The AC-electricity demand $\dot{E}_{el_{AC},t}$ is supplied by the electric output power of the organic

Rankine cycle (ORC) $\dot{V}_{ORC,t}$, the combined heat and power engine (CHP) $\dot{V}_{CHP_{el},t}$ and the wind turbines (WIND) $\dot{V}_{WIND,t}$. Furthermore, the net energy output of the inverter $\Delta\dot{V}_{INV,t}$ consumes or supplies additional AC-electricity. Excess or lack of AC-electricity can be exchanged with the public electricity grid. The DC-electricity demand $\dot{E}_{el_{DC},t}$ is supplied by photovoltaic (PV). Moreover, the net energy output $\Delta\dot{V}_{Bat,t}$ of the battery (BAT) and the net energy input of the inverter $\Delta\dot{U}_{INV,t}$ consume or supply additional DC-electricity. For the real-world case study, no DC-electricity demand is present, $\dot{E}_{el_{DC},t} = 0$.

A.5 General energy-conversion unit model

Investment costs

The investment costs I_n^{N} of the units follow the capacity power law (Smith, 2005):

$$I_n^{\mathrm{N}} = I_n^{\mathrm{B}} \cdot \left(\frac{\dot{V}_n^{\mathrm{N}}}{\dot{V}_n^{\mathrm{B}}}\right)^{\mathrm{K}_n} \qquad \forall n \in \mathcal{C}. \tag{A.4}$$

The costs of a unit I_n^{N} are defined as the reference cost for the unit I_n^{B} multiplied by the ratio of the capacity $\dot{V}_n^{\mathrm{N}}$ and a reference value $\dot{V}_n^{\mathrm{B}}$ to the power of K_n, Table A.2. The reference value is set to $\dot{V}_n^{\mathrm{B}} = 1\mathrm{kW}$ for all units except the heat exchanger, Section A.8.

Table A.2: Parameter of the energy-conversion units: Reference costs of a unit I_n^B, investment exponent K_n, maintenance factor c_n^m, nominal efficiency η_n^N for every unit, and thermal efficiency $\eta^{N,th}$ of CHPs and ORCs. Data sources are given as footnotes. Data sources for unit parameters based on previous work of our group are not shown.

unit ($n \in \mathcal{C}$)	I_n^B /€	K_n	c_n^m	η_n^N	η_{th}^N
absorption chiller[a]	8847.5	0.4345	0.01	0.67	–
battery[b]	2116.1	0.8382	0.025	0.92	–
boiler[a]	2701.6	0.4502	0.015	0.9	–
CHP_1[a]	9332.6	0.539	0.1	0.87	0.4604
CHP_2[a]	9332.6	0.539	0.1	0.87	0.4244
CHP_3[a]	9332.6	0.539	0.1	0.87	0.3884
compression chiller[a]	444.3	0.8732	0.04	5.54	–
electrode boiler[c]	18307	0.464	0.01	0.99	–
heat exchanger[d]	3600	0.65	0.025	1	–
heat pump[e]	1654.7	0.6611	0.01	0.5	–
inverter[f]	486.8	0.8687	0.01	0.95	–
ORC[g]	5974.9	0.876	0.025	0.98	0.8
photovoltaic[h]	4264.3	0.9592	0.01	1	–
solar thermal[i]	2215.4	0.735	0.01	1	–
storage system[j]	83.83	0.8663	0.01	0.95	–
wind turbine[k]	Section A.13		0.0257	1	–

[a]model of Voll et al. (2013)
[b]based on industrial data
[c]costs from Eller (2015), efficiencies based on industrial data
[d]data from Timmerman et al. (2017)
[e]based on industrial data
[f]based on industrial data
[g]costs from Quoilin et al. (2013), efficiencies based on industrial data
[h]model of Bahl et al. (2017a), updated costs based on Theo et al. (2017)
[i]costs from IEA-ETSAP and IRENA (2015)
[j]model of Bahl et al. (2018b), costs from industrial data, efficiencies from Facci et al. (2014)
[k]model of Bahl et al. (2017a)

In the MILP model, the non-linear investment cost functions is approximated by piecewise linear functions h. The following constraints, Eq. (A.5-A.10), model the investment costs I_n^N ($\forall n \in \mathcal{C}$):

$$I_n^N = \sum_h \left(\gamma_{h,n} \cdot I_{h,n} + \left(\dot{V}_{h,n}^N - \dot{V}_{h,n}^0 \right) \cdot \left(\frac{\mathrm{d}I_n}{\mathrm{d}\dot{V}_n^0} \right)_h \right), \tag{A.5}$$

$$\gamma_{h,n} \cdot \dot{V}_{h,n}^0 \leq \dot{V}_{h,n}^N \leq \gamma_{h,n} \cdot \dot{V}_{h+1,n}^0 \qquad \forall h \in \mathcal{H}, \tag{A.6}$$

$$\gamma_n = \sum_h \gamma_{h,n} \leq 1, \tag{A.7}$$

$$\dot{V}_n^{\mathrm{N}} = \sum_h \dot{V}_{h,n}^{\mathrm{N}}. \tag{A.8}$$

$\gamma_{h,n}$ is a binary variable representing which part h of the linearized investment cost curve is active (active: $\gamma_{h,n} = 1$, inactive: $\gamma_{h,n} = 0$). Eq. (A.6) ensures that the nominal power $\dot{V}_{h,n}^{\mathrm{N}}$ lies within the boundaries of $\dot{V}_{h,n}^{0}$ and $\dot{V}_{h+1,n}^{0}$ with the supporting points h and $h+1$, values for all units $\forall n \in \mathcal{C}$ are presented in Eq. (A.3). If the unit is not installed ($\gamma_n = 0$) the nominal power $\dot{V}_n^{\mathrm{N}}$ is zero, because no part of the linearized investment costs curve can be chosen, Eq. (A.7). In Eq. (A.8), a single continuous variable is introduced representing the installed nominal power $\dot{V}_n^{\mathrm{N}}$.
In Eq. (A.5), $I_{h,n}$ and $\left(\frac{\mathrm{d}I_n}{\mathrm{d}\dot{V}_n^{\mathrm{N}}}\right)_h$ are coefficients representing the intercepts and the slopes of the piecewise linear approximation of the investment cost curves:

$$I_{h,n} = \mathrm{I}_n^{\mathrm{B}} \cdot \left(\frac{\dot{V}_{h,n}^{0}}{\dot{V}_n^{\mathrm{B}}}\right)^{\mathrm{K}_n} \qquad \forall h \in \mathcal{H}, \forall n \in \mathcal{C}, \tag{A.9}$$

$$\left(\frac{\mathrm{d}I_n}{\mathrm{d}\dot{V}_n^{\mathrm{N}}}\right)_h = \gamma_{h,n} \cdot \frac{I_{h+1,n} - I_{h,n}}{\dot{V}_{h+1,n}^{0} - \dot{V}_{h,n}^{0}} \qquad \forall h \in \mathcal{H}, \forall n \in \mathcal{C}. \tag{A.10}$$

Table A.3: Capacity $\dot{V}^{\mathrm{N}}_{h,n}$ at the supporting points $h \in \mathcal{H}$ for the linearization of the investment cost curve for each unit

unit ($n \in \mathcal{C}$)	**supporting points** $\dot{V}^{0}_{h,n}$/**kW**			
	h_1	h_2	h_3	h_4
absorption chiller	50	750	6500	–
battery $V^0_{h,n}$/kWh	40	60	2000	6000
boiler	100	14000	–	–
CHP_1	500	711.56	1400	–
CHP_2	1400	2300	–	–
CHP_3	2300	3200	–	–
compression chiller	400	10000	–	–
electrode boiler	100	500	2000	5000
heat exchanger	20	200	1000	5000
heat pump	20	200	1000	5000
inverter	220	330	1400	2000
ORC	50	1000	4000	–
photovoltaic	100	1000	5000	20000
solar thermal	50	500	5000	10000
storage unit (heat) $V^0_{h,n}$/kWh	115	5000	25000	115000
storage unit (cold) $V^0_{h,n}$/kWh	115	1000	5000	25000
wind turbine	Section A.13			

Efficiencies

The efficiency $\eta_{n,t}$ relates the input power $\dot{U}_{n,t}$ of a unit n to its output power $\dot{V}_{n,t}$:

$$\dot{U}_{n,t} = \frac{\dot{V}_{n,t}}{\eta_{n,t}} \qquad \forall n \in \mathcal{C}, \forall t \in \mathcal{T}. \tag{A.11}$$

To follow a more linear trend, η_n is converted into output/input curves ($\dot{U}_{n,t}/\dot{V}_{n,t}$). Thereby, the operating performance curves can be modeled by fewer piecewise linear functions g ($\forall n \in \mathcal{C}, \forall t \in \mathcal{T}$):

$$\dot{U}_{n,t} = \sum_g \left(\delta_{g,n,t} \cdot \dot{U}_{g,n} + \left(\dot{V}_{n,t} - \dot{V}_{g,n}\right) \cdot \left(\frac{d\dot{U}_{g,n}}{d\dot{V}_{g,n}}\right)_g \right), \tag{A.12}$$

$$\delta_{g,n,t} \cdot \dot{V}_{g,n} \leq \dot{V}_{n,t} \leq \delta_{g,n,t} \cdot \dot{V}_{g+1,n} \qquad \forall g \in \mathcal{G}, \tag{A.13}$$

$$\sum_g \delta_{g,n,t} \leq \gamma_n. \tag{A.14}$$

$\delta_{g,n,t}$ is the binary decision variable representing the discrete on/off status of the unit n for every time step t and line segment g (active $\sum_g \delta_{g,n,t} = 1$, inactive $\sum_g \delta_{g,n,t} = 0$). Eq. (A.13) ensures that the supplied power $\dot{V}_{n,t}$ lies within the limits of its line segment load $\dot{V}_{n,g}$ and $\dot{V}_{n,g+1}$. Eq. (A.14) ensures that if a unit is not installed ($\gamma_n = 0$), the supplied power $\dot{V}_{n,t}$ is zero.
To enable continuous sizing, the characteristic performance model is employed, assuming that the part-load performance of a certain unit is valid regardless of their nominal capacity $\dot{V}_n^{\mathrm{N}}$. For CHP engines, three size classes are defined to consider the size-dependency of the ratio between nominal electric and thermal efficiencies. With this assumption, the coefficients in Eq. (A.12) and Eq. (A.13) can be expressed as functions of the nominal capacity $\dot{V}_n^{\mathrm{N}}$ as follows:

$$\dot{U}_{g,n} = u_{g,n} \cdot \dot{U}_n^{\mathrm{N}} = u_{g,n} \cdot \frac{\dot{V}_n^{\mathrm{N}}}{\eta_n^{\mathrm{N}}} \qquad \forall g \in \mathcal{G}, \forall n \in \mathcal{C}, \tag{A.15}$$

$$\left(\frac{d\dot{U}_{g,n}}{d\dot{V}_{g,n}}\right)_g = \delta_{g,n,t} \cdot \frac{u_{g+1,n} - u_{g,n}}{v_{g+1,n} - v_{g,n}} \cdot \frac{1}{\eta_n^{\mathrm{N}}} \qquad \forall g \in \mathcal{G}, \forall n \in \mathcal{C}, \tag{A.16}$$

$$\dot{V}_{g,n} = \dot{V}_n^{\mathrm{N}} \cdot v_{g,n} \qquad \forall g \in \mathcal{G}, \forall n \in \mathcal{C}. \tag{A.17}$$

In Eq. (A.15)-(A.17), $u_{g,n}$ and $v_{g,n}$ are constant coefficients representing the supporting points of the linear approximation of the characteristic performance model, Table A.4. The minimum relative output power, i.e., the minimum load fraction, of a unit is given by the supporting point g_1. The nominal input power $\dot{U}_n^{\mathrm{N}}$ is expressed in terms of the output power $\dot{V}_n^{\mathrm{N}}$ and the nominal efficiency η^{N}, Eq. (A.15). The ratio of the constant coefficients $\frac{u_{g+1,n}-u_{g,n}}{v_{g+1,n}-v_{g,n}}$ represents the slope of the linear approximation of the characteristic performance model, Eq. (A.16).

Table A.4: Relative thermal input power u_g and output power v_g at the supporting points $g \in \mathcal{G}$ for the linearization of the part-load behavior for each unit.

unit ($n \in \mathcal{C}$)	**supporting points** u_g **and** v_g		
	u_{g_1}/v_{g_1}	u_{g_2}/v_{g_2}	u_{g_3}/v_{g_3}
absorption chiller	0.272/0.2	0.483/0.6	1/1
battery	0/0	1/1	–
boiler	0.218/0.2	1/1	–
$\text{CHP}_{th,1/2/3}$	0.479/0.5	1/1	–
$\text{CHP}_{el,1/2/3}$	0.194/0.1	1/1	–
compression chiller	0.318/0.2	0.594/0.7	1/1
electrode boiler	0.205/0.2	1/1	–
heat exchanger	0.05/0.05	1/1	–
heat pump	0.216/0.2	0.509/0.5	1/1
inverter	0.048/0.05	0.503/0.5	1/1
ORC_{th}	0.3/0.3	1/1	–
ORC_{el}	0.386/0.3	0.635/0.6	1/1
photovoltaic	0/0	1/1	–
solar thermal	0/0	1/1	–
storage unit (heat)	0/0	1/1	–
storage unit (cold)	0/0	1/1	–
wind turbine	0/0	1/1	–

Linearization

The linearization of the investment cost curve and the characteristic performance model leads to bilinear products, e. g., $\delta_{g,n,t} \cdot \dot{V}_n^{\mathrm{N}}$. Bilinear products are the product of a binary ($\delta_{g,n,t}$) and a continuous decision variable ($\dot{V}_n^{\mathrm{N}}$), and thus represent a non-linearity for the optimization problem. This non-linearity is circumvented by applying the reformulation strategy developed by Petersen (Petersen, 1971) and Glover (Glover, 1975): The bilinear product is substituted by a single continuous decision variable whose value is determined by linear constraints guaranteeing that the behavior of the bilinear product is reproduced correctly. For the presented formulation, e. g., the bilinear product $\delta_{g,n,t} \cdot \dot{V}_n^{\mathrm{N}}$ is substituted by a continuous auxiliary variable $\xi_{n,t}$. To guarantee that $\xi_{n,t}$ reproduces the behavior of the bilinear product $\delta_{g,n,t} \cdot \dot{V}_n^{\mathrm{N}}$, two linear constraints are added, $\forall g \in \mathcal{G}, \forall n \in \mathcal{C}, \forall t \in \mathcal{T}$:

$$\sum_g (\delta_{g,n,t} \cdot \dot{V}^{\mathrm{N}}_{h\,\min,n}) \leq \xi_{n,t} \leq \sum_h (\delta_{g,n,t} \cdot \dot{V}^{\mathrm{N}}_{h\,\max,n}), \tag{A.18}$$

$$\sum_h \dot{V}^{\mathrm{N}}_{h,n} + \sum_h ((\delta_{g,n,t} - 1) \cdot \dot{V}^{\mathrm{N}}_{h\max,n}) \leq \xi_{n,t} \leq \sum_h \dot{V}^{\mathrm{N}}_{h,n}. \tag{A.19}$$

By this general unit model, the complete model formulation is given for the absorption chiller, boiler, compression chiller, and electrode boiler. In the following, additional constraints and model formulations are given for the other units.

A.6 Battery and thermal storage systems

For the model of batteries and thermal storage systems, the general unit model contributes capacity-dependent investment costs, maintenance costs, maximal and minimal capacity limits, and storage efficiencies. Thus, here, we complete the unit model by adding the stored energy, minimal and maximal storage limits, charging and discharging limits, and storage losses. The storage level $V_{n,t}$ is given by $(\forall BAT/TS \in \mathcal{C}, \forall t \in \mathcal{T})$:

$$V_{n,t+1} = V_{n,t} + \Delta \mathrm{t}_t \cdot \left(-c_n^{\mathrm{loss}} \cdot V_{n,t} + \eta_n \cdot \left(\dot{V}^{\mathrm{in}}_{n,t} - \dot{V}^{\mathrm{out}}_{n,t}\right)\right). \tag{A.20}$$

Based on the storage level $V_{n,t}$, the next storage level $V_{n,t+1}$ is calculated. The next time step $t+1$ is given by a index vector. This index vector includes a cycle condition. Thus, the storage level $V_{n,t=T}$ at the end of the time series $t = T$ is coupled with the beginning of the time series $t = 1$. Eq. (A.21) limits the stored energy $V_{n,t}$ to a maximum and minimum level, proportional to the capacity V_n^{N} with a factor c_n^{Vmax} and c_n^{Vmin},

$$c_n^{\mathrm{Vmin}} \cdot V_n^{\mathrm{N}} \leq V_{n,t} \leq c_n^{\mathrm{Vmax}} \cdot V_n^{\mathrm{N}} \qquad \forall BAT/TS \in \mathcal{C}, \forall t \in \mathcal{T}. \tag{A.21}$$

The input power $\dot{V}^{\mathrm{in}}_{n,t}$ and the output power $\dot{V}^{\mathrm{out}}_{n,t}$ are limited by

$$\dot{V}^{\mathrm{in}}_{n,t} \leq \dot{c}_n^{\dot{\mathrm{V}}\mathrm{in,max}} \cdot V_n^{\mathrm{N}} \qquad \forall BAT/TS \in \mathcal{C}, \forall t \in \mathcal{T}, \tag{A.22}$$

$$\dot{V}^{\mathrm{out}}_{n,t} \leq \dot{c}_n^{\dot{\mathrm{V}}\mathrm{out,max}} \cdot V_n^{\mathrm{N}} \qquad \forall BAT/TS \in \mathcal{C}, \forall t \in \mathcal{T}. \tag{A.23}$$

At one time step t, the binary variables $\epsilon^{\mathrm{in}}_{n,t}$ and $\epsilon^{\mathrm{out}}_{n,t}$ ensure that only charging or discharging is possible:

$$0 \leq \dot{V}^{\mathrm{in}}_{n,t} \leq \epsilon^{\mathrm{in}}_{n,t} \cdot V_n^{\mathrm{N}} \qquad \forall BAT/TS \in \mathcal{C}, \forall t \in \mathcal{T}, \tag{A.24}$$

$$0 \leq \dot{V}^{\mathrm{out}}_{n,t} \leq \epsilon^{\mathrm{out}}_{n,t} \cdot V_n^{\mathrm{N}} \qquad \forall BAT/TS \in \mathcal{C}, \forall t \in \mathcal{T}, \tag{A.25}$$

$$1 \geq \epsilon^{\mathrm{in}}_{n,t} + \epsilon^{\mathrm{out}}_{n,t} \qquad \forall BAT/TS \in \mathcal{C}, \forall t \in \mathcal{T}. \tag{A.26}$$

The additional parameters for the battery and the thermal storage systems are given in Table A.5.

Table A.5: Parameters for battery and thermal storage systems

parameter	symbol	value
battery charge losses	c_{BAT}^{loss}	$4.2 \times 10^{-5}\,\text{kW/kWh}$
battery rate limits	$c_{BAT}^{\dot{V}\text{in,max}}, c_{BAT}^{\dot{V}\text{out,max}}$	$0.36\,\text{kW/kWh}$
thermal storage losses	c_{TS}^{loss}	$0.005\,\text{kW/kWh}$
thermal storage rate limits	$c_{TS}^{\dot{V}\text{in,max}}, c_{TS}^{\dot{V}\text{out,max}}$	$1\,\text{kW/kWh}$

A.7 CHP engine

CHP engines produce electricity and low-temperature heat as output power $\dot{V}_{CHP,t}$. The power-to-heat ratio depends on the capacity class and the current load. The electric efficiency $\eta_{el,CHP,t}^{\text{N}}$ is defined by the difference of the nominal efficiency η_{CHP}^{N} and the thermal efficiency $\eta_{th,CHP,t}^{\text{N}}$:

$$\eta_{el,CHP,t}^{\text{N}} = \eta_{CHP}^{\text{N}} - \eta_{th,CHP,t}^{\text{N}} \qquad \forall CHP \in \mathcal{C}, \forall t \in \mathcal{T}. \tag{A.27}$$

The gas input power $\dot{U}_{CHP,t}$ is defined in the general unit model. The electric output power $\dot{V}_{CHP_{el},t}$ is computed by

$$\dot{V}_{CHP_{el},t} = \dot{U}_{CHP,t} \cdot \eta_{el,CHP,t} \qquad \forall CHP \in \mathcal{C}, \forall t \in \mathcal{T}. \tag{A.28}$$

A.8 Heat exchanger

Heat exchangers (HEX) transfer thermal power from the steam level to the low-temperature level. Here, heat exchangers are assumed to be adiabatic, i.e., the heat transfer is free of losses.

The parameters $\text{I}_{HEX}^{\text{B}}$ and K_{HEX} of the investment cost curve for the heat exchanger, Eq. (A.4), are given in square meters. Thus, the reference value $\dot{V}_{HEX}^{\text{B}}$ is used to transfer the capacity $\dot{V}_{HEX}^{\text{N}}$ to an area. For given temperature levels of steam and low-temperature heat demand, we use a constant ratio of nominal heat transfer rate to area $\text{k}\Delta\text{T} = 6.9881\text{kW}\,\text{m}^{-2}$ as reference value. This reference value results from the multiplication of the logarithmic temperature difference between the steam demand

and the heating demand and a heat-transfer coefficient of $50\mathrm{W\,K^{-1}\,m^{-2}}$. The heat exchanger area A_{HEX} is given by,

$$A_{HEX} = \frac{\dot{V}^{textN}_{HEX}}{\mathrm{k}\Delta\mathrm{T}} \qquad \forall HEX \in \mathcal{C}. \tag{A.29}$$

A.9 Heat pump

The efficiency of a heat pump from the general unit system model $\eta_{HP,t}$ is extended by a temperature influence $\eta^{\text{temperature}}_{HP,t}$ by

$$\eta^{\text{temperature}}_{HP,t} = \eta_{HP,t} \cdot \eta^{\text{Carnot}}_{t} \qquad \forall HP \in \mathcal{C}, \forall t \in \mathcal{T}, \tag{A.30}$$

with

$$\eta^{\text{Carnot}}_{t} = 1 - \frac{T^{\text{ambient}}_{t}}{T_{HP,t}} \qquad \forall HP \in \mathcal{C}, \forall t \in \mathcal{T}. \tag{A.31}$$

The ambient temperature T^{ambient}_{t} is included as time series , Fig. 6b. For the heat pump temperature, the low-temperature heat demand temperature is used $T_{HP,t} = 90\,^{\circ}\mathrm{C}$.

A.10 Inverter

The inverter (INV) is modeled as a bi-directional device capable to convert DC to AC electricity and vice versa. The electricity generation and consumption of the inverter are considered both from the perspective of the DC network and from the perspective of the AC network. The electricity consumption of inverters is denoted as $\dot{U}^{\text{el,AC}}_{INV,t}$ and $\dot{U}^{\text{el,DC}}_{INV,t}$ while the generation is denoted as $\dot{V}^{\text{el,AC}}_{INV,t}$ and $\dot{V}^{\text{el,DC}}_{INV,t}$ for the AC and DC perspective, respectively. The linearized part-load behavior is applicable for both directions of conversion. The variable $\dot{U}^{el}_{INV,t}$ denotes the input power and the variable $\dot{V}^{\text{el}}_{INV,t}$ denotes the output power:

$$\dot{U}^{\text{el}}_{INV,t} = \dot{U}^{\text{el,AC}}_{INV,t} + \dot{U}^{\text{el,DC}}_{INV,t} \qquad \forall INV \in \mathcal{C}, \forall t \in \mathcal{T}, \tag{A.32}$$

$$\dot{V}^{\text{el}}_{INV,t} = \dot{V}^{\text{el,AC}}_{INV,t} + \dot{V}^{\text{el,DC}}_{INV,t} \qquad \forall INV \in \mathcal{C}, \forall t \in \mathcal{T}. \tag{A.33}$$

The bi-directional inverter allows an electricity flow in only one direction at a time. The binary variables $\delta^{\text{AC/DC}}_{INV,t}$ and $\delta^{\text{DC/AC}}_{INV,t}$ indicate if electricity is flowing from AC to DC or vice versa:

$$\delta^{\text{AC/DC}}_{INV,t} + \delta^{\text{DC/AC}}_{INV,t} \leq 1 \qquad \forall INV \in \mathcal{C}, \forall t \in \mathcal{T}, \tag{A.34}$$

$$\dot{V}^{\text{el,AC}}_{INV,t} \leq \dot{V}^{\text{N}}_{INV,t} \cdot \delta^{\text{DC/AC}}_{INV,t} \qquad \forall INV \in \mathcal{C}, \forall t \in \mathcal{T}, \tag{A.35}$$

$$\dot{V}^{\text{el,DC}}_{INV,t} \leq \dot{V}^{\text{N}}_{INV,t} \cdot \delta^{\text{AC/DC}}_{INV,t} \qquad \forall INV \in \mathcal{C}, \forall t \in \mathcal{T}. \tag{A.36}$$

A.11 Organic Rankine cycle

The organic Rankine cycle (ORC) consumes steam $\dot{U}_{ORC,t}$ and produces electricity and low-temperature heat as output power $\dot{V}_{ORC,t}$. The ratio of electricity and low-temperature heat is part-load dependent. The electric efficiency $\eta^{\text{N}}_{el,ORC,t}$ is defined by the difference between the nominal efficiency η^{N}_{ORC} and the thermal efficiency $\eta^{\text{N}}_{th,ORC,t}$,

$$\eta^{\text{N}}_{el,ORC,t} = \eta^{\text{N}}_{ORC} - \eta^{\text{N}}_{th,ORC,t} \qquad \forall ORC \in \mathcal{C}, \forall t \in \mathcal{T}. \tag{A.37}$$

The operating efficiencies $\eta_{el,ORC,t}$ and $\eta_{th,ORC,t}$ are extended by a thermal correction term $\frac{\eta^{\text{Carnot}}_t}{\eta^{\text{N,Carnot}}}$. In the real-world case study, the temperatures of the steam and low temperature heat demand are considered as constant and thus $\frac{\eta^{\text{Carnot}_t}}{\eta^{\text{N,Carnot}}} = 1$.

$$\eta^{\text{temperature}}_{el,ORC,t} = \eta_{el,ORC,t} \cdot \frac{\eta^{\text{Carnot}}_t}{\eta^{\text{N,Carnot}}} \qquad \forall ORC \in \mathcal{C}, \forall t \in \mathcal{T}, \tag{A.38}$$

$$\eta^{\text{temperature}}_{th,ORC,t} = \eta_{th,ORC,t} \cdot \frac{\eta^{\text{N,Carnot}}}{\eta^{\text{Carnot}}_t} \qquad \forall ORC \in \mathcal{C}, \forall t \in \mathcal{T}, \tag{A.39}$$

$$\eta^{\text{Carnot,t}} = 1 - \frac{T_{ORC,t}}{T_{steam,t}} \qquad \forall ORC \in \mathcal{C}, \forall t \in \mathcal{T}. \tag{A.40}$$

The electric output power $\dot{V}_{ORC_{el},t}$ and the thermal output power $\dot{V}_{ORC_{el},t}$ are defined by

$$\dot{V}_{ORC_{el},t} = \dot{U}_{ORC,t} \cdot \eta^{\text{temperature}}_{el,ORC,t} \qquad \forall ORC \in \mathcal{C}, \forall t \in \mathcal{T}, \tag{A.41}$$

$$\dot{V}_{ORC_{th},t} = \dot{U}_{ORC,t} \cdot \eta^{\text{temperature}}_{th,ORC,t} \qquad \forall ORC \in \mathcal{C}, \forall t \in \mathcal{T}. \tag{A.42}$$

A.12 Photovoltaic and solar thermal

For photovoltaic, the electric output power $\dot{V}_{\mathrm{PV},t}$ is limited by

$$\dot{V}_{PV,t} \leq v_{PV,t} \cdot \dot{V}^{\mathrm{N}}_{PV} \qquad \forall PV \in \mathcal{C}, \forall t \in \mathcal{T}, \tag{A.43}$$

$$v_{PV,t} = \frac{G_t}{G^{\mathrm{STC}}} \cdot \left[1 + \gamma \left(T^{\mathrm{M}}_t - T^{\mathrm{M,\,STC}}\right)\right] \qquad \forall PV \in \mathcal{C}, \forall t \in \mathcal{T}. \tag{A.44}$$

The specific electric power $v_{PV,t}$ depends on the irradiation flux G_t, the module temperature T^{M}_t and the standard testing conditions ($G^{\mathrm{STC}} = 1000\,\mathrm{W\,m^{-2}}$ and $T^{\mathrm{M,\,STC}} = 25\,°\mathrm{C}$). The electric power of the module decreases with increasing temperature with the temperature coefficient $\gamma = -0.0045\,°\mathrm{C}^{-1}$. The irradiation G_t is calculated based on the diffuse irradiation model for tilted surfaces (Perez et al., 1987). The module temperature T^{M}_t is calculated based on the temperature model by Eicker (Eicker, 2005). The rooftop area A^{rooftop} can be used for either photovoltaic A_{PV} or solar thermal A_{SH} units. The rooftop area in the given case study is limited to $A^{\mathrm{rooftop}} = 20{,}000\,\mathrm{m}^2$.

$$A_{PV} = \dot{V}^{\mathrm{N}}_{PV} \cdot a_{PV} \qquad \forall PV \in \mathcal{C}, \tag{A.45}$$

$$A_{SH} = \dot{V}^{\mathrm{N}}_{SH} \cdot a_{SH} \qquad \forall SH \in \mathcal{C}, \tag{A.46}$$

$$A^{\mathrm{rooftop}} \geq \sum_{PV} A_{PV} + \sum_{SH} A_{SH}. \tag{A.47}$$

In Eq. (A.45) and (A.46), the coefficients to calculate the area for the installed capacities are $a_{PV} = 7\,\mathrm{m^2\,kW^{-1}}$ and $a_{SH} = 1.667\,\mathrm{m^2\,kW^{-1}}$. The low-temperature heat output $\dot{V}_{SH,t}$ is limited by ($\forall SH \in \mathcal{C}, \forall t \in \mathcal{T}$):

$$\dot{V}_{SH,t} \leq v_{SH,t} \cdot \dot{V}^{\mathrm{N}}_{SH}, \tag{A.48}$$

$$v_{SH,t} = \frac{G_t}{a_{SH}} \cdot \left(\eta^{\mathrm{opt}}_{SH} - k_1 \left(\frac{(T_{SH} - T^{\mathrm{ambient}}_t)}{G_t}\right) - k_2 \left(\frac{(T_{SH} - T^{\mathrm{ambient}}_t)^2}{G_t}\right)\right). \tag{A.49}$$

The specific heating output $v_{SH,t}$ depends on the irradiation flux G_t, the optical efficiency $\eta^{\mathrm{opt}}_{SH} = 0.8$, the linear and quadratic temperature loss factor $k_1 = 1.5\,\mathrm{W\,K^{-1}\,m^{-2}}$ and $k_2 = 0.005\,\mathrm{W\,K^{-2}\,m^{-2}}$ and the temperature difference between panel and ambient ($T_{SH} - T^{\mathrm{ambient}}_t$). The panel temperature T_{SH} is set to the temperature of the low-heat temperature demand.

A.13 Wind turbine

The electricity output power $\dot{V}_{WIND,t}$ of a wind turbine mainly depends on the wind speed at hub height u_t^{H} and the installed capacity $\dot{V}_{WIND}^{\mathrm{N}}$. Thus, two decision variables represent the wind turbines (WIND) in the model: the installed capacity $\dot{V}_{WIND}^{\mathrm{N}}$ and the hub height H_{WIND}. The performance $\dot{V}_{WIND,t}$ of a wind turbine is described by a generic performance curve. A generic performance curve is employed based on the mean of seven (1-7) normalized performance curves by data sheets of real-world wind turbines, Fig. A.1a. The real-world performance curves are normalized by the maximum of the provided electric power output, i.e., the installed capacity $\dot{V}_{WIND}^{\mathrm{N}}$, Fig. A.1b:

$$v_{WIND,t} = \frac{\dot{V}_{WIND,t}}{\dot{V}_{WIND}^{\mathrm{N}}} \qquad \forall WIND \in \mathcal{C}, \forall t \in \mathcal{T}. \tag{A.50}$$

To consider different operation ranges of wind speed at hub height H_{WIND}, the wind speed at hub height u_t^{H} is normalized relative to the reference wind speed $\hat{u}^{\mathrm{H}}$ with $v_{WIND}\left(\hat{u}^{\mathrm{H}}\right) = 0.8$, Fig. A.1c,

$$\widetilde{u}_t^{\mathrm{H}} = \frac{u_t^{\mathrm{H}}}{\hat{u}^{\mathrm{H}}} \qquad \forall t \in \mathcal{T}. \tag{A.51}$$

The mean of all resulting normalized performance curves represents the generic performance curve $v_{WIND,t}$, Fig. A.1d. Hence, the electricity output power $\dot{V}_{WIND,t}$ is given by,

$$\dot{V}_{WIND,t} \leq v_{WIND,t} \cdot \dot{V}_{WIND}^{\mathrm{N}} \qquad \forall WIND \in \mathcal{C}, \forall t \in \mathcal{T}. \tag{A.52}$$

To model about 2000 h full-load hours per year, the reference wind speed $\hat{u}^{\mathrm{H}}$, Eq. (A.51), is set to $\hat{u}^{\mathrm{H}} = 11.8\,\mathrm{m/s}$.
The actual wind speed u_t^{WS} is measured at a specific measuring height H^{WS}. To calculate the wind speed u_t^{H} at hub height H_{WIND}, we use the atmospheric boundary layer

$$u_{\mathrm{H},t} = u_{\mathrm{WS},t} \cdot \frac{\ln H_{\mathrm{WIND}}/Z_0}{\ln H_{\mathrm{WS}}/Z_0} \qquad \forall t \in \mathcal{T}. \tag{A.53}$$

Based on reports of the German Weather Service (Christoffer and Ulbricht-Eissing, 1989), the roughness of the ground for near-urban regions is $Z_0 = 0.3\,\mathrm{m}$.
The investment costs I_{WIND}^{N} are modeled differently than in the generic unit model,

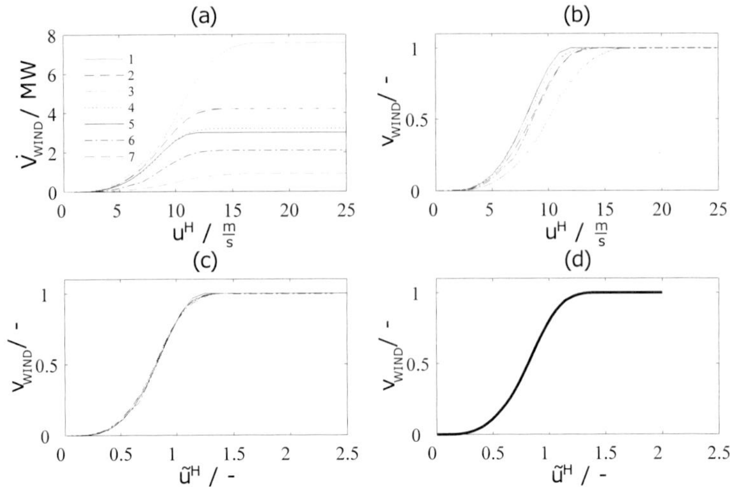

Figure A.1: Wind turbine performance curves (a), normalization of output power (b), normalization of wind speed (c) and final generic performance curve (d)

Section A.5. The supporting points for the investment costs depend on the installed capacity $\dot{V}^{N}_{WIND}$ and the hub height H_{WIND}. Thus, the supporting points are defined over two dimensions, Table A.6. Thus, the standard piecewise linearization of the generic unit model cannot be applied. We use the generalized convex combination

Table A.6: The investment costs I_{WIND} for a wind turbine at the supporting points h depending on the installed capacity of the wind turbine $\dot{V}^{N}_{WIND}$ and the hub height H_{WIND}

installed capacity	**hub height H_{WIND}**		
	80 m	110 m	140 m
1.25 MW	1.709 M€	1.934 M€	2.146 M€
2.75 MW	3.770 M€	4.213 M€	4.593 M€
5 MW	6.885 M€	7.535 M€	7.960 M€

method to describe a multi-dimensional piecewise linearization (Geißler, 2011; Geißler et al., 2012).

$$I_{WIND} = \sum_{h \in \mathcal{H}} \lambda_{WIND,h} \cdot I_{WIND,h} \qquad \forall WIND \in \mathcal{C}, \tag{A.54}$$

$$\dot{V}^{\mathrm{N}}_{WIND} = \sum_{h\in\mathcal{H}} \lambda_{WIND,h} \cdot \dot{V}^{\mathrm{N}}_{WIND,h} \qquad \forall WIND \in \mathcal{C}, \tag{A.55}$$

$$H_{WIND} = \sum_{h\in\mathcal{H}} \lambda_{WIND,h} \cdot H_{WIND,h} \qquad \forall WIND \in \mathcal{C}, \tag{A.56}$$

$$\dot{V}_{WIND,t} = \sum_{h\in\mathcal{H}} \lambda_{WIND,h} \cdot \dot{V}_{\mathrm{WIND},h,t} \qquad \forall WIND \in \mathcal{C}, \forall t \in \mathcal{T}, \tag{A.57}$$

$$\gamma_{WIND} = \sum_{h\in\mathcal{H}} \lambda_{WIND,h} \qquad \forall WIND \in \mathcal{C}, \tag{A.58}$$

$$\gamma_{WIND} = \sum_{s\in\mathcal{S}} \kappa_{WIND,s} \qquad \forall WIND \in \mathcal{C}, \tag{A.59}$$

$$\lambda_{WIND,h} \leq \sum_{s\in\mathcal{S}} \kappa_{WIND,s} \Lambda_{h,s} \qquad \forall WIND \in \mathcal{C}, \forall h \in \mathcal{H}. \tag{A.60}$$

In Eq. (A.54) -(A.57), the investment costs I_{WIND}, the installed capacity $\dot{V}^{\mathrm{N}}_{WIND}$, the hub height H_{WIND}, and the electric power output $\dot{V}_{WIND,t}$ are calculated based on the weighting coefficient $\lambda_{WIND,h}$ of the supporting points $h \in \mathcal{H}$. The linearization is represented by Eq. (A.58)-(A.60), where the weighting coefficient $\lambda_{WIND,h}$ of every supporting point h is calculated. The binary variable $\kappa_{WIND,s}$ describes which simplex is active. Simplices are a multi-dimensional equivalent of a line segment in one-dimensional piecewise linearization. Only supporting points h of the active simplex have a weight ($\lambda_{WIND,h} \geq 0$), Eq. (A.60). $\Lambda_{h,s}$ is an assignment matrix between simplex $s \in \mathcal{S}$ and supporting point h.

A.14 Model adaptations

The industrial energy system model presented in this appendix is used in Chapter 3 as operational model, in Chapter 4, as single objective synthesis model, and in Chapter 5, as multi-objective synthesis model. In Chapter 5, we use the energy system model as presented here. For Chapter 4, we omit Eq. (A.2) as objective and further adapt the superstructure and the time resolution as described in Chapter 4. In Chapter 3, we use the industrial energy system model as a long-term operational model. For this purpose, we add long-term variables and constraints to the energy system model as described in Chapter 3. Further, to render the synthesis model to an operational model, we use a fixed design given in Table A.7. Table A.7 presents both energy-conversion units and storage units in alphabetical order.

Table A.7: Parameter of the energy-conversion units n: unit capacity $\dot{V}_n^{\mathrm{N}}$ and V_n^{N}, maintenance costs of a unit $\mathrm{M}_n^{\mathrm{B}}$, nominal efficiency η_n^{N} for every unit, and thermal efficiency $\eta_n^{\mathrm{N,th}}$ of the combined heat and power engines

unit ($n \in \mathcal{C}$)	$\dot{V}_n^{\mathrm{N}}$/**MW**	M_n^{N} /**€**	η_n^{N}	$\eta_n^{\mathrm{N,th}}$
absorption $\text{chiller}_1 \in \mathcal{A}\text{b}\mathcal{C}$	2.03	2421	0.67	–
absorption $\text{chiller}_2 \in \mathcal{A}\text{b}\mathcal{C}$	1.79	2292	0.67	–
absorption $\text{chiller}_3 \in \mathcal{A}\text{b}\mathcal{C}$	1.02	1795	0.67	–
absorption $\text{chiller}_4 \in \mathcal{A}\text{b}\mathcal{C}$	0.61	1436	0.67	–
$\text{boiler}_1 \in \mathcal{B}$	3.14	1648	0.9	–
$\text{boiler}_2 \in \mathcal{B}$	2.95	1573	0.9	–
$\text{boiler}_3 \in \mathcal{B}$	1.57	1063	0.9	–
CHP$\in \mathcal{CHP}$	3.2	85056	0.87	0.4604
compression $\text{chiller}_1 \in \mathcal{CC}$	5.62	35821	5.54	–
compression $\text{chiller}_2 \in \mathcal{CC}$	2.01	13510	5.54	–
compression $\text{chiller}_3 \in \mathcal{CC}$	1.36	9493	5.54	–
compression $\text{chiller}_4 \in \mathcal{CC}$	1.11	7948	5.54	–
compression $\text{chiller}_5 \in \mathcal{CC}$	0.42	3683	5.54	–
heat $\text{exchanger}_{1/2/3/4/5/6/7} \in \mathcal{HEX}$	0.715	1823	1	–
$\text{inverter}_1 \in \mathcal{INV}$	1.9	3432	0.95	–
$\text{inverter}_{2/3} \in \mathcal{INV}$	0.64	1334	0.95	–
photovoltaic $\in \mathcal{PV}$	0.5	16546	1	–
unit ($n \in \mathcal{C}$)	V_n^{N}/**MWh**	M_n^{N} /**€**	η_n^{N}	$\eta_n^{\mathrm{N,th}}$
$\text{battery}^{c} \in \mathcal{BAT}$	8.4	83503	0.92	–
heat storage system$\in \mathcal{TS}$	210	35789	0.95	–
cold storage system$\in \mathcal{TS}$	120	20196	0.95	–

Appendix B

MILP model for the pump system

In this section, we provide the complete MILP model of the pump system synthesis problem used in case study ***PS***. The MILP model and model description is from our previous work Bahl et al. (2018a) and is only restated for completeness of this work.

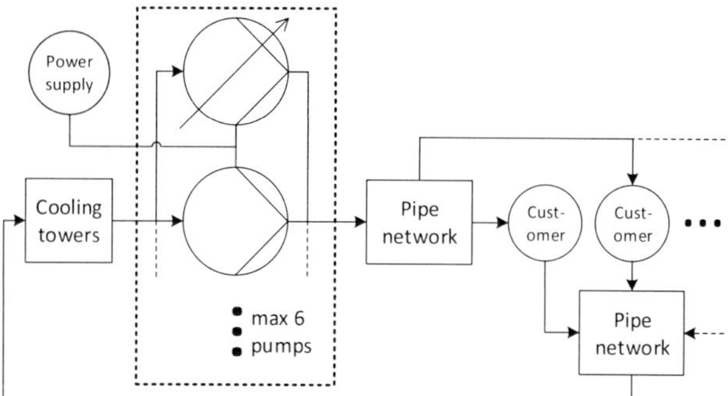

Figure B.1: The cooling network of real-world case study ***PS*** with the superstructure of the pump system (dotted rectangle), which is connected to several customers and cooling towers via a piping network. All pumps are connected to an external power supply.

The pump network (B.1) has to provide the total volume flow $\dot{V}_t^{\text{total}}$ and to generate a minimum pressure difference $\Delta p^{\min}$ demanded by the consumer. The total volume flow and the pressure difference can be expressed as

$$\sum_{p \in \mathcal{P}} \dot{V}_{p,t} = \dot{V}_t^{\text{total}}, \qquad \forall t \in \mathcal{T}, \tag{B.1}$$

$$\epsilon_{p,t} \Delta p_{p,t} \geq \Delta p^{\min}, \qquad \forall t \in \mathcal{T}, \forall p \in \mathcal{P}. \tag{B.2}$$

A maximum number of six pumps p can be built at the plant. Each pump provides a volume flow $\dot{V}_{p,t}$ while the sum needs to fulfill the total volume flow demand,

Eq. (B.1)). The binary variable $\epsilon_{p,t}$ describes if pump p is operated in time step t. All pumps are set up in parallel connection in the superstructure, Fig. B.1. Thus, Eq. (B.2) requires that, if a pump is operated, the pressure difference of the pump $\Delta p_{p,t}$ needs to provide at least the demanded pressure difference. Both equations are stated for all time steps $t \in \mathcal{T}$.

The superstructure has six positions for possible pump installation. For each of these six positions, a pump can be selected from the set of pumps $\mathcal{P}$. Three types of fixed-speed pumps and three types of variable-speed pumps are considered. However, not all positions have to be selected by the synthesis:

$$\sum_{p \in \mathcal{P}} \gamma_p \leq 6 \tag{B.3}$$

Throughout this section, bi-linear terms (e.g., $\epsilon_{p,t} \Delta p_{p,t}$ in Eq. (B.2)) will not be further discussed, but in the final implementation, bi-linear products are linearized (Glover, 1975), as in Appendix A.
First, in Section B.1, we introduce the cost objective function considered in the synthesis problem. Details of the pump models are provided in Section B.2 and B.3. In Section B.4, we systematically list all parameter values considered in the real-world case study.

B.1 Cost objective function

The synthesis problem has to consider both capital expenditure $CAPEX$ and operational expenditure $OPEX$. Thus, we minimize total annualized cost as objective function:

$$\min_{variables} \; CAPEX + OPEX \tag{B.4}$$

In the following, we introduce and explain the constraints of the synthesis problem. For readability, we use Greek letters and upper case Latin letters for variables and parameters are described by lower case letters. The capital expenditure and operational expenditure are calculated by

$$CAPEX = \sum_{p \in \mathcal{P}} \gamma_p c_p^{\text{invest}}, \tag{B.5}$$

$$OPEX = \sum_{t \in \mathcal{T}} \sum_{p \in \mathcal{P}} \Delta t_t c^{\text{o}} P_{p,t} + c^{\text{p}} P^{\max} + \sum_{p \in \mathcal{P}} \gamma_p c_p^{\text{maint}}. \tag{B.6}$$

In Eq. (B.5), the capital expenditure of each pump p is calculated by the binary decision variable γ_p to describe if a pump p is built and the annualized investment cost c_p^{invest}.

In Eq. (B.6), the operational expenditure is the sum of the operating cost, the peak-power cost and maintenance cost. The operating cost for each time step t and each pump p is calculated by the power consumption variable $P_{p,t}$ multiplied with the duration of each time step Δt_t and the electricity price c^{o}. The peak-power cost are calculated by the peak-power variable $P^{\max}$ multiplied with the specific peak-power cost c^{P}. The maintenance cost are considered by a fixed yearly amount c_p^{maint} if a pump is build (γ_p).
The peak-power variable is constrained by

$$\sum_{p\in\mathcal{P}} P_{p,t} \le P^{\max}, \quad \forall t \in \mathcal{T}. \tag{B.7}$$

Moreover, the binary variable for the on/off status of a pump ($\epsilon_{p,t}$) is constrained by the existence of a pump:

$$\epsilon_{p,t} \le \gamma_p, \quad \forall t \in \mathcal{T}, \forall p \in \mathcal{P}. \tag{B.8}$$

In Section B.2 and B.3, we describe the operation conditions of the pumps, namely the relation between $\dot{V}_{p,t}$, $\Delta p_{p,t}$ and $P_{p,t}$. The operation conditions are separately stated for the fixed-speed pumps $\mathcal{P}^{\text{fixed}}$ and variable-speed pumps $\mathcal{P}^{\text{var}}$ and it holds:

$$\mathcal{P}^{\text{fixed}} \sqcup \mathcal{P}^{\text{var}} = \mathcal{P}. \tag{B.9}$$

B.2 Fixed-speed pumps

In this section, we describe the operation conditions of a fixed-speed pump $p \in \mathcal{P}^{\text{fixed}}$. Throughout, we state the index t and p in the equations and all equations are equal and valid for all time steps $t \in \mathcal{T}$ and all pumps $p \in \mathcal{P}^{\text{fixed}}$.
The pressure difference $\Delta p_{p,t}$ and the power $P_{p,t}$ required to provide this pressure difference are non-linear functions of the volume flow $\dot{V}_{p,t}$. Here, we linearize the non-linear equations using the linear segmentation method (Floudas, 1995; Misener et al., 2009). We use the general function $\Phi_{p,t}$ to describe the linearization of $\Delta p_{p,t}$ and $P_{p,t}$. The linearization uses n equidistant breakpoints $a_{1,p}, \dots, a_{n,p}$ of $\dot{V}_{p,t}$, resulting in $n-1$ line segments $i \in \mathcal{I}$. The function values of $\Phi_{p,t}$ at the breakpoints are denoted

by $\phi_p(a_{i,p})$. Thus, using the linear segmentation method, we can state the following linearized fixed-speed pump model:

$$\Phi_{p,t} = \sum_{i \in \mathcal{I}} \delta_{i,p,t}\phi_p(a_{i,p}) + c_{i,p}^{\text{grad}}\left(\dot{V}_{i,p,t}^{\text{lin}} - \delta_{i,p,t}a_{i,p}\right), \quad \forall t \in \mathcal{T}, \forall p \in \mathcal{P}^{\text{fixed}}, \tag{B.10}$$

$$\dot{V}_{p,t} = \sum_{i \in \mathcal{I}} \dot{V}_{i,p,t}^{\text{lin}}, \quad \forall t \in \mathcal{T}, \forall p \in \mathcal{P}^{\text{fixed}}, \tag{B.11}$$

$$\sum_{i \in \mathcal{I}} \delta_{i,p,t} = \epsilon_{p,t}, \quad \forall t \in \mathcal{T}, \forall p \in \mathcal{P}^{\text{fixed}}, \tag{B.12}$$

$$\delta_{i,p,t}a_{i,p} \leq \dot{V}_{i,p,t}^{\text{lin}} \leq \delta_{i,p,t}a_{i+1,p}, \quad \forall t \in \mathcal{T}, \forall p \in \mathcal{P}^{\text{fixed}}, \forall i \in \mathcal{I}, \tag{B.13}$$

$$\delta_{i,p,t} \in \{0,1\}, \quad \forall t \in \mathcal{T}, \forall p \in \mathcal{P}^{\text{fixed}}, \forall i \in \mathcal{I}.$$

The newly introduced binary variable $\delta_{i,p,t}$ is 1, if the operating point is on the line segment i. The difference $c_{i,p}^{\text{grad}}$ describes the linear relation between $\Phi_{p,t}$ and $\dot{V}_{i,p,t}^{\text{lin}}$ in each line segment i. In Eq. (B.10), the linearized function value of $\Phi_{p,t}$ is calculated as the sum of all line segments. Each segment consists of a supporting value $\phi_p(a_{i,p})$ plus a linear dependency on the actual value of $\dot{V}_{i,p,t}^{\text{lin}}$. The difference of each line segment $c_{i,p}^{\text{grad}}$ is calculated by:

$$c_{i,p}^{\text{grad}} = \frac{\phi_p(a_{i+1,p}) - \phi_p(a_{i,p})}{a_{i+1,p} - a_{i,p}}, \quad \forall i \in \mathcal{I}, \forall p \in \mathcal{P}^{\text{fixed}}. \tag{B.14}$$

Only one line segment can be active at each time step, Eq. (B.12). In Eq. (B.11), the volume flow $\dot{V}_{p,t}$ is calculated as the sum of all linearized volume flow rates $\dot{V}_{i,p,t}^{\text{lin}}$ of the segments i. The linearized volume flow variable $\dot{V}_{i,p,t}^{\text{lin}}$ describes the volume flow at each line segment and is zero, if the binary variable $\delta_{i,p,t}$ is zero, Eq. (B.13).

The equation of the linear segmentation method (Eq. B.10-B.13) are separately stated for $\Delta p_{p,t}$ and $P_{p,t}$ by replacing $\Phi_{p,t}$. More precisely, we calculate

$$\Delta p_{p,t} = \rho \cdot g \cdot \Phi_{p,t}^{\text{head}}, \tag{B.15}$$

$$P_{p,t} = \frac{\Phi_{p,t}^{\text{mechPower}}}{e_p^{\text{motor}} \cdot e_p^{\text{fc}}}. \tag{B.16}$$

Thus, the pressure $\Delta p_{p,t}$ is described by the hydraulic head expressed in units of length multiplied with the density of water ρ and the gravity g. The electrical power consumption $P_{p,t}$ is described by the mechanical power required by the pump divided

by the motor efficiency e_p^{motor} and the frequency converter efficiency e_p^{fc}.
The linear segmentation method requires the introduction of one binary variable $\delta_{i,p,t}$, one continuous variable $\dot{V}_{i,p,t}^{\text{lin}}$ plus two inequalities per line segment i. Thus, the complexity of the optimization problem increases with the number of line segments $|\mathcal{I}| = n - 1$.
The volume flow provided by a fixed-speed pump is limited between

$$\epsilon_{p,t}\dot{V}_p^{\min} \leq \dot{V}_p, t \leq \epsilon_{p,t}\dot{V}_p^{\max}, \quad \forall t \in \mathcal{T}, \forall p \in \mathcal{P}^{\text{fixed}}. \tag{B.17}$$

Next, we describe the model of a variable-speed pump.

B.3 Variable-speed pumps

In this section, we describe the operation conditions of a variable-speed pump $p \in \mathcal{P}^{\text{var}}$. We throughout state the index t and p in the equations and all equations are equal and valid for all time steps $t \in \mathcal{T}$ and all pumps $p \in \mathcal{P}^{\text{fixed}}$.
The pressure difference $\Delta p_{p,t}$ and the power $P_{p,t}$ are non-linear functions of the volume flow $\dot{V}_{p,t}$ and the rotational speed $\omega_{p,t}$. This two-dimensional non-linear dependency cannot be linearized by the linear segmentation method employed for the fixed-speed pumps, Section B.2. However, the generalized convex-combination method (Geißler, 2011; Geißler et al., 2012) can handle multivariate functions.
We use the general function $\widetilde{\Phi}_{p,t}$ as placeholder for the two-dimensional functions $\Delta p_{p,t}$ and $P_{p,t}$. The generalized convex-combination method requires the introduction of vertices. The domain between the vertices - a so-called simplex - is triangulated by a linear combination of the function values at the vertices surrounding a simplex, Fig. B.2. The grid of vertices $\left(n^{\dot{V}}, n^{\omega}\right)$ is addressed with the sets of one-dimensional nodes $n^{\dot{V}} \in \mathcal{N}^{\dot{V}}$ and $n^{\omega} \in \mathcal{N}^{\omega}$. Using the one-dimensional nodes, we can describe equidistant breakpoints $a_{1,p}^{\dot{V}}, \ldots, a_{|\mathcal{N}^{\dot{V}}|,p}^{\dot{V}}$ of $\dot{V}_{p,t}$ and $a_{1,p}^{\omega}, \ldots, a_{|\mathcal{N}^{\omega}|,p}^{\omega}$ of $\omega_{p,t}$. The function values of $\widetilde{\Phi}_{p,t}$ at the vertices are denoted by $\widetilde{\phi}_p\left(a_{n^{\dot{V}},p}^{\dot{V}}, a_{n^{\omega},p}^{\omega}\right)$. A simplex is addressed by $\left(i^{\dot{V}}, i^{\omega}\right)$ with

$$i^{\dot{V}} \in \mathcal{I}^{\dot{V}} = \left\{1, 2, \ldots, 2\left|\mathcal{N}^{\dot{V}}\right| - 1, 2\left|\mathcal{N}^{\dot{V}}\right|\right\}, \tag{B.18}$$

$$i^{\omega} \in \mathcal{I}^{\omega} = \left\{1, 2, \ldots, |\mathcal{N}^{\omega}| - 1, |\mathcal{N}^{\omega}|\right\}. \tag{B.19}$$

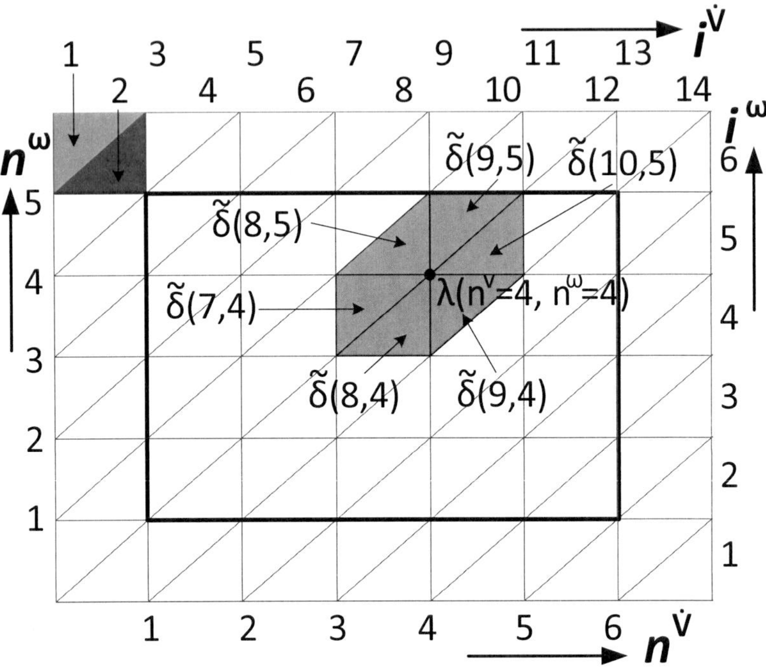

Figure B.2: Generalized convex-combination: domain (black rectangle) of volume flow $\dot{V}_{p,t}$ and the rotational speed $\omega_{p,t}$ addressed by vertices $\left(n^{\dot{V}},n^{\omega}\right)$ and simplices $\left(i^{\dot{V}},i^{\omega}\right)$, the binary variable $\widetilde{\delta}$ is stated for each simplex, the continuous variable λ is stated for each vertex.

Please note, as we use triangles, the index $i^{\dot{V}}$ has more entries than the corresponding node index $n^{\dot{V}}$, Fig. B.2. Using theses definitions for the generalized convex-combination method, we can state the following linearized variable-speed pump model:

$$\widetilde{\Phi}_{p,t} = \sum_{n^{\dot{V}} \in \mathcal{N}^{\dot{V}}} \sum_{n^{\omega} \in \mathcal{N}^{\omega}} \lambda_{n^{\dot{V}},n^{\omega},p,t} \cdot \widetilde{\phi}_p \left(a^{\dot{V}}_{n^{\dot{V}},p}, a^{\omega}_{n^{\omega},p} \right), \qquad \forall t \in \mathcal{T}, \forall p \in \mathcal{P}^{\text{var}}, \tag{B.20}$$

$$\dot{V}_{p,t} = \sum_{n^{\dot{V}} \in \mathcal{N}^{\dot{V}}} \sum_{n^{\omega} \in \mathcal{N}^{\omega}} \lambda_{n^{\dot{V}},n^{\omega},p,t} \cdot a^{\dot{V}}_{n^{\dot{V}},p}, \qquad \forall t \in \mathcal{T}, \forall p \in \mathcal{P}^{\text{var}}, \tag{B.21}$$

$$\omega_{p,t} = \sum_{n^{\dot{V}} \in \mathcal{N}^{\dot{V}}} \sum_{n^{\omega} \in \mathcal{N}^{\omega}} \lambda_{n^{\dot{V}},n^{\omega},p,t} \cdot a^{\omega}_{n^{\omega},p}, \qquad \forall t \in \mathcal{T}, \forall p \in \mathcal{P}^{\text{var}}, \tag{B.22}$$

$$\epsilon_{p,t} = \sum_{n^{\dot{V}} \in \mathcal{N}^{\dot{V}}} \sum_{n^{\omega} \in \mathcal{N}^{\omega}} \lambda_{n^{\dot{V}},n^{\omega},p,t}, \qquad \forall t \in \mathcal{T}, \forall p \in \mathcal{P}^{\text{var}}, \tag{B.23}$$

$$0 \leq \lambda_{n^{\dot{V}},n^{\omega},p,t} \leq 1, \qquad \forall t \in \mathcal{T}, \forall p \in \mathcal{P}^{\text{var}}, \; \forall n^{\dot{V}} \in \mathcal{N}^{\dot{V}}, \forall n^{\omega} \in \mathcal{N}^{\omega}, \tag{B.24}$$

$$\lambda_{n^{\dot{V}},n^{\omega},p,t} \leq \widetilde{\delta}_{2n^{\dot{V}}-1,n^{\omega}} + \widetilde{\delta}_{2n^{\dot{V}},n^{\omega}} + \widetilde{\delta}_{2n^{\dot{V}}+1,n^{\omega}} \tag{B.25}$$

$$+ \widetilde{\delta}_{2n^{\dot{V}},n^{\omega}+1} + \widetilde{\delta}_{2n^{\dot{V}}+1,n^{\omega}+1} + \widetilde{\delta}_{2n^{\dot{V}}+2,n^{\omega}+1}, \qquad \forall t \in \mathcal{T}, \forall p \in \mathcal{P}^{\text{var}}, \; \forall n^{\dot{V}} \in \mathcal{N}^{\dot{V}}, \forall n^{\omega} \in \mathcal{N}^{\omega}, \tag{B.26}$$

$$\epsilon_{p,t} = \sum_{i^{\dot{V}} \in \mathcal{I}^{\dot{V}}} \sum_{i^{\omega} \in \mathcal{I}^{\omega}} \widetilde{\delta}_{i^{\dot{V}},i^{\omega},p,t}, \qquad \forall t \in \mathcal{T}, \forall p \in \mathcal{P}^{\text{var}}, \tag{B.27}$$

$$\widetilde{\delta}_{i^{\dot{V}},i^{\omega},p,t} \in \{0,1\}, \qquad \forall t \in \mathcal{T}, \forall p \in \mathcal{P}^{\text{var}}, \; \forall i^{\dot{V}} \in \mathcal{I}^{\dot{V}}, \forall i^{\omega} \in \mathcal{I}^{\omega}. \tag{B.28}$$

$$\widetilde{\delta}_{i^{\dot{V}},i^{\omega},p,t} = 0 \quad \forall t \in \mathcal{T}, \forall p \in \mathcal{P}^{\text{var}}, \; \forall i^{\dot{V}} \in \left\{1, 2, 2\left|\mathcal{N}^{\dot{V}}\right| + 1, 2\left|\mathcal{N}^{\dot{V}}\right| + 2\right\}, \forall i^{\omega} \in \{1, |\mathcal{N}^{\omega}| + 1\}. \tag{B.29}$$

The binary variable $\widetilde{\delta}_{i^{\dot{V}},i^{\omega},p,t}$ is one, if the operating point is on the simplex $\left(i^{\dot{V}},i^{\omega}\right)$ and zero if not. The continuous variable $\lambda_{n^{\dot{V}},n^{\omega},p,t}$ is the linear combination variable stated for each vertex $\left(n^{\dot{V}},n^{\omega}\right)$.

In Eq. (B.23), the sum of all linear combination variables must be 1, if the pump is operated. The linearized function $\widetilde{\Phi}_{p,t}$, the volume flow rate $\dot{V}_{p,t}$ and the rotational speed $\omega_{p,t}$ are calculated as a linear combination of the function values at the vertices (Eq. (B.20)) and the values of the breakpoints (Eq. (B.21) and (B.22)). Eq. (B.24) limits the continuous variable $\lambda_{n^{\dot{V}},n^{\omega},p,t}$ between 0 and 1.

Eq. (B.26) describes that the vertex can only be part of the linear combination $\lambda_{n^{\dot{V}},n^{\omega},p,t} > 0$, if the operating point is on one of the bordering simplices ($\widetilde{\delta}_{i^{\dot{V}},i^{\omega},p,t} = 1$). Eq. (B.29) ensures that the binary variable of the simplices outside the domain are all set to 0, Fig. B.2, the domain is the area within the thick line). However, the variables have to exist for the inequality Eq. (B.26).
The equation of the linear segmentation method (Eq. B.20-B.29) are separately stated for $\Delta p_{p,t}$ and $P_{p,t}$ instead of $\widetilde{\Phi}_{p,t}$ using Eq. (B.15) and (B.16). The generalized convex-combination method requires the introduction of one binary variable $\widetilde{\delta}_{i^{\dot{V}},i^{\omega},p,t}$ per simplex and one continuous variable per vertex $\lambda_{n^{\dot{V}},n^{\omega},p,t}$. Thus, the complexity of the optimization problem increases with the number of vertices and simplices.
For variable-speed pumps, the minimal and maximal volume flow depends on the rotational speed. The dependency of the allowed volume flow rate can be described by the following constraints:

$$\dot{V}_p^{\min}\left(\omega_p^{\min}\right)\epsilon_{p,t} + c_p^{\mathrm{V},\min}\left(\omega_{p,t} - \omega_p^{\min}\epsilon_{p,t}\right) \leq \dot{V}_{p,t}, \quad \forall t \in \mathcal{T}, \forall p \in \mathcal{P}^{\mathrm{var}}, \quad \text{(B.30)}$$

$$\dot{V}_{p,t} \leq \dot{V}^{\max}\left(\omega_p^{\max}\right)_p \epsilon_{p,t} - c_p^{\mathrm{V},\max}\left(\omega_p^{\max}\epsilon_{p,t} - \omega_{p,t}\right), \quad \forall t \in \mathcal{T}, \forall p \in \mathcal{P}^{\mathrm{var}}. \quad \text{(B.31)}$$

B.4 Parameters

In this section, we list the values of all parameters used in case study ***PS***. Please note that the original parameter values used in the project together with InfraServ GmbH & Co. Knapsack KG were modified for this publication due to confidentiality requirements on the price data as well as pump data sheets of suppliers. In the case study, we consider three fixed-speed pumps p_{f1}, p_{f2}, p_{f3} with the nominal volume flow rates of $\dot{V}^{\mathrm{nom}} = 1000\,\mathrm{m^3/h}, 2000\,\mathrm{m^3/h}, 3000\,\mathrm{m^3/h}$. Similarly, we consider three variable-speed pumps p_{v1}, p_{v2}, p_{v3} with the nominal volume flow rates $\dot{V}^{\mathrm{nom}} = 1000\,\mathrm{m^3/h}, 2000\,\mathrm{m^3/h}, 3000\,\mathrm{m^3/h}$ and a nominal frequency $\omega^{\mathrm{nom}} = 50\,\mathrm{Hz}$.
Pump specific parameters on the minimal and maximal volume flow rate and the rotation speed limits for variable-speed pumps are listed in Table B.1 and B.2. The

Table B.1: Parameters of fixed-speed pumps: limits of volume flow $\dot{V}$

Pump	$\dot{V}_p^{\min}$	$\dot{V}_p^{\max}$
p_{f1}	$420.5\,\mathrm{m^3/h}$	$1400.0\,\mathrm{m^3/h}$
p_{f2}	$841.0\,\mathrm{m^3/h}$	$2800.0\,\mathrm{m^3/h}$
p_{f3}	$1261.5\,\mathrm{m^3/h}$	$4200.0\,\mathrm{m^3/h}$

Table B.2: Parameters of variable-speed pumps: limits of volume flow $\dot{V}$ and rotation speed ω

Pump	$\dot{V}_p^{\min}\left(\omega_p^{\min}\right)$	$\dot{V}_p^{\min}\left(\omega_p^{\max}\right)$	$\dot{V}_p^{\max}\left(\omega_p^{\max}\right)$	$\dot{V}_p^{\max}\left(\omega_p^{\min}\right)$	$\omega_p^{\min}$	$\omega_p^{\max}$
p_{v1}	220.0 m³/h	490.0 m³/h	1430.0 m³/h	650.0 m³/h	25 Hz	55 Hz
p_{v2}	440.0 m³/h	980.0 m³/h	2860.0 m³/h	1300.0 m³/h	25 Hz	55 Hz
p_{v3}	660.0 m³/h	1470.0 m³/h	4290.0 m³/h	1950.0 m³/h	25 Hz	55 Hz

coefficient for the volume flow limits (Eq. B.30 and B.31) of the variable-speed pumps are calculated based on the data in Table B.2, by

$$c_p^{\mathrm{V,min}} = \frac{\dot{V}_p^{\min}\left(\omega_p^{\max}\right) - \dot{V}_p^{\min}\left(\omega_p^{\min}\right)}{\omega_p^{\max} - \omega_p^{\min}}, \qquad \forall p \in \mathcal{P}^{\mathrm{var}}, \tag{B.32}$$

$$c_p^{\mathrm{V,max}} = \frac{\dot{V}_p^{\max}\left(\omega_p^{\max}\right) - \dot{V}_p^{\max}\left(\omega_p^{\min}\right)}{\omega_p^{\max} - \omega_p^{\min}}, \qquad \forall p \in \mathcal{P}^{\mathrm{var}}. \tag{B.33}$$

The coefficient for the investment cost c_p^{invest} is the annualized cost of the total investment cost c_p^{total} of the pump. We considered the present value factor with an interest rate of 7.5 % over a time horizon of 5 years, thus

$$c_p^{\mathrm{invest}} = c_p^{\mathrm{total}} \frac{(1+0.075)^5\, 0.075}{(1+0.075)^5 - 1}. \tag{B.34}$$

The total investment cost c_p^{total} of the pump accounts for the cost of the pump itself, the cost of the motor and installation cost. The values for the total investment cost are given in Table B.3.

Table B.3: Total investment cost c_p^{total} of all pumps

Pump	c_p^{total}
p_{f1}	75,198 €
p_{f2}	90,238 €
p_{f3}	131,598 €
p_{v1}	112,797 €
p_{v2}	135,357 €
p_{v3}	197,397 €

The maintenance cost c_p^{maint} are calculated relative to the total investment cost of each pump (Table B.3) by

$$c_p^{\text{maint}} = 0.1 \cdot c_p^{\text{total}}, \quad \forall p \in \mathcal{P}. \tag{B.35}$$

The cost coefficients for Eq. (B.6) are listed in Table B.4 together the efficiency for pump and frequency converter and the values for the density of water and gravity used in Eq. (B.15) and (B.16).

Table B.4: Cost coefficients and efficiencies for pump system synthesis

Parameter	**Description**	**Value**
c^{o}	electricity price	0.15 €/kWh
c^{P}	peak-power price	77.71 €/kW
e_p^{motor}	motor efficiency	$0.98, \forall p \in \mathcal{P}$
e_p^{fc}	frequency converter efficiency	$1, \forall p \in \mathcal{P}^{\text{fixed}}$
e_p^{fc}	frequency converter efficiency	$0.97, \forall p \in \mathcal{P}^{\text{var}}$
ρ	density of water	$998\,\text{kg/m}^3$
g	gravity	$9.81\,\text{m/s}^2$

The coefficients of the model for the fixed-speed pump (Eq. B.10-B.13) are listed in Table B.5.

Table B.5: Node values $a_{n,p}$ of the volume flow rate $\dot{V}_{p,t}$ and function values $\phi_p(a_{n,p})$ for pressure difference $\Delta p_{p,t}$ and power $P_{p,t}$ (Coefficients of the linear segmentation method for the fixed-speed pump model)

Pump p	n	$a_{n,p}$	$\phi_p^{\text{head}}(a_{n,p})$	$\phi_p^{\text{mechPower}}(a_{n,p})$
p_{f1}	1	$420.5\,\text{m}^3/\text{h}$	77.78 m	137.04 kW
	2	$910.25\,\text{m}^3/\text{h}$	66.30 m	191.09 kW
	3	$1400\,\text{m}^3/\text{h}$	44.64 m	221.93 kW
p_{f2}	1	$841\,\text{m}^3/\text{h}$	77.78 m	272.23 kW
	2	$1820.5\,\text{m}^3/\text{h}$	66.30 m	379.62 kW
	3	$2800\,\text{m}^3/\text{h}$	44.64 m	440.88 kW
p_{f3}	1	$1261.5\,\text{m}^3/\text{h}$	77.78 m	392.71 kW
	2	$2730.75\,\text{m}^3/\text{h}$	66.30 m	547.63 kW
	3	$4200\,\text{m}^3/\text{h}$	44.64 m	636.00 kW

The coefficients of the model for the variable-speed pump (Eq. B.20-B.29) are listed in Table B.6, Table B.7 and B.8.

Table B.6: Breakpoint values $a^{\dot{V}}_{n^{\dot{V}},p}$ of the volume flow rate $\dot{V}_{p,t}$ and breakpoint values $a^{\omega}_{n^{\omega},p}$ of the rotational speed $\omega_{p,t}$ (Coefficients of the generalized convex-combination method for the variable-speed pump model)

Pump p	$a^{\dot{V}}_{n^{\dot{V}},p} / \mathrm{m^3/h}$			$a^{\omega}_{n^{\omega},p} / \mathrm{min^{-1}}$		
p_{v1}	220	825	1430	1500	2400	3300
p_{v2}	440	1650	2860	1500	2400	3300
p_{v3}	660	2475	4290	1500	2400	3300
Nodes $n^{\dot{V}}, n^{\omega}$	1	2	3	1	2	3

Table B.7: Functional values $\widetilde{\phi}^{\text{head}}_{p}\left(a^{\dot{V}}_{n^{\dot{V}},p}, a^{\omega}_{n^{\omega},p}\right)$ for the pressure $\Delta_{p,t}$ of all pumps p_{v1},p_{v2},p_{v3} at vertices $\left(n^{\dot{V}}, n^{\omega}\right)$ (Coefficients of the generalized convex-combination method for the variable-speed pump model)

n^{ω} \ $n^{\dot{V}}$	1	2	3
1	20.48 m	17.1 m	10.8 m
2	51.26 m	44.22 m	27.26 m
3	97.26 m	83.27 m	52.35 m

Table B.8: Functional values $\widetilde{\phi}^{\text{mechPower}}_{p}\left(a^{\dot{V}}_{n^{\dot{V}},p}, a^{\omega}_{n^{\omega},p}\right)$ for the power $P_{p,t}$ of pumps p_{v1},p_{v2},p_{v3} at vertices $\left(n^{\dot{V}}, n^{\omega}\right)$ (Coefficients of the generalized convex-combination method for the variable-speed pump model)

Pump p	n^{ω} \ $n^{\dot{V}}$	1	2	3
p_{v1}	1	32.24 kW	45.73 kW	56.84 kW
	2	73.37 kW	117.75 kW	142.77 kW
	3	134.33 kW	220.5 kW	270.31 kW
p_{v2}	1	64.04 kW	90.85 kW	112.92 kW
	2	145.75 kW	233.92 kW	283.62 kW
	3	266.85 kW	438.03 kW	537 kW
p_{v3}	1	92.38 kW	131.06 kW	162.89 kW
	2	210.25 kW	337.45 kW	409.14 kW
	3	384.95 kW	631.89 kW	774.66 kW

APPENDIX C

LP model for a national energy system

The sector model (SecMOD) established in this thesis is a synthesis and operational optimization model. SecMOD includes the power and the heating sector, both including private, commercial, and industry demands, and the private transportation sector. In the synthesis optimization (SO), the capacity is expanded and in the operational optimization (OP), the existing and the expanded capacities are dispatched. SecMOD maps multiple investment periods using a rolling-horizon approach with variable foresight. To holistically evaluate the environmental impact of current and future energy systems, SecMOD incorporates a life-cycle assessment (LCA) (International Organization for Standardization, 2006a,b).

In this work, we parametrize SecMOD for the German energy transition from year 2016 until year 2050. The German high-voltage grid and power sector formulation are based on the open-source electricity model ELMOD-DE by Egerer (2016) published by DIW (German Institute for Economic Research) Berlin. ELMOD-DE is an optimal dispatch model for the power sector (Egerer, 2016). SecMOD considers high spatial resolution based on the high-voltage grid represented by 416 nodes. To reduce the computational burden, the resolution can be simplified to the 18 regions defined by Deutsche Energie-Agentur (2018). The electricity grid, the existing conventional capacities, and the installed renewable energy capacities, all of which are included in ELMOD-DE for the year 2012, are updated to the year 2016, Section C.2.

For transparency, we present a complete model description, including all novelties of SecMOD compared to ELMOD-DE. First, the general optimization method for the transition pathway is presented in Section C.1. The description of the general optimization method is split up into the greenhouse gas (GHG) emission targets (Section C.1.1), the rolling-horizon method (Section C.1.2), and the general equations of the synthesis optimization (Section C.1.3) as well as of the operational optimization (Section C.1.4). Second, the respective technologies are described and specific equations are presented if not already covered within the general set of equations (Section C.2, Section C.3 and Section C.4). Additionally, the data sources for cost, capacity and life-cycle inventory (LCI) data are listed for each sector.

C.1 Transition path

In SecMOD, the transition towards a sustainable energy system is computed by an economic optimization constrained by greenhouse gas emission (GHG) limits resulting from GHG-reduction targets. In this section, the basics of determining the transition path are presented. Presentation includes the GHG-emission targets set in SecMOD, the rolling-horizon approach as well as the general set of equations of the optimization problems.

C.1.1 Greenhouse gas emission targets

The German GHG-reduction targets are set by the German federal government and are described in detail in a document from Federal Ministry for the Environment, Nature Conservation and Nuclear Safety (2018). The document provides a detailed insight into the projected and recorded German GHG emissions originating from the following sectors: power, industrial, transportation, commercial, household, waste, and agriculture. In SecMOD, the GHG-emission targets of the federal government are included for the power, industrial incineration, private vehicle transportation, household as well as the commercial sector. The given target is to reduce GHG emissions by 55 % until 2030, by 70 % until 2040 and by 80-95 % by 2050 compared to the level of 1990. In SecMOD, a linear reduction path is imposed between the given target years. For this purpose, we linearly interpolate between the current state represented by 2016 and the target of 2030, the target of 2030 and the target of 2040, and finally the target of 2040 and the target of 2050. In this German version of SecMOD, 77 % (690 $\mathrm{Mt_{CO_{2eq}}}$ of 899 $\mathrm{Mt_{CO_{2eq}}}$) of the total GHG emissions for the year 2016 are accounted for. The targets apply to the 77 % of total emissions (Federal Ministry for the Environment, Nature Conservation and Nuclear Safety, 2018) . Table C.1 shows the linearly interpolated emission targets implemented in SecMOD. The reduction target of the GHG emissions for the year 2050 is set to 85 % compared to 1990.

Table C.1: Emission targets in $Mt_{CO_{2eq}}$ from 2016 to 2050 assuming a total reduction goal of 85 % until 2050 compared to 1990.

Year	Emission target ($Mt_{CO_{2eq}}$)	Reduction to 1990 (%)	and 2016 (%)
2016	690	28 %	0 %
2020	617	37 %	13 %
2025	526	46 %	25 %
2030	435	55 %	38 %
2035	362	63 %	48 %
2040	290	70 %	59 %
2045	217	78 %	69 %
2050	145	85 %	79 %

C.1.2 Rolling horizon

The transition path of the German energy system is determined in a multi-period investment problem starting with a synthesis optimization in 2016. In the synthesis optimization, the existing infrastructure in the power, heating and transportation sector is expanded to cover given demands for electricity, heat and transportation at each node n for each time step t. As objective function, we choose the total annualized costs.
To render the synthesis optimization computationally feasible, all time-dependent input parameters are aggregated according to the method presented in Bahl et al. (2018a). The result of the synthesis optimization is an energy system at minimal cost for the invest period y which complies with the given GHG-reduction targets.

After the synthesis optimization, an operational optimization for the first investment period, i. e., the hand-over period, is conducted. If there the demands for electricity, heat and transportation at each node n for each time step t can be supplied, the feasibility of the hand-over energy system is proven. The described optimization sequence of synthesis and operational optimization is repeated until $y = 2050$.
For the rolling-horizon approach, an optimization horizon consists of linked investment periods. The number of linked investment periods per optimization horizon can be varied as a parameter. The combination of all optimization horizons results in the transition path. Figure C.1 visualizes the rolling-horizon optimization in SecMOD. For each optimization horizon, the synthesis optimization is carried out resulting in investment decisions for the hand-over period.

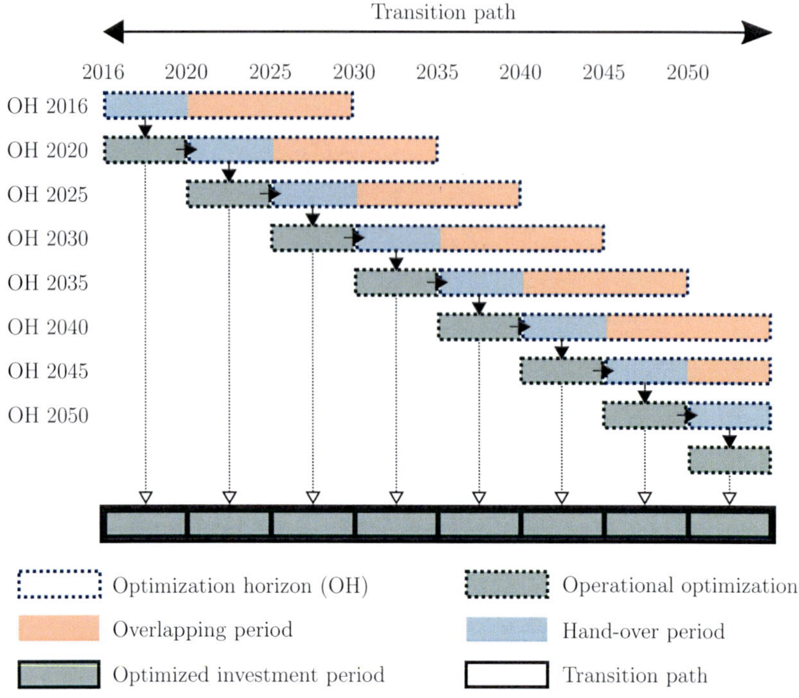

Figure C.1: Sequence of the synthesis and operational optimization by rolling-horizon method. SecMOD ueses an optimization horizon of 3 investment periods, resulting in a transition path.

Investment decisions of overlapping periods are not included in the transition path, as these investment decisions are recalculated in the following optimization horizons. The optimized energy system serves as input energy system for the next optimization horizon. If the optimization horizon ends after the end of the observation period - the year 2050 - the overlapping period is reduced accordingly.

C.1.3 General equations for the synthesis optimization

For the synthesis optimization, the equations for the investment and operational costs, the energy balance as well as the emission balance are valid for all sectors. Variables are typeset in italic upper case letters (e. g., $INVCOST$ for the investment costs), parameters in non-italic lower case letters (e. g., pvf for the present value factor). Non-italic upper case letters indicate sets (e. g., TECH for the set of technologies), while italic lower case letters indicate running indices (e. g., yex for the existing investment periods). To refer to a specif running index of a set, we use quotation marks as indicators (e. g., 'cc' for the climate change impact of the set IMPACT). Superscripts are used to specify variables or parameters (e. g., ^{SO} for the synthesis optimization).

Investment and operational costs

The objective function of an optimization horizon with the investment periods $y \in \text{Y}$ is the total costs $TOTALCOST^{\text{SO}}$ of all technologies $tech \in \text{TECH}$ given by

$$TOTALCOST^{\text{SO}} = \sum_{y \in \text{Y}} \sum_{tech \in \text{TECH}} COST^{\text{SO}}_{tech,y}. \tag{C.1}$$

In Eq. C.1, the discounted net present value $COST^{\text{SO}}_{tech,y}$ contains all payments in € /a for a technology $tech$ in investment period y. The discounted net present value is discounted at the initial investment period by discount_y and multiplied by $\text{pvf}^{\text{Investtime}}$ to incorporating all costs a technology generates over the five years of an investment period. The discounted net present value $COST^{\text{SO}}_{tech,y}$is calculated using

$$COST^{\text{SO}}_{tech,y} = \text{discount}_y \cdot \text{pvf}^{\text{Investtime}} \cdot \left(INVCOST^{\text{SO}}_{tech,y} + OPCOST^{\text{SO}}_{tech,y} \right), \tag{C.2}$$

where $INVCOST^{\text{SO}}_{tech,y}$ are the annualized investment and $OPCOST^{\text{SO}}_{tech,y}$ the annualized operational costs. The annualized costs of all investments in future periods are discounted using the discount factor discount_y to determine the present value of all costs. Further, The annualized costs are multiplied with the present value factor $\text{pvf}^{\text{Investtime}}$ to account for all costs arising within each investment period.

$$\text{pvf}^{\text{Investtime}} = \frac{(1+\text{i})^{\text{Investtime}} - 1}{(1+\text{i})^{\text{Investtime}} \cdot \text{i}}, \tag{C.3}$$

where i is the interest rate and Investtime is the investment period duration. As the investment period incorporates 5 years, we use pvf^5. The discount factor discount_y is used to discount all costs to the initial investment period. The discount factor discount_y is calculated using the interest rate i and the time difference to the initial investment period $y^{\text{OH}}_{\text{initial}}$:

$$\text{discount}_y = \frac{1}{(1+\text{i})^{y - y^{\text{OH}}_{\text{initial}}}}. \tag{C.4}$$

In Eq. C.3 and C.4, we assume a constant interest rate of $\mathrm{i} = 5\,\%$, as in the JRC-EU-TIMES model (Simoes et al., 2013). The annualized investment costs $INVCOST^{\mathrm{SO}}_{tech,y}$ for a technology $tech$ in an investment period y consist of the costs for newly built capacity as well as the costs for existing capacity. They are described by

$$INVCOST^{\mathrm{SO}}_{tech,y} = \sum_{n \in \mathrm{N}} \Big[\sum_{\substack{yy \in \mathrm{Y} \\ yy \leq y}} (CNEW_{n,yy,tech} \cdot \mathrm{invcost}_{tech,yy}) + \sum_{\substack{yex \in \mathrm{YEX} \\ yex \leq y}} (\mathrm{cex}_{n,y,yex,tech} \cdot \mathrm{invcost}_{tech,yex}) \Big] \quad \forall\, tech \in \mathrm{TECH}, y \in \mathrm{Y}. \tag{C.5}$$

The specific investment costs $\mathrm{invcost}^0_{tech,y}$ for a technology $tech$ built in y are annualized using the minimum of technology lifetime and economic investment time (30 years) according to

$$\mathrm{invcost}_{tech,y} = \mathrm{invcost}^0_{tech,y} \cdot \frac{(1+\mathrm{i})^{\min(\mathrm{Lifetime}_{tech},\, 30\,\mathrm{a})} \cdot \mathrm{i}}{(1+\mathrm{i})^{\min(\mathrm{Lifetime}_{tech},\, 30\,\mathrm{a})} - 1} \tag{C.6}$$

where $\mathrm{invcost}^0_{tech,y}$ are the total investment costs for a capacity of technology $tech$ built in y. For specific technologies within the power sector, the annualized investment costs are computed according to their lifetime, which is longer than 30 years (Section C.2). The capacity $CNEW_{n,yy,tech}$ is the newly built capacity in the investment period yy at a node n within the set of nodes N. The newly built capacity is multiplied by the specific investment costs $\mathrm{invcost}_{tech,yy}$ for a technology $tech$ in the investment period yy. The sum over the optimization horizon Y if $yy \leq y$ accounts for all new capacity built in the investment period y. The index yy is introduced to differentiate between the current investment period y and the past investment periods yy and to enable the expression of the condition $yy \leq y$. The capacity $\mathrm{cex}_{n,y,yex,tech}$ is the capacity of technology $tech$ in the investment period y from the building year yex at a node n which is multiplied by the specific investment costs $\mathrm{invcost}_{tech,yex}$ for a technology $tech$ from a year yex. The sum over all existing capacity built in a year yex before the investment period y yields the costs paid for existing capacity in this respective investment period, which are not yet fully paid off.

The operational costs $OPCOST^{\text{SO}}_{tech,y}$ for a technology $tech$ in an investment period y are calculated using

$$OPCOST^{\text{SO}}_{tech,y} = \sum_{n \in \text{N}} \sum_{t \in \text{T}} \sum_{\substack{yex \in \text{YEX} \\ yex \leq y}} GENLOAD^{\text{SO}}_{n,t,y,yex,tech} \cdot \Delta \text{t}_t \cdot \text{opcost}_{yex,tech} \qquad \forall\, tech \in \text{TECH}, y \in \text{Y}. \tag{C.7}$$

The load $GENLOAD^{\text{SO}}_{n,t,y,yex,tech}$ of a technology $tech$ built in year yex at a time step t and a node n in the investment period y is multiplied by the duration of the time step Δt_t and the marginal operational costs $\text{opcost}_{yex,tech}$ of a technology $tech$ built in year yex . The operational variable $GENLOAD^{\text{SO}}_{n,t,y,yex,tech}$ is restricted to the sum of the existing capacity $\text{cex}_{n,y,yex,tech}$ and the newly built capacity $CNEW_{n,yex=y,tech}$ in the respective investment period y by

$$\begin{aligned} GENLOAD^{\text{SO}}_{n,t,y,yex,tech} &\leq cex_{n,y,yex,tech} + CNEW_{n,yex=y,tech} \\ &\forall\, n \in \text{N}, t \in \text{T}, y \in \text{Y}, yex \in \text{YEX}, tech \in \text{TECH}. \end{aligned} \tag{C.8}$$

These general Eq.(C.5, C.7, and C.8) are valid for most technologies used in SecMOD, especially in the power and heating sector. For some technologies, special equations are needed which are described in the respective subsection.

Energy balance

The set of different energy demands E includes power, heat on different temperature levels, and vehicle kilometers for the private transportation sector. For the sake of readability, the running index for the energy demand e is neglected in all parameters as well as capacity and load variables, if possible. Further information on the demand structure for the different energy demands is presented in Section C.2, Section C.3, and Section C.4, respectively.

The load generated by a technology $tech$ from the construction year yex at a node n and a time step t in the investment period y is $GENLOAD^{\text{SO}}_{n,t,y,yex,tech}$. The exogenously given demand is imposed as energy load $\text{energyload}_{n,t,e}$ of the energy demand e at a node n and a time step t. The demand needs to be fulfilled by the total supply of all generation technologies of the respective energy demand:

$$\sum_{tech \in \text{TECH}} \sum_{\substack{yex \in \text{YEX} \\ yex \leq y}} GENLOAD^{\text{SO}}_{n,t,y,yex,tech} = \text{energyload}_{n,t,e} \; \forall\, n \in \text{N}, t \in \text{T}, y \in \text{Y}, e \in \text{E}. \tag{C.9}$$

Environmental Impact Balance
Building capacities for a technology *tech* causes environmental impacts $INVEMIT^{\mathrm{SO}}_{tech,y,impact}$ in categories according to the Environmental Footprint (EF) 2.5 (European Commission, 2018) method:

$$INVEMIT^{\mathrm{SO}}_{tech,y,impact} = \sum_{n \in \mathrm{N}} \Big[\sum_{\substack{yy \in \mathrm{Y} \\ yy \leq y}} (CNEW_{n,yy,tech} \cdot \mathrm{invemit}_{tech,yy,impact}) + \sum_{\substack{yex \in \mathrm{YEX} \\ yex \leq y}} (\mathrm{cex}_{n,y,yex,tech} \cdot \mathrm{invemit}_{tech,yex,impact}) \Big] \quad \forall\, tech \in \mathrm{TECH}, y \in \mathrm{Y}, impact \in \mathrm{IMPACT}. \tag{C.10}$$

The specific environmental impacts $\mathrm{invemit}_{tech,yy,impact}$ and $\mathrm{invemit}_{tech,yex,impact}$ are annualized impacts for the construction of technology *tech* in year *yy* and *yex*. They are annualized using the minimum out of the technology lifetime and the economic investment time (30 years), assuming that a technology is depreciated at the latest after 30 years. Unlike the costs, the impacts are not discounted, but distributed evenly over the respective number of years.
The annual environmental impacts $OPEMIT^{\mathrm{SO}}_{tech,y,impact}$ of a category *impact* in the investment period y resulting from the operation of a technology *tech* are calculated by

$$OPEMIT^{\mathrm{SO}}_{tech,y,impact} = \sum_{n \in \mathrm{N}} \sum_{t \in \mathrm{T}} \sum_{\substack{yex \in \mathrm{YEX} \\ yex \leq y}} GENLOAD^{\mathrm{SO}}_{n,t,y,yex,tech} \cdot \Delta \mathrm{t}_t \cdot \mathrm{opemit}_{yex,tech,impact} \quad \forall\, tech \in \mathrm{TECH}, y \in \mathrm{Y}, impact \in \mathrm{IMPACT}, \tag{C.11}$$

where $\mathrm{opemit}_{yex,tech,impact}$ are the specific environmental impacts of a category *impact* for a technology *tech*. As above, $GENLOAD^{\mathrm{SO}}_{n,t,y,yex,tech}$ is the operational variable describing the generated load by the respective technology at a node n and a time step t. Analogous to the environmental impacts resulting from the investment into a capacity, also the operational impacts are not discounted at the initial investment period.
In SecMOD, the CO_{2eq} emissions resulting from the operation of the energy system is the sum of all impacts in the category climate change impact ('cc'). For each investment period y, the operational CO_{2eq} emissions are limited by an annual limit $\mathrm{maxco2}_y$:

$$\sum_{tech \in \mathrm{TECH}} OPEMIT^{\mathrm{SO}}_{y,tech,\text{'cc'}} \leq \mathrm{maxco2}_y \;\; \forall\, y \in \mathrm{Y}. \tag{C.12}$$

The annual limit of CO_{2eq} emissions $\mathrm{maxco2}_y$ is exogenously given by the targets set by the German federal government and described in Section C.1.1. The values are listed in Table C.1.

C.1.4 General equations for the operational optimization

After the synthesis optimization, an operational optimization with an adjustable resolution up to 1 hour is conducted for the hand-over energy system as described in Section C.1.2. The operational optimization results in an economically optimal operation and is a feasible solution of the full problem. Due to the aggregated load used for the synthesis optimization, the resulting optimal design might be infeasible in the operational optimization. Thus, a blackout might occur. A blackout occurs if the hand-over energy system cannot meet the demand in every time step as required by the energy balance, Eq. (C.9). The operational optimization is computationally efficient since the capacities of the energy system are fixed in the parameter $\mathrm{cex}_{n,tech,yex}$. If a blackout is detected, additional conventional capacity in the power, heating and transportation sector is added after the operational optimization. We identify nodes with heat and transportation blackouts and add gas boilers and diesel vehicles to the capacity at each of these nodes. For the electricity sector, we identify the largest power blackout of all time steps and add gas turbines to the node with the largest blackout. Then, the operational optimization is carried out again. This scheme is repeated until the energy system is able to cover the demand for the whole year. In the presented transition path, no blackout occurred.

Although the synthesis variables are fixed, the operational optimization with the full time resolution might still not be computationally feasible due to memory limits. Thus, to decrease the complexity of the full resolution, the operational optimization is split into an adjustable number of operational optimization units (OP-units) U (e. g., 10 OP-units of 876 h duration each). In the operational optimization, every single OP-unit $u \in \mathrm{U}$ is optimized. This decomposition of the year prohibits long-term time coupling constraints such as the limit for CO_{2eq} operational emissions (Eq. (C.12)) or year-long storage cycles. Still, storage cycles within one OP-unit are considered. The storage level at the beginning equals the storage level at the end of one OP-unit. As storage is used predominately for daily demand shifts, we expect the influence of the storage cycle limitation to be small. The reduction target for CO_{2eq} emissions is substituted by the specification of a CO_2 decomposition price to indirectly punish emissions within the objective function as described in Eq. (C.17). Within SecMOD, the CO_2 limits and the decomposition price are adapted iteratively, leading to a deviation of less than 1 % compared to the given CO_2 limit in Table C.1 and required by Eq. (C.12). In the following, the equations for the operational optimization not already covered within Section C.1.3 are presented.

Investment and Operational Costs

All operational optimizations in a year *yop* result in the total annualized costs (TAC) of the system and are calculated by

$$TAC = \sum_{u \in \mathrm{U}} TAC_u^{\mathrm{OP}}, \tag{C.13}$$

where TAC_u^{OP} are the total annualized costs resulting from the optimization of the OP-unit u. For the OP-unit, the objective function consists of the annualized investment, operational and CO_{2eq} emission costs which are minimized:

$$\min TAC_u^{\mathrm{OP}} = \min \sum_{tech \in \mathrm{TECH}} INVCOST_{tech,u}^{\mathrm{OP}} + OPCOST_{tech,u}^{\mathrm{OP}} + CO2COST_{tech,u}^{\mathrm{OP}} \qquad \forall\, u \in \mathrm{U} \tag{C.14}$$

In the operational optimization, the infrastructure is fixed by the decisions made in the synthesis optimization.

$$INVCOST_{tech,u}^{\mathrm{OP}} = \frac{1}{|\mathrm{U}|} \cdot \sum_{n \in \mathrm{N}} \sum_{\substack{yex \in \mathrm{YEX} \\ yex \leq yop}} \mathrm{cex}_{n,yex,tech} \cdot \mathrm{invcost}_{tech,yex}, \tag{C.15}$$

Thus, the investment costs $INVCOST_{tech,u}^{\mathrm{OP}}$ are fixed. The operational costs $OPCOST_{tech,u}^{\mathrm{OP}}$ and the CO_{2eq} emission costs resulting from the decomposition $CO2COST_{tech,u}^{\mathrm{OP}}$ are determined by the operational optimization:

$$OPCOST_{tech,u}^{\mathrm{OP}} = \sum_{t \in \mathrm{T}} \sum_{n \in \mathrm{N}} \sum_{\substack{yex \in \mathrm{YEX} \\ yex \leq yop}} GENLOAD_{n,t,yex,tech,u}^{\mathrm{OP}} \cdot \Delta \mathrm{t}_t \cdot \mathrm{opcost}_{tech,yex}, \tag{C.16}$$

$$CO2COST_{tech,u}^{\mathrm{OP}} = \mathrm{decompprice} \cdot \sum_{t \in \mathrm{T}} \sum_{n \in \mathrm{N}} \sum_{\substack{yex \in \mathrm{YEX} \\ yex \leq yop}} GENLOAD_{n,t,yex,tech,u}^{\mathrm{OP}} \cdot \Delta \mathrm{t}_t \cdot \mathrm{opemit}_{tech,yex,\text{'cc'}} \qquad \forall\, u \in \mathrm{U}, tech \in \mathrm{TECH}. \tag{C.17}$$

|U| represents the number of OP-units for a year *yop*. The total annual investment costs are divided by |U| to split them up equally to the OP-units. The operational costs $OPCOST^{\mathrm{OP}}_{tech,u}$ in Eq. (C.16) result from the sum of the generated operational costs over all time steps t, all nodes n, and all technologies *tech* with their respective construction year *yex*. In Eq. (C.17), the CO_{2eq} emission costs $CO2COST^{\mathrm{OP}}_{tech,u}$ are dependent on the CO_2 decomposition price decompprice. The decomposition price is endogenously derived from the synthesis optimization as the marginal CO_2 emission price corresponding to the given CO_2 emission limit and handed over to the operational optimization. For the real costs of the energy system, only the investment and operational costs are considered. The CO_{2eq} emission costs are virtual costs for the optimization and not arising for the economy in reality.

Energy Balance

Analogously to (C.9) in the synthesis optimization, the demand must be fulfilled in every time step of the operational optimization.

The operational variable $GENLOAD^{\mathrm{OP}}_{n,t,yex,tech,u}$ is restricted by the existing capacity of the technology *tech* at each node n and each time step t according to

$$\begin{aligned} &GENLOAD^{\mathrm{OP}}_{n,t,yex,tech,u} \leq \mathrm{cex}_{n,tech,yex} \\ &\qquad \forall n \in \mathrm{N}, t \in \mathrm{T}, yex \in \mathrm{YEX}, tech \in \mathrm{TECH}, u \in \mathrm{U}. \end{aligned} \tag{C.18}$$

Environmental Impact Balance

Analogously to the investment and operational costs, the operational environmental impact $OPEMIT^{\mathrm{OP}}_{tech,u,impact}$ for a category *impact* in a specific OP-unit u can be derived from

$$\begin{aligned} &OPEMIT^{\mathrm{OP}}_{tech,u,impact} = \\ &\quad \sum_{t \in \mathrm{T}} \sum_{n \in \mathrm{N}} \sum_{\substack{yex \in \mathrm{YEX} \\ yex \leq yop}} GENLOAD^{\mathrm{OP}}_{n,t,yex,tech,u} \cdot \Delta t_t \cdot \mathrm{opemit}_{tech,yex,impact} \\ &\qquad \forall u \in \mathrm{U}, tech \in \mathrm{TECH}, impact \in \mathrm{IMPACT}. \end{aligned} \tag{C.19}$$

The total CO_{2eq} emissions $TOTALCO2$ of the year *yop* are calculated by summing over the CO_{2eq} emissions of every OP-unit u:

$$\begin{aligned} &TOTALCO2 = \sum_{u \in \mathrm{U}} OPEMIT^{\mathrm{OP}}_{tech,u,\text{'cc'}} \\ &\forall n \in \mathrm{N}, t \in \mathrm{T}, yex \in \mathrm{YEX}, tech \in \mathrm{TECH}, u \in \mathrm{U}. \end{aligned} \tag{C.20}$$

If the CO_{2eq} emission target is not met by the operational optimization, the targets for the synthesis optimization are iteratively adapted. For this purpose, we first adapt the emission targets in the synthesis optimization based on the deviation between target and operational optimization. Then we repeat the synthesis and the operational optimization. If the emission target is still not met by the operational optimization, we (extra-)interpolate between the targets for the synthesis optimization and the emissions of the operational optimization, till the operational optimization satisfies the target within a user-defined deviation.

After this description of the general equations for the synthesis and operational optimization, the following sections present the power, heating and private transportation sector in further detail, including data documentation as well as equations for specific technologies.

C.1.5 Lify-cycle inventory data

In the following section, we describe the life-cycle inventory (LCI) data used for the implementation of the considered technologies across the sectors in SecMOD. The LCI data of all technologies are divided into infrastructural and operational processes. The infrastructural processes cover the construction and subsequent dismantling, potential recycling, treatment and final disposal of wastes for each technology. Based on the infrastructural processes, the environmental impacts of capacities are estimated according to Eq. C.10 and are based on the functional unit of 1 MW or 1 vehicle. The operational LCI data results from the operation of a technology due to the consumption and combustion of fuels, maintenance etc. The operational processes are based on the output of the technology depending on the considered sector: 1 MWh_{power}, 1 MWh_{heat} and 1 vkm. The operational environmental impacts in SecMOD are calculated based on Eq. C.19. All LCI datasets are based on the LCI database ecoinvent version 3.5 APOS (Wernet et al., 2016). The infrastructural LCI data are directly used from ecoinvent and scaled by the process quantity processquantity to derive the LCI of technology $\text{invemit}_{tech,y,impact}$ based on the functional unit:

$$\text{invemit}_{tech,y,impact} = \text{ecoinv}_{tech,impact} \cdot \text{processquantity}. \tag{C.21}$$

For the operation of the technologies, we use the LCI data for maintenance and direct emissions solely, and, if applicable, the fuel consumption. Since ecoinvent provides solely aggregated processes including the operation and infrastructure, we estimate the specific operational impact $\text{opemit}_{tech,y,impact}$ by subtracting the infrastructure:

$$\begin{aligned} \text{opemit}_{tech,y,impact} = \text{ecoinv}_{tech,impact} \cdot \text{processquantity} \\ -\text{invemit}_{tech,y,impact} \cdot \text{invshare}. \end{aligned} \quad (C.22)$$

The aggregated ecoinvent process $\text{ecoinv}_{tech,impact}$ is scaled by the required processquantity, while the subtracted $\text{invemit}_{tech,y,impact}$ is scaled by the share of the required capacity invshare. Technologies are modeled individually based on literature data if ecoinvent data is unavailable. The models can be found for each sector in Sections C.2.4, C.3.4 and C.4.4.

For the end-of-life of the technologies, we assume recycling of metals (Table C.3), and treatment and disposal of all other materials (Table C.4). Technology-specific end-of-life processes can also be found in Sections C.2.4, C.3.4, and C.4.4 for each sector.

Ecoinvent provides LCI data for several geographical regions which are documented in Table C.2. We use German datasets if available. If not, we use datasets from the next closest region. A detailed description of the LCIs for the single technologies can also be found in Sections C.2.4, C.3.4, and C.4.4 for each sector.

Table C.2: Location codes of LCI datasets according to ecoinvent

Location code	**Full name**
CH	Switzerland
DE	Germany
EFTA	European Free Trade Association (Iceland, Liechtenstein, Norway, and Switzerland)
EU27	EU member states except United Kingdom
GLO	Global
IAI	International Aluminum Institute
IAI Area, EU27 & EFTA	Aluminum producing area, EU27 and EFTA countries
RER	Europe
RoW	Rest-of-World
SE	Sweden

Table C.3: Life-cycle inventory of end-of-life technologies part 1. Metal recycling rate according to Reuter et al. (2013). Plastics recycling rate according to (Lindner et al., 2017). Processes without explicit source are based on ecoinvent 3.5 APOS (Wernet et al., 2016)

$ecoinv_{tech,impact}$	Location	Unit	Rate
Aluminum			
treatment of waste aluminum, sanitary landfill	CH	kg	10 %
aluminum recycling		kg	90 %
Cobalt			
treatment of inert waste, inert material landfill	CH	kg	32 %
cobalt recycling		kg	68 %
Copper			
treatment of inert waste, inert material landfill	CH	kg	4 %
treatment of copper scrap by electrolytic refining	RER	kg	26 %
copper recycling		kg	70 %
Iridium			
treatment of inert waste, inert material landfill	CH	kg	75 %
iridium recycling		kg	25 %
Nickel			
treatment of inert waste, inert material landfill	CH	kg	41 %
nickel recycling		kg	59 %
PET			
treatment of waste polyethylene terephthalate, sanitary landfill	CH	kg	2.5 %
market for waste polyethylene terephthalate, for recycling, sorted	Europe w/o CH	kg	27.7 %
treatment of waste polyethylene terephthalate, municipal incineration	CH	kg	69.8 %
Plastic			
treatment of waste plastic, mixture, sanitary landfill	CH	kg	2.5 %
treatment of waste plastic, mixture, municipal incineration	CH	kg	97.5 %

Table C.4: Life-cycle inventory of end-of-life technologies part 2. Metal recycling rate according to Reuter et al. (2013). Plastics recycling rate according to (Lindner et al., 2017). Processes without explicit source are based on ecoinvent 3.5 APOS (Wernet et al., 2016)

$ecoinv_{tech,impact}$	Location	Unit	Rate
Platinum			
treatment of inert waste, inert material landfill	CH	kg	35 %
platinum recycling		kg	65 %
Polyethylene			
treatment of waste polyethylene, sanitary landfill	CH	kg	2.5 %
treatment of waste polyethylene, municipal incineration	CH	kg	69.8 %
market for waste polyethylene, for recycling, sorted	Europe w/o CH	kg	27.7 %
Polypropylene			
treatment of waste polypropylene, sanitary landfill	CH	kg	2.5 %
treatment of waste polypropylene, municipal incineration	CH	kg	97.5 %
Polystyrene			
treatment of waste polystyrene, sanitary landfill	CH	kg	2.5 %
treatment of waste polystyrene, municipal incineration	CH	kg	97.5 %
Polyvinylchloride			
treatment of waste polyvinylchloride, sanitary landfill	CH	kg	2.5 %
treatment of waste polyvinylchloride, municipal incineration	CH	kg	97.5 %
Steel			
treatment of scrap steel, inert material landfill	CH	kg	15 %
steel recycling		kg	85 %
Titanium			
treatment of inert waste, inert material landfill	CH	kg	9 %
titanium recycling		kg	91 %
Zinc			
treatment of inert waste, inert material landfill	CH	kg	58 %
zinc recycling		kg	42 %

C.2 Power sector

The German high-voltage grid and power sector formulation in SecMOD is based on the model of the electricity grid in 2012 in ELMOD-DE by Egerer (2016). We updated the electricity grid based on all realized projects between 2012 and 2016 published by the Federal Network Agency (2018). The new projects included are shown in Figure C.2.

Figure C.2: Additional grid projects included in SecMOD according to Federal Network Agency (2018).

The generation system was also updated to the year 2016. The conventional generation portfolio is based on the power plant list of the Federal Network Agency (Federal Network Agency, 2018). The heat supply of the conventional power plants is added using the Umweltbundesamt's list of power plants in Germany (Umweltbundesamt, 2018). The conventional plants are matched with the closest of the 416 nodes by their ZIP code. The renewable generation capacity (biomass, geothermal, run of river, PV, wind on and offshore) is based on the renewable network transparency data provided by the transmission system operators (Netztransparenz.de, 2018). The capacities are aggregated to the closest node by their ZIP code.

C.2.1 Technologies in the power sector

Table C.5 lists the technologies in the power sector with their respective investment cost, including data sources. For the investment costs, learning curves are implemented to model a cost degression for technologies with for example greater ongoing research and development efforts. The original infrastructure incorporated in ELMOD-DE considers combined cycle gas turbine (CCGT), coal, gas, lignite, nuclear, oil, others, waste incineration, biomass, run of river and geothermal plants. For lignite, CCGT and gas, the construction of new capacities is enabled in SecMOD. All other technologies in the list have been added to SecMOD. For the technologies which have been added, the construction of new capacities is enabled as well. In SecMOD, all capacities denoted 'other' in ELMOD-DE are replaced by oil power plants as a conservative estimate for the GHG-emission intensity. The technology 'waste' is assumed as 'biomass' due to data availability. Life-cycle inventory (LCI) data for all technologies can be found in Section C.2.4.

Table C.5: Investment costs for technologies in the power sector included in SecMOD. CCGT: Combined Cycle Gas Turbine

Technology	Investment cost **2016** (k€/MW_{el})	Investment cost **2050** (k€/MW_{el})	Maintenance cost (k€/($MW_{el} \cdot a$))	Source
Biomass	3297	3287	165	A
CCGT	700	700	24	A
Coal	1467	1467	38	A
Gas	400	400	13	A
Geothermal	10504	9026	380	A
H_2 electrolysis	800	300	14 (2016) - 5 (2050)	A
H_2 fuel-cell	1425	575	22.8 (2016) - 9.2 (2050)	B
Lignite	1596	1596	46	A
Lithium-ion battery	550	350	12	A
New power line	73.53/100km	73.53/100km	included in invcost	E
Nuclear	3323	3323	102	A
Oil	483	483	6.25	C
Other	483	483	6.25	C
Power-to-gas	1438	407	18.5(2016) – 6.8(2050)	A, D
Pumped hydro storage	1218	1218	12	A
Photovoltaics	927	547	17	A
Run of River	500	500	12	A
Switching	41.32/100km	41.32/100km	included in invcost	E
Waste incineration	3297	3287	165	A, F
Wind onshore	1113	938	13	A
Wind offshore	2590	1285	93	A

[A] Deutsche Energie-Agentur (2018)
[B] Palzer (2016)
[C] Chatzimouratidis and Pilavachi (2009)
[D] Breyer et al. (2018)
[E] Section C.2.3.1
[F] The technology 'waste' is assumed as 'biomass'

For all power plants p, except from nuclear, pumped hydro storage plants and the renewable technologies, a lifetime lifetime_{tech} is added to the start-up year Y_p^0, so that the year of shut-down Y_p^{end} is calculated by

$$Y_p^{\text{end}} = Y_p^0 + \text{lifetime}_{tech}. \tag{C.23}$$

The start-up year Y_p^0 is obtained from the list of power plants published by the Federal Network Agency (2018). The shut-down year is defined as the first investment period where the power plant is not available anymore. Thus, the shut-down year is rounded in five-year steps. As the renewable generation capacity is aggregated to the nodes, no explicit start-up year is considered. For the renewable generation capacity, generic shut-down curves are implemented at all nodes also using the data from Federal Network Agency (2018). The shut-down curves are documented in Section C.2.3.5. The shut-down year for all nuclear power plants is set at the end of 2022 according to Federal Ministry of Justice and Consumer Protection and Federal Office of Justice (1985).

The annualized investment costs $\text{invcost}_{tech,y}$ are calculated according to Eq. (C.6). Here, the minimum of the technology lifetime and the economic investment time (30 years) is considered. An exception is made for the grid technologies "new power line" and "switching" as well as for the pumped hydro storage plants. We assume that these technologies are maintained and renewed after their lifetime. Thus, for these 3 technologies, the investment costs are annualized using the technology lifetime. Table C.6 lists the lifetimes considered for the technologies in the power sector.

While maintenance costs are considered as percentage of the investment costs and added to the investment costs, the specific operational costs $\text{opcost}_{yex,tech,t}$ consist only of the fuel costs. They are calculated by

$$\text{opcost}_{yex,tech,t} = \frac{c_{tech,t}^{\text{fuel}}}{\eta_{tech,yex}} \;\forall\; tech \in \text{TECH},\; yex \in \text{YEX},\; t \in \text{T}. \tag{C.24}$$

Analogous to ELMOD-DE, the fuel price $c_{tech,t}^{\text{fuel}}$ is exogenously given as an hourly time series. Table C.7 lists the average fuel prices. The fuel prices, like all other prices, do not include taxes. Additional fuel shipping costs are considered for power plants powered by hard coal from Egerer et al. (2014). The shipping costs are measured in €/MWh and depend on the plant's location (dena zone). The costs are considered as additional operational costs in €/MWh. Figure C.3 shows the spatial shipping costs depending on the dena zone. The efficiencies of the conventional power plant portfolio

Table C.6: Lifetime of the technologies in the power sector.

Technology	Lifetime $lifetime_{tech}$	Source
Biomass	30	A
CCGT	30	A
Coal	45	A
Gas	25	A
Geothermal	30	A
H_2 electrolysis	11 (2016) – 16 (2050)	A
H_2 fuel-cell	16	B
Lignite	45	A
Lithium-ion battery	20	A
New power line	80	B
Nuclear	End of 2022	C
Oil	40	B
Other	40	B
Power-to-gas	20	A
Pumped hydro	100	A
Photovoltaics	25	A
Run of River	100	A
Switching voltage	80	B
Waste incineration	65	D
Wind offshore	25	A
Wind onshore	25	A

[A] Deutsche Energie-Agentur (2018)
[B] Own assumption.
[C] Federal Ministry of Justice and Consumer Protection and Federal Office of Justice (1985)
[D] Markewitz and Stein (2003)

Table C.7: Average fuel price considered in SecMOD, from Egerer (2016).

Fuel	Average fuel price (€/MWh_{th})
Nuclear	3.01
Lignite	4.01
Coal	11.45
Gas	32.46
Oil	52.04
Waste	0

$\eta_{tech,yex}$ depend on their respective start-up year yex. The efficiencies for the start-up years from 1950 to 2020 are taken from Egerer et al. (2014). For start-up years after 2020, the efficiencies are assumed to remain constant.

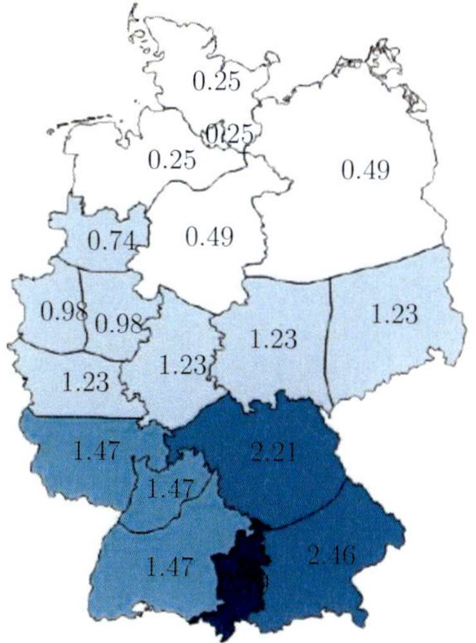

Figure C.3: Transportation costs in €/MWh for hard coal, from Egerer et al. (2014).

C.2.2 Demand structure

In ELMOD-DE, the electricity demand is given as an hourly time series. The total electricity demand of Germany is split up into the nodes using an allocation method based on the population density and the gross domestic product (GDP). However, this allocation method used in ELMOD-DE underestimates significant industrial consumers which cause locally high electricity demands. Therefore, SecMOD integrates the electricity demand of 217 industrial sites (>100MW) which are most significant concerning their CO_{2eq} emissions according to the "European Pollutant Release and Transfer Register" (E-PRTR) (European Environment Agency, 2019). For the electricity demand of the industrial sites, industry-wide data for electricity, heating and fuel demand from Fraunhofer ISE (2018) is used. As an estimate for a single industrial site's share of the energy demand in its sub-industry, the site's share of CO_{2eq} emissions in its sub-industry is taken. The data required for the share of CO_{2eq} emissions is derived from E-PRTR (European Environment Agency, 2019). Thus, the electricity demand of the sub-industry is split up to the industrial sites included in E-PRTR. The electricity demand resulting from industrial sites is allocated to the closest node by ZIP code. The industrial electricity demand is assumed to remain constant in each time step t.

The total electricity demand of the German households and business, commerce, services sector $\mathrm{h}_t^{\mathrm{dem,HH\&GHD}}$ is derived by subtracting the total industrial demand $\mathrm{h}^{\mathrm{dem,ind}}$ from all industrial site from the original electricity demand $\mathrm{h}_t^{\mathrm{dem,ELMOD\text{-}DE}}$ for each time step t:

$$\mathrm{h}_t^{\mathrm{dem,HH\&GHD}} = \mathrm{h}_t^{\mathrm{dem,ELMOD\text{-}DE}} - \mathrm{h}_t^{\mathrm{dem,ind}} \tag{C.25}$$

For the spatial allocation, the electricity of the households and the business, commerce, services sector $\mathrm{h}_t^{\mathrm{dem,HH\&GHD}}$ is split up to the nodes based on the population density and GDP as in Egerer et al. (2014). The electricity demand $\mathrm{h}_{n,t}^{\mathrm{dem,SecMOD}}$ at a node n and a time step t is calculated by summing up the nodal industry and household, business, commerce, services and the industry demand:

$$\mathrm{h}_{n,t}^{\mathrm{dem,SecMOD}} = \mathrm{h}_{n,t}^{\mathrm{dem,HH\&GHD}} + \mathrm{h}_{n,t}^{\mathrm{dem,ind}} \forall n \in \mathrm{N}, t \in \mathrm{T} \tag{C.26}$$

For the synthesis optimization, time series aggregation is used to render the optimization model computationally feasible. Time series aggregation averages the load and causes extreme load points to be filtered out. Thus, generation capacities are likely to be underestimated which may lead to blackouts in case of extreme situations such as dark doldrums. To guarantee coverage of these extreme situations, the guaranteed capacity must always be greater than the sum of the annual peak load and the endogenously generated power demand. This constraint is described by

$$\begin{aligned} \max_t \sum_{n \in \mathrm{N}} \left(\mathrm{h}_{n,t}^{\mathrm{dem,SecMOD}} + \mathrm{power}_{n,t}^{\mathrm{ind}} \right) + \mathrm{endogpower}_{y-1} \leq \\ \sum_{tech \in \mathrm{TECH}} \left[\mathrm{cf}_{tech} \cdot \mathrm{cex}_{tech,y} + \sum_{\substack{yy \in \mathrm{YY} \\ yy \leq y}} \mathrm{cf}_{tech} \cdot CNEW_{tech,yy}^{\mathrm{SO}} \right] \\ \forall y \in \mathrm{Y}. \end{aligned} \tag{C.27}$$

Here, $\max_t \sum_{n \in \mathrm{N}} \left(\mathrm{h}_{n,t}^{\mathrm{dem,SecMOD}} + \mathrm{power}_{n,t}^{\mathrm{ind}} \right)$ is the sum of the annual peak load resulting from the electricity demand calculated in Eq. C.26 and the loads caused by industrial sites at node n and time step t. The term endogpower$_{y-1}$ sums up the electricity demand endogenously generated by electricity-consuming technologies within SecMOD. Therefore, the value is zero for the first investment period. After each investment period, endogpower$_{y-1}$ is calculated as the sum of the electricity demand of newly built electricity-consuming technologies (Power-to-Gas, H_2 electrolysis, heat pump, electrode boiler, electric vehicle, hybrid vehicle, Power-to-Diesel). The presumed endogenously generated power demand is the calculated value of the last investment period. To calculate the guaranteed capacity, the existing and newly built capacities of each technology $tech$ are multiplied by the technology's capacity factor cf$_{tech}$ and summed over all investment periods in the optimization horizon and all technologies. The capacity factors of the power-generating technologies incorporate the relation between guaranteed capacity and installed capacity and are listed in Table C.9. The guaranteed capacity includes all capacities available for supply minus not available capacities due to ,e. g., revision and reserves for system services. The available capacities are based on a report of the German transmission net operators (50Hertz et al., 2012).

Table C.9: Capacity factors assumed in SecMOD for power-generation technologies based on (Deutsche Energie-Agentur, 2010).

Technology	**Capacity factor** cf$_{tech}$ (%)	**Source**
Biomass	88	A
CCGT	86	A
Coal	86	A
Gas	42	A
Geothermal	90	A
Lignite	92	A
Lithium	90	A
Nuclear	93	A
Oil	92	B
Others	92	B
Photovoltaics	1	A
Pumped storage	90	A
Run of river	40	A
Waste	90	B
Wind offshore	5	A
Wind onshore	5	A

[A] Lower bound according to Deutsche Energie-Agentur (2010).
[B] Chatzimouratidis and Pilavachi (2009)

C.2.3 Description of specific technologies

In the following section, all technologies are described in greater detail, for which extensions to the general equations are necessary.

C.2.3.1 Grid - switching and new power line

The two grid technologies include switching the voltage level ($\mathrm{v}(l)$) of a transmission line from 220 kV to 380 kV (switching) and building a new power line with the voltage level 380 kV between two nodes (new power line). Switching the voltage level increases the power flow limit. Thus, higher amounts of electricity may be transported by the power line. Due to the high complexity of a connection between all of the 416 nodes, new power lines may only be built between nodes with an existing connection. This constraint is also implemented if the grid and nodes are aggregated according to the dena zones. A grid technology does not have a shut-down year, as it is assumed to be renewed after its lifetime. Thus, the investment costs are annualized over the complete lifetime of the installation.

Following ELMOD-DE (Egerer, 2016), SecMOD represents network flows with the DC-load-flow approximation by Overbye et al. (2004) which is a linearization of the AC power flow. The linearization assumes small differences in voltage angles and in voltage levels. The DC-load-flow approximation is described in detail by Egerer (2016).
The real power flow $PF_{l,t,y}$ on a line $l \in \mathrm{L}^{\mathrm{i}} = \{l \in L : v(l) = i\}$ at a time step t in a calculation period y is described by

$$PF_{l,t,y} = \sum_{n \in \mathrm{N}} \mathrm{h}_{l,n} \cdot DELTA_{n,t,y} \; \forall \; l \in \mathrm{L}^{i}, \; t \in \mathrm{T}, \; y \in \mathrm{Y}. \tag{C.28}$$

The network transfer matrix $\mathrm{h}_{l,n}$ aggregates the physical line parameters such as reactance and resistance and the grid topology. The network transfer matrix is derived from the incidence matrix $\mathrm{incidence}_{l,n}$ which indicates the grid topology by taking the value $+1$ for the start node and -1 for the end node of a specific line l. For this purpose, the network transfer matrix has entries for the start and end nodes of all lines $l \in \mathrm{L}$. The optimization variable $DELTA_{n,t,y}$ describes the difference in voltage angles at a time step t in a year y between a node n and the reference node n^0. The voltage angle of the reference node is defined as 0° to enforce unique solutions for the voltage angles at the other nodes.

Furthermore, the absolute value of the power flow limit is restricted leading to equations for the positive and for the negative end of the restriction:

$$\begin{aligned} PF_{l,t,y} \leq\ & \text{pflimit}_l \\ & + (1 - \text{trm}) \cdot \Big[\sum_{\substack{yy \in \mathrm{Y} \\ yy \leq y}} CNEW_{l,yy,\text{'switching'}} + CNEW_{l,yy,\text{'newpowerline'}} \\ & + \sum_{\substack{yex \in \mathrm{Y} \\ yex \leq y}} \text{cex}_{l,yy,\text{'switching'}} + \text{cex}_{l,yy,\text{'newpowerline'}} \Big] \end{aligned} \tag{C.29}$$

$$\begin{aligned} PF_{l,t,y} \geq\ & -\text{pflimit}_l \\ & - (1 - \text{trm}) \cdot \Big[\sum_{\substack{yy \in \mathrm{Y} \\ yy \leq y}} CNEW_{l,yy,\text{'switching'}} + CNEW_{l,yy,\text{'newpowerline'}} \\ & + \sum_{\substack{yex \in \mathrm{Y} \\ yex \leq y}} \text{cex}_{l,yy,\text{'switching'}} + \text{cex}_{l,yy,\text{'newpowerline'}} \Big] \\ & \forall\, l \in \mathrm{L^i},\, t \in \mathrm{T},\, y \in \mathrm{Y}. \end{aligned} \tag{C.30}$$

The capacity of a line l consists of the power flow limit pflimit_l for the original infrastructure of the base year and the additional capacity from the synthesis optimization. The physical line parameters (power flow limit, reactance, resistance, number of circuits, voltage rating and length of the line) are included in the infrastructure of the base year and thus taken from Egerer (2016) and updated accordingly to Figure C.2. The power-flow limit includes a thermal reliability margin (trm) of 20 % to approximate the n–1 criterion. The n–1 criterion ensures that the grid is able to transmit the demanded electricity even if 1 power line fails. The power flow limit pflimit_l is computed by:

$$\text{pflimit}_l = \text{maxpower}_{\mathrm{v}(l)} \cdot \text{circuits}_l \cdot (1 - \text{trm}) \qquad \forall\, l \in \mathrm{L^i}, \tag{C.31}$$

where $\text{maxpower}_{\mathrm{v}(l)}$ is the maximum power of one circuit of voltage level $\mathrm{v}(l)$ and circuits_l the number of parallel circuits on the power line l.

The general cost equations presented in Eq. C.5 are not valid for switching or for a new power line, because the total investment costs are dependent on the length length_l of the respective line. Further, for the switching costs, the costs are summed over all lines with the voltage level $\mathrm{v}(l) = 220\,\mathrm{kV}$ instead of all nodes. Thus, the investment costs for switching the voltage level are calculated by

$$INVCOST^{\mathrm{SO}}_{\text{'switching'},y} = \sum_{l \in \mathrm{L}^{220}} \Bigg[\sum_{\substack{yy \in \mathrm{Y} \\ yy \leq y}} (\mathrm{length}_l \cdot CNEW_{n,yy,\text{'switching'}} \cdot \mathrm{invcost}_{\text{'switching'},yy}) + \sum_{\substack{yex \in \mathrm{YEX} \\ yex \leq y}} (\mathrm{length}_l \cdot \mathrm{cex}_{n,y,yex,\text{'switching'}} \cdot \mathrm{invcost}_{\text{'switching'},yex}) \Bigg] \; \forall\, y \in \mathrm{Y}. \tag{C.32}$$

$CNEW_{n,y,\text{'switching'},l}$ is the newly built capacity of a line l in year y, whereas the existing capacity of a line l switched in the year yex is described by $\mathrm{cex}_{n,y,yex,\text{'switching'},l}$. The capacities are multiplied by the length of the power line length_l and the specific investment cost for switching the voltage level in an investment period y, $\mathrm{invcost}_{\text{'switching'},y}$. Table C.10 lists the total investment costs of the two grid technologies. For the grid technologies, we assume the investment costs to remain constant over the investment periods. The investment costs are calculated based on standardized costs from Rippel et al. (2017) and the power limits from Kießling et al. (2001).

Table C.10: Investment costs for grid technologies included in SecMOD. Source: Calculations based on prices from Rippel et al. (2017) and power limits from Kießling et al. (2001).

Technology	**Investment cost** (k€/(MW · 100 km))
New power line	73.53
Switching	41.32

The power limits for 380 kV and 220 kV from Kießling et al. (2001) are used to calculate the limit of allowed switches from a 220 kV transmission line to a 380 kV transmission line. The power limit for a transmission line of 380 kV or of 220 kV are 1700 MW or 490 MW respectively. As a result, the limit of allowed switches is set to $\mathrm{switchlimit}_y = 1700\,\mathrm{MW} - 490\,\mathrm{MW} = 1210\,\mathrm{MW}$ for each circuit in each investment period y:

$$CNEW_{l,y,\text{'switching'}} \leq \mathrm{switchlimit}_y \cdot \mathrm{circuits}_l - \sum_{\substack{yex \in \mathrm{YEX} \\ yex \leq y}} \mathrm{cex}_{n,y,yex,\text{'switching'}} - \sum_{\substack{yy \in \mathrm{Y} \\ yy < y}} CNEW_{n,yy,\text{'switching'}} \quad \forall\, y \in \mathrm{Y}, l \in \mathrm{L}^{220} \tag{C.33}$$

Additionally, switching is not permitted for lines which already are on the 380 kV voltage level:

$$CNEW_{l,y,\text{'switching'}} = 0 \;\forall\, y \in \mathrm{Y}, l \in \mathrm{L}^{380} \tag{C.34}$$

The investment costs for building a new 380 kV power line $INVCOST^{\mathrm{SO}}_{\text{'newpowerline'},y}$ are calculated by

$$\begin{aligned} & INVCOST^{\mathrm{SO}}_{\text{'newpowerline'},y} = \\ & \sum_{l \in \mathrm{L}^{\mathrm{i}}} \Bigg[\sum_{\substack{yy \in \mathrm{Y} \\ yy \leq y}} \left(\mathrm{length}_l \cdot CNEW_{n,yy,\text{'newpowerline'}} \cdot \mathrm{invcost}_{\text{'newpowerline'},yy}\right) \\ & + \sum_{\substack{yex \in \mathrm{YEX} \\ yex \leq y}} \left(\mathrm{length}_l \cdot \mathrm{cex}_{n,y,yex,\text{'new powerline'}} \cdot \mathrm{invcost}_{\text{'new powerline'},yex}\right) \\ & + \mathrm{maxpower}_{\mathrm{v}(l)} \cdot \mathrm{circuits}_l \cdot \mathrm{invcost}_{\text{'new powerline'},y} \Bigg] \quad \forall\, y \in \mathrm{Y}. \end{aligned} \tag{C.35}$$

$CNEW_{n,y,\text{'new powerline'},l}$ is the newly built capacity of a line l in year y, whereas $\mathrm{cex}_{n,y,yex,\text{'new powerline'},l}$ is the existing capacity of a line l newly built in the year yex. The capacities are multiplied by the length of the powerline length_l and the specific investment cost for building a new power line in an investment period y, $\mathrm{invcost}_{\text{'switching'},y}$.

The investment costs of the base year grid infrastructure $INVCOST^{\mathrm{SO}}_{\text{'basegrid'},y}$ are considered by summing over the maximum power of a circuit multiplied by the number of circuits $\mathrm{circuits}_l$ as well as over all lines with their respective voltage levels ($\forall\, y \in \mathrm{Y}$):

$$INVCOST^{\mathrm{SO}}_{\text{'basegrid'},y} = \sum_{\substack{l \in \mathrm{L}^{\mathrm{i}} \\ yex(l) \leq y_{\mathrm{initial}}}} \mathrm{maxpower}_{\mathrm{v}(l)} \cdot \mathrm{circuits}_l \cdot \mathrm{invcost}_{\text{'newpowerline'},y}, \tag{C.36}$$

while the specific investment costs $\mathrm{invcost}_{\text{'newpowerline'},y}$ equal the specific investment costs of a newly built power line.

The operational costs for the switching and building new power lines are fixed to zero as there are no fuel costs involved. Maintenance costs are included in the annualized investment costs and no further operational costs are considered.

C.2.3.2 Power-storage technologies

In order to store electricity in SecMOD, 3 technologies contained in the set STORAGES are implemented. The 3 technologies are lithium-ion batteries, pumped-hydro-storage plants and hydrogen storage. For the considered German, case no capacity expansions for pumped-hydro-storage plants is allowed as the potential is assumed to be fully exploited. For hydrogen storage, produced hydrogen can be stored in underground caverns and converted to electricity using fuel-cells. The potential is limited to the availability of storage caverns.

The general equations, presented in Section C.1.3, do not account for storage technologies. Thus, an additional equation for the storage level for all storage systems is necessary. This storage level equation is valid for all technologies, while the remaining storage-specific equations are presented in the following subsections.

The net power of a storage *storage* at a node n and in a time step t results from

$$\begin{aligned} GENLOAD^{\mathrm{SO}}_{n,t,y,storage} = \sum_{yex \in \mathrm{YEX}} \Big[& GENPOWER^{\mathrm{SO}}_{n,t,y,yex,storage} \\ & - STOREDPOWER^{\mathrm{SO}}_{n,t,y,yex,storage} \Big] \\ & \forall\ storage \in \mathrm{STORAGE},\ n \in \mathrm{N},\ t \in \mathrm{T},\ y \in \mathrm{Y}, \end{aligned} \tag{C.37}$$

where the generated power by discharging each technology *storage* is described by $GENPOWER^{SO}_{n,t,y,yex,storage}$ and the power for charging by $STOREDPOWER^{SO}_{n,t,y,yex,storage}$. Restrictions for the charging and discharging power are following for each technology in the corresponding sections.

The power from charging $STOREDPOWER^{SO}_{n,t,y,yex,\text{'storage'}}$ and discharging $GENPOWER^{SO}_{n,t,y,yex,\text{'storage'}}$ a storage are restricted by the available capacity at the node including newly built and existing capacity, and by non-negativity:

$$0 \leq GENPOWER^{\mathrm{SO}}_{n,t,y,yex,\text{'storage'}} \leq CNEW_{n,y=yex,\text{'storage'}} + \mathrm{cex}_{n,y,yex,\text{'storage'}} \tag{C.38}$$

$$\begin{aligned} & 0 \leq STOREDPOWER^{\mathrm{SO}}_{n,t,y,yex,\text{'storage'}} \leq CNEW_{n,y=yex,\text{'storage'}} + \mathrm{cex}_{n,y,yex,\text{'storage'}} \\ & \forall\ n \in \mathrm{N},\ t \in \mathrm{T},\ y \in \mathrm{Y}. \end{aligned} \tag{C.39}$$

We assume that the potential of pumped-hydro storage is estimated to be already exploited. Therefore, $CNEW_{n,y=yex,\text{'storage'}}$ is neglected for this storage technology.

The amount of energy stored in a storage at a time step $t+1$ following the current time step t in the specific storage capacity from the construction year yex at a node n is described by the storage level $STORAGELEVEL^{\mathrm{SO}}_{n,t+1,y,yex,storage}$. The storage level at a time step $t+1$ is calculated using the storage level at the current time step t and the energy from charging and discharging discretized with the length of a time step $\Delta \mathrm{t}_t$:

$$\begin{aligned}&STORAGELEVEL^{\mathrm{SO}}_{n,t+1,y,yex,storage} = STORAGELEVEL^{\mathrm{SO}}_{n,t,y,yex,storage}\\&+ \left[STOREDPOWER^{\mathrm{SO}}_{n,t,y,yex,storage} \cdot \eta_{storage,yex} - GENPOWER^{\mathrm{SO}}_{n,t,y,yex,storage}\right] \cdot \Delta \mathrm{t}_t\\&\forall\ storage \in \mathrm{STORAGE},\ n \in \mathrm{N},\ t \in \mathrm{T},\ yex \in \mathrm{YEX}.\end{aligned} \tag{C.40}$$

In the synthesis optimization, we use the time series aggregation method from Bahl et al. (2018a). Therefore, the storage level at the beginning of each typical period is coupled to the end of that period. Thus, for the synthesis optimization the maximal storage cycle is limited by the length of the typical periods. The charging efficiency $\eta_{storage,yex}$ for the pumped-hydro storage is set to 75 % according to Egerer (2016).

Lithium-ion Batteries

The storage level of lithium-ion batteries is restricted to the maximum energy amount storable with the existing and newly built capacity:

$$\begin{aligned}&STORAGELEVEL^{\mathrm{SO}}_{n,t,y,yex,\text{'battery'}} \leq\\&\left[CNEW_{n,y=yex,\text{'battery'}} + \mathrm{cex}_{n,y,yex,\text{'battery'}}\right] \cdot \mathrm{etop}_{\text{'battery'},y}\end{aligned} \tag{C.41}$$

Capacity and power of the lithium-ion battery can be built in the specific ratio of $1.25\,\frac{\mathrm{MWh}}{\mathrm{MW}}$ based on industrial data. The parameter $\mathrm{etop}_{\text{'battery'},y}$ is the energy-to-power ratio and converts the built power capacities $CNEW_{n,y,\text{'battery'}}$ and $\mathrm{cex}_{n,y,yex,\text{'battery'}}$ to energy capacity.

Pumped-hydro storage

Pumped-hydro-storage plants 'psp' are assumed to be renewed after their lifetime. Thus, the investment costs are annualized over the complete lifetime of the plant.

The storage level is restricted to the maximum energy amount $\text{energycapacity}_{n,yex,\text{'psp'}}$ storable within the capacity from the year yex at the node n:

$$STORAGELEVEL^{\text{SO}}_{n,t,y,yex,\text{'psp'}} \leq \text{energycapacity}_{n,t,y,yex,\text{'psp'}} \quad \forall\, n \in \text{N},\, t \in \text{T},\, yex \in \text{YEX},\, y \in \text{Y}. \tag{C.42}$$

The maximum energy amount $\text{energycapacity}_{n,yex,\text{'psp'}}$ is part of the base year energy system updated using the data from Federal Network Agency (2018).

Power-to-Hydrogen & Hydrogen-to-Power

Hydrogen with a lower heating value of $\text{LHV}^{\text{H2}} = 33.33\,\frac{\text{kWh}_{\text{H2}}}{\text{kg}_{\text{H2}}}$ is produced by the proton exchange membrane (PEM) electrolysis using electric power. The efficiency of PEM-electrolysis $\eta_{\text{'pem'},yex}$ regarding the LHV is 67 % for 2016 and increases linearly to 80 % until 2050 (Agora Verkehrswende et al., 2018).

Produced hydrogen can be stored in underground caverns. The potential of hydrogen storage is limited due to the availability of storage caverns. As a conservative estimate, this work assumes a storage potential of 26.5 TWh_{H2} referencing the lower heating value (Noack et al., 2015). The potential is split up into the nodes which are matching the regions with suitable salt caverns identified by Noack et al. (2015). All other nodes remain without the possibility to store hydrogen. Thus, the storage technology of producing and storing hydrogen can only be deployed at the nodes with storage potential. Table C.12 lists the nodes with potential as well as the actual potential. Figure C.4 visualizes the potential.

Table C.12: Nodes n with a cavern storage potential for the storage of hydrogen according to Noack et al. (2015).

Node n	**Latitude**	**Longitude**	**Cavern potential** (MWh_{H2})
35	53.543	9.502	7,015
41	53.342	7.207	6,476
42	53.333	8.053	204,612
49	53.190	8.409	18,168
50	53.126	7.312	166,028
55	53.093	8.696	13,581
100	52.203	7.034	205,511
117	51.909	11.644	118,720
144	51.619	6.614	10,793
188	51.405	11.893	44,430

Figure C.4: Cavern storage potential for the storage of hydrogen according to Noack et al. (2015).

The charging of the cavern storage systems is described by the technology Power-to-Hydrogen 'p2h,' while the discharging is described by the technology Hydrogen-to-Power 'h2p.' Other than for the storage technologies above, the generated load of both technologies is limited analogous to Eq. C.8 for the nodes with cavern potential. The storage level $STORAGELEVEL^{\mathrm{SO}}_{n,t,y,\text{'hydrogen'}}$ at a node n with cavern potential is limited to the cavernpotential$_n$:

$$STORAGELEVEL^{\mathrm{SO}}_{n,t,y,\text{'hydrogen'}} \leq \mathrm{cavernpotential}_n \quad \forall\, n \in \mathrm{N},\, t \in \mathrm{T},\, yex \in \mathrm{YEX},\, y \in \mathrm{Y}. \tag{C.43}$$

If there is no cavern potential at a node n, the storage level at the respective node is fixed to zero. The storage level at a time step $t+1$ is calculated using the storage level at the current time step t and the energy from charging and discharging discretized with the time step length $\Delta \mathrm{t}_t$:

$$\begin{aligned} & STORAGELEVEL^{\mathrm{SO}}_{n,t+1,y,\text{'hydrogen'}} = STORAGELEVEL^{\mathrm{SO}}_{n,t,y,\text{'hydrogen'}} \\ & + \sum_{\substack{yex \in \mathrm{YEX} \\ yex \leq y}} \left[GENLOAD^{\mathrm{SO}}_{n,t,y,yex,\text{'p2h'}} \cdot \eta_{\text{'p2h'},yex} \cdot \Delta \mathrm{t}_t \right] \\ & - \sum_{\substack{yex \in \mathrm{YEX} \\ yex \leq y}} \left[GENLOAD^{\mathrm{SO}}_{n,t,y,yex,\text{'h2p'}} \cdot \frac{1}{\eta_{\text{'h2p'},yex}} \cdot \Delta \mathrm{t}_t \right] \\ & \forall\, n \in \mathrm{N},\, t \in \mathrm{T},\, yex \in \mathrm{YEX}. \end{aligned} \tag{C.44}$$

The efficiency of the electrolysis is $\eta_{\text{'p2h'},yex}$ is 67 % in 2016 and 80.7 % in 2050 Agora Verkehrswende et al. (2018). The efficiency of the fuel-cell is $\eta_{\text{'h2p'},yex}$ 66.9,% in 2016 and 79.9,% in 2050 each referring to the upper heating value of hydrogen (Deutsche Energie-Agentur, 2018).

C.2.3.3 Power-to-methane

The technology "Power-to-Methane" 'p2g' produces synthetic natural gas from electricity through CO_2 and H_2. H_2 is produced via proton exchange membrane (PEM) electrolysis, while CO_2 is captured by direct air capture, both using electric power. In SecMOD, integrated production capacities are considered incorporating all 3 subcomponents as 1 technology (see Table C.13).
The equations for the investment costs and environmental impacts are implemented as presented in Section C.1.3. During operation (gate-to-gate), the utilization of CO_2 abates CO_2 emissions. Thus, the operational environmental impacts include a credit for utilized CO_2. The annual operational CO_{2eq} emissions $OPEMIT_{\text{'p2g'},y,\text{'cc'}}$ in the investment period y are calculated by

$$\begin{aligned} OPEMIT_{\text{'p2g'},y,\text{'cc'}} = -\sum_{n \in \mathrm{N}} \sum_{t \in \mathrm{T}} \sum_{\substack{yex \in \mathrm{YEX} \\ yex \leq y}} \Big[& GENLOAD_{n,t,y,yex,\text{'p2g'}} \cdot \Delta \mathrm{t}_t \\ & \cdot \eta_{\text{'p2g'},yex} \cdot \mathrm{co2benefit}_{\text{'p2g'}} \Big] \quad \forall\, y \in \mathrm{Y}, \end{aligned} \tag{C.45}$$

where $\eta_{\text{'p2g'},yex}$ converts the generated power load to the amount of produced natural gas in MWh and $\mathrm{co2benefit}_{\text{'p2g'}}$ indicates the amount of CO_2 (in t) captured from the

air to produce 1 MWh of natural gas.
The operational costs $OPCOST_{\text{'p2g'},y}$ in investment period y are calculated by

$$OPCOST_{\text{'p2g'},y} = -\sum_{n \in \mathrm{N}} \sum_{t \in \mathrm{T}} \sum_{\substack{yex \in \mathrm{YEX} \\ yex \leq y}} \Big[GENLOAD_{n,t,y,yex,\text{'p2g'}} \cdot \Delta \mathrm{t}_t \cdot \eta_{\text{'p2g'},yex} \cdot \mathrm{fuelprice}_{\text{'gas'},t} \Big] \quad \forall y \in \mathrm{Y}, \tag{C.46}$$

where $\mathrm{fuelprice}_{\text{'gas'},t}$ is the fuel price for natural gas introduced in Section C.2.1. The Eq. C.45 and C.46 are valid for both synthesis and operational optimization. For the sake of readability, the superscripts SO and OP at the variables are neglected.

SecMOD assumes that the produced synthetic natural gas is directly fed into the gas grid. Hence, the operational costs include a cost credit for the replacement of natural gas by the synthetic natural gas produced. Each of the nodes in the model is connected to the gas grid, as the spatial resolution of the German gas grid is high. The production of synthetic natural gas is also not restricted by the storage capacity of the gas grid, because the storage capacity of the grid is assumed to be sufficient with a storage capacity of 217 TWh (Zentrum für Energieforschung, 2012). Therefore, the production of synthetic natural gas is only restricted by the amount of gas consumed:

$$\sum_{n \in \mathrm{N}} \sum_{t \in \mathrm{T}} \sum_{\substack{yex \in \mathrm{YEX} \\ yex \leq y}} \Big[GENLOAD^{\mathrm{SO}}_{n,t,y,yex,\text{'p2g'}} \cdot \Delta \mathrm{t}_t \cdot \eta_{\text{'p2g'},yex} \Big] \leq \sum_{n \in \mathrm{N}} \sum_{t \in \mathrm{T}} \sum_{\substack{yex \in \mathrm{YEX} \\ yex \leq y}} \sum_{tech \in \mathrm{GAS}} \Big[GENLOAD^{\mathrm{SO}}_{n,t,y,yex,tech} \cdot \Delta \mathrm{t}_t \cdot \frac{1}{\eta_{tech,yex}} \Big] \quad \forall y \in \mathrm{Y}. \tag{C.47}$$

The set GAS is a subset of the set TECH and includes all technologies consuming gas: gas power plants, combined cycle gas plants, centralized and decentralized gas boilers, and natural gas vehicles. The equation is only valid for the synthesis optimization to avoid the coupling of all time steps in the operational optimization. Therefore, the produced and consumed synthetic natural gas are compared after the operational optimization. If the produced amount exceeds the consumption, the optimization is repeated with individual constraints for the OP-units. In practice, these individual constraints have not been necessary.

The investment costs $\text{invcost}^0_{\text{'p2g'},yex}$ for the integrated production capacity are derived by the sum over the set SUB:

$$\text{invcost}^0_{\text{'p2g'},yex} = \sum_{sub \in \text{SUB}} \text{x}_{sub,yex} \cdot \text{invcost}^0_{sub,yex} \cdot \frac{\text{pvf}^{\text{lifetime 'p2g'},yex}}{\text{pvf}^{\text{lifetime } sub,yex}}. \tag{C.48}$$

The set SUB contains all 3 sub-components sub incorporated in the integrated production capacity: direct air capture ('dac'), PEM-electrolysis ('pem') and methanation ('met'). Within Eq. C.48, both the different lifetimes $\text{lifetime}_{sub,yex}$ as well as the shares $\text{x}_{sub,yex}$ of each subcomponent are taken into account. Similar to all other technologies $\text{invcost}^0_{sub,yex}$ additionally contains the maintenance costs of each sub-component.

The shares of the sub-components are calculated based on the required mass $\text{m}_{sub,yex}$ and electricity demand $\text{w}_{sub,yex}$ to produce 1 MWh synthetic natural gas:

$$\text{x}_{sub,yex} = \frac{\text{w}_{sub,yex} \cdot \text{m}_{sub,yex}}{\sum_{sub \in \text{SUB}} \text{w}_{sub,yex} \cdot \text{m}_{sub,yex}} \quad \forall\, sub \in \text{SUB}. \tag{C.49}$$

All data for modeling the technology “Power-to-Methane” and the corresponding sources are specified in Table C.13.

Table C.13: Modeling assumptions for "Power-to-Methane" ('p2g').

Specification	Unit	2016	2030	2040	2050	Source
System efficiency: $\eta_{\text{'p2g'},yex}$	$\text{MWh}_{\text{CH4}}/\text{MWh}_{\text{el,tot}}$	0.498	0.525	0.555	0.584	A,B,C
CO_2 benefit: $\text{co2benefit}_{\text{'p2g'}}$	$\text{t}_{\text{CO2}}/\text{MWh}_{\text{CH4}}$	0.197				B
Direct air capture electricity demand: $\text{w}^{\text{el,'dac'}}_{\text{yex}}$	$\text{MWh}_{\text{el,'dac'}}/\text{t}_{\text{CO2}}$	0.500	0.500	0.500	0.500	B
PEM-electrolysis efficiency: $\eta_{\text{'pem'},yex}$	$\text{MWh}_{\text{H2}}/\text{MWh}_{\text{el,'pem'}}$	0.670	0.710	0.755	0.807	C
Inputs:						
$\text{m}_{CO2,yex}$	$\text{t}_{\text{CO2}}/\text{t}_{\text{CH4}}$	2.755				B,E
$\text{m}_{H2,yex}$	$\text{t}_{\text{H2}}/\text{t}_{\text{CH4}}$	0.504				B,E
Total investment costs: $\text{invcost}^{0}_{\text{'p2g'},yex}$	$\text{k€}/\text{MW}_{\text{el}}$	1105	784	526	312	
Investment costs:	$\text{k€}/\text{MW}_{\text{el}}$	1140	810	545	325	
Direct air capture	$\text{k€}/\text{MW}_{\text{el}}$	19	10	8	6	D
PEM electrolysis	$\text{k€}/\text{MW}_{\text{el}}$	800	633	467	300	A
Methanation	$\text{k€}/\text{MW}_{\text{CH4}}$	400	310	220	130	A
Maintenance costs:	$\text{k€}/(\text{MW}_{\text{el}} \cdot \text{a})$	18.5	14.6	10.7	6.8	
Direct air capture	$\text{k€}/(\text{MW}_{\text{el}} \cdot \text{a})$	0.08	0.04	0.03	0.02	D
PEM electrolysis	$\text{k€}/(\text{MW}_{\text{el}} \cdot \text{a})$	14	11	8	5	A
Methanation	$\text{k€}/(\text{MW}_{\text{CH4}} \cdot \text{a})$	8.0	6.3	4.7	3.0	A
Lifetime: $\text{lifetime}_{\text{'p2g'},yex}$	a	20	20	20	20	
Direct air capture ($\text{lifetime}_{\text{'dac'},yex}$)	a	20	25	30	30	D
PEM electrolysis ($\text{lifetime}_{\text{'pem'},yex}$)	a	11	13	14	16	A
Methanation ($\text{lifetime}_{\text{'meth'},yex}$)	a	16	19	22	25	A

[A] Deutsche Energie-Agentur (2018)
[B] Deutz and Bardow (2020)
[C] Agora Verkehrswende et al. (2018)
[D] Breyer et al. (2018)
[E] Bongartz et al. (2018)

C.2.3.4 Power-to-diesel

The technology "Power-to-Diesel" 'p2fts' produces synthetic diesel via the Fischer-Tropsch synthesis (FTS). The FTS process is modeled based on König et al. (2015) and includes the reverse water gas shift to generate syngas from CO_2 and H_2. Due to high temperatures, the FTS process is coupled with a solid oxide electrolysis (SOE) cell to produce H_2. CO_2 is obtained by a DAC unit. In SecMOD, integrated production capacities are considered incorporating all 3 subcomponents: FTS process, SOE cell and DAC unit, Table C.14.
The "Power-to-Diesel" 'p2fts' technology is modeled analogously to the technology "Power-to-Methane." Therefore, we state no additional equations but refer to the equations of the "Power-to-Methane" section. The equations for the investment costs and environmental impacts are implemented according to Section C.1.3. During operation (gate-to-gate), the utilization of CO_2 avoids CO_2 emissions, analogous to Eq. C.45 with the efficiency $\eta_{\text{'p2fts'},yex}$. The efficiency $\eta_{\text{'p2fts'},yex}$ converts the generated power load to the amount of produced synthetic diesel in MWh.
Due to the replacement of fossil by synthetic diesel, the operational costs include a cost credit analogously to Eq. C.46 with the $\text{fuelprice}_{\text{'diesel'},t}$ of fossil diesel introduced in Section C.4.3.1.

The production of synthetic diesel is analogous to Eq. C.47 restricted by the amount of diesel consumed by all technologies consuming diesel: conventional diesel vehicles and plug-in hybrid electric vehicles (PHEVs) (see Section C.4). If the produced amount exceeds the consumption, the optimization is repeated with individual constraints for the OP-units. As for the "Power-to-Methane" technology, in practice, these individual constraints have not been necessary.

The investment costs $\text{invcost}^0_{\text{'p2fts'},yex}$ for the integrated production capacity are derived by the sum over all 3 subcomponents: FTS process, SOE cell and DAC unit (see Table C.14, analogous to Eq. C.48 and C.49).

All data for modeling the technology "Power-to-Methane" and the corresponding sources are specified in Table C.13.

Table C.14: Modeling assumptions for "Power-to-Diesel" ('p2fts').

Specification	Unit	2016	2030	2040	2050	Source
System efficiency: $\eta_{\text{'p2fts'},yex}$	$\text{MWh}_{\text{Diesel}}/\text{MWh}_{\text{el,tot}}$	0.398	0.412	0.4261	0.440	A,B,C
CO_2 benefit: $\text{co2benefit}_{\text{'p2fts'}}$	$\text{t}_{\text{CO2}}/\text{MWh}_{\text{Diesel}}$			0.265		
Direct air capture electricity demand: $\text{w}_{\text{yex}}^{\text{el,'dac'}}$	$\text{MWh}_{\text{el,'dac'}}/\text{t}_{\text{CO2}}$	0.500	0.500	0.500	0.500	B
SOEC-electrolysis efficiency: $\eta_{\text{'soec'},yex}$	$\text{MWh}_{\text{H2}}/\text{MWh}_{\text{el,'soec'}}$	0.810	0.840	0.870	0.900	C
Inputs:						
$\text{m}_{CO2,yex}$	$\text{t}_{\text{CO2}}/\text{t}_{\text{Diesel}}$			3.111		E
$\text{m}_{H2,yex}$	$\text{t}_{\text{H2}}/\text{t}_{\text{Diesel}}$			0.691		E
Total investment costs: $\text{invcost}^{0}_{\text{'fts'},yex}$	$\text{k€}/\text{MW}_{\text{el}}$	3091	1508	926	369	
Investment costs:	$\text{k€}/\text{MW}_{\text{el}}$	2129	1507	926	369	
Direct air capture	$\text{k€}/\text{MW}_{\text{el}}$	19	10	8	6	D
SOEC electrolysis	$\text{k€}/\text{MW}_{\text{el}}$	1700	1223	747	270	A
FTS synthesis	$\text{k€}/\text{MW}_{\text{Diesel}}$	600	413	284	195	A
Maintenance costs:	$\text{k€}/(\text{MW}_{\text{el}} \cdot \text{a})$	99.5	76.08	53.06	29.86	
Direct air capture	$\text{k€}/(\text{MW}_{\text{el}} \cdot \text{a})$	0.08	0.04	0.03	0.02	D
SOEC electrolysis	$\text{k€}/(\text{MW}_{\text{el}} \cdot \text{a})$	68	49	30	10.8	A
FTS synthesis	$\text{k€}/(\text{MW}_{\text{CH4}} \cdot \text{a})$	31	27	23	19	A
Lifetime: $\text{lifetime}_{\text{'fts'},yex}$	a	20	20	20	20	E
Direct air capture ($\text{lifetime}_{\text{'dac'},yex}$)	a	20	25	30	30	B
SOEC electrolysis ($\text{lifetime}_{\text{'soec'},yex}$)	a	16	16	16	16	A
FTS synthesis ($\text{lifetime}_{\text{'meth'},yex}$)	a	16	19	22	25	A

[A] Deutsche Energie-Agentur (2018)
[B] Deutz and Bardow (2020)
[C] Agora Verkehrswende et al. (2018)
[D] Breyer et al. (2018)
[E] König et al. (2015)

C.2.3.5 Renewable energy generation

The renewable energy generation of the set REN includes wind offshore 'windof,' wind onshore 'windon' and PV 'pv.' The renewable generation do not need fuel, so the fuel costs equal 0. Maintenance costs are included in the annual investment costs and thus no further operational costs are considered. Analogous to ELMOD-DE, this work includes the availability of wind and solar power by using an hourly time series for the local full load percentage, depending on the dena zone of the respective node. The energy balance therefore results in

$$\sum_{\substack{yex \in \mathrm{YEX} \\ yex \leq y}} GENLOAD^{\mathrm{SO}}_{n,t,y,yex,tech} + CUR_{n,t,y,tech} = \Big[\sum_{\substack{yex \in \mathrm{YEX} \\ yex \leq y}} \mathrm{cex}_{n,y,yex,tech} + \sum_{\substack{yy \in \mathrm{Y} \\ yy \leq y}} CNEW_{n,yy,tech} \Big] \cdot \mathrm{h}_{n,t,tech} \quad \forall\, n \in \mathrm{N},\, t \in \mathrm{T},\, y \in \mathrm{Y},\, tech \in \mathrm{REN}. \tag{C.50}$$

The variable $CUR^{\mathrm{SO}}_{n,t,y,tech}$ represents the curtailment and describes the energy produced by the renewable capacity which is curtailed and not fed into the grid. The local full load percentage of an hour $\mathrm{h}_{n,t,tech}$ are taken from Egerer (2016).

The newly built capacity $CNEW_{n,y,tech}$ is restricted by the potential of the renewable energy source. The data on the regional potentials for onshore wind energy (McKenna et al., 2015) and roof-PV (Mainzer et al., 2014) kindly have been made available by Russell McKenna from Karlsruher Institute of Technology. The data are based on an analysis of the available area considering political restrictions. From the data, a limit of the annual energy from onshore wind and PV for each municipality is derived. Figures C.5 and C.6 show the potential of onshore wind and photovoltaics included in this work. The restriction of the newly built capacity for onshore wind and PV is described by

$$CNEW_{n,y,tech} \leq \max\{0,\, \mathrm{potlimit}_{n,tech} - \sum_{\substack{yex \in \mathrm{YEX} \\ yex \leq y}} \mathrm{cex}_{n,y,yex,tech} - \sum_{\substack{yy \in \mathrm{Y} \\ yy \leq y}} CNEW_{n,yy,tech}\} \quad \forall\, n \in \mathrm{N},\, y \in \mathrm{Y},\, tech \in \{\text{'windon,' 'pv'}\}. \tag{C.51}$$

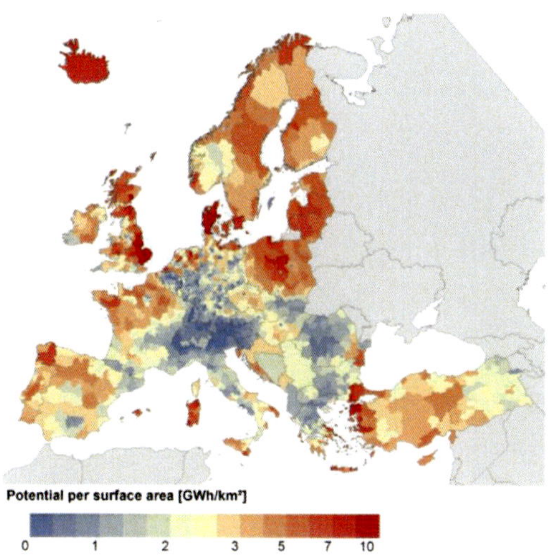

Figure C.5: Potential of onshore wind energy. Source: McKenna et al. (2015).

The limit for the annual energy $\text{maxannualenergy}_{n,tech}$ from McKenna et al. (2015) and Mainzer et al. (2014) for each municipality is matched to the closest node n using geographic coordinates. The limit of potential $\text{potlimit}_{n,tech}$ is calculated using the annual full load hours $\text{flh}_{n,tech}$ for the respective dena zone from Egerer (2016):

$$\text{potlimit}_{n,tech} = \frac{\text{maxannualenergy}_{n,tech}}{\text{flh}_{n,tech}} \tag{C.52}$$

In this work, offshore wind capacity can only be built at the 3 nodes linked to the main offshore platforms. The 3 nodes connected are node 9 at Bentwisch, node 19 at Büttel and node 50 at Dörpen. The limit of potential for offshore wind energy, 38 GW, is based on Gerhardt et al. (2015). The limit is equally split up to the 3 nodes above. Thus, the potential in the North Sea is approximated as $\frac{2}{3}$ of the maximum potential,

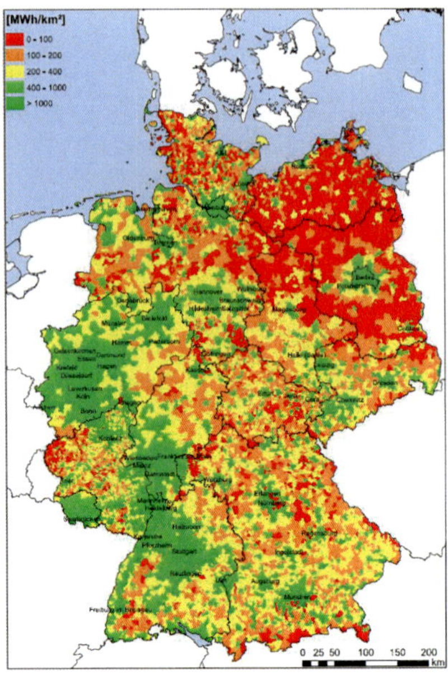

Figure C.6: Technical potential for photovoltaics on residential buildings roofs in Germany. Source: Mainzer et al. (2014).

whereas the potential in the Baltic Sea is approximated as $\frac{1}{3}$ of the maximum potential. The restriction for offshore wind is modeled by

$$CNEW_{n,y,\text{'windof'}} \leq \text{potlimit}_{n,\text{'windof'}} - \sum_{\substack{yex \in \text{YEX} \\ yex \leq y}} \text{cex}_{n,y,yex,\text{'windof'}} - \sum_{\substack{yy \in \text{Y} \\ yy \leq y}} CNEW_{n,yy,\text{'windof'}}$$
$$\forall\, n \in \{\text{n009, n019, n050}\},\, y \in \text{Y} \qquad \text{(C.53)}$$

and at all other nodes by

$$CNEW_{n,y,\text{'windof'}} = 0 \qquad \forall\, n \in \text{N} \setminus \{\text{n009, n019, n050}\},\, y \in \text{Y}. \qquad \text{(C.54)}$$

Table C.15 shows the total potential for the fluctuating renewables included in this work. The potentials for onshore wind and PV are given in annual energy amounts and are converted to maximum capacity limits using Eq. C.52. Because renewable

Table C.15: Limits of potential for fluctuating renewables included in SecMOD.

Technology	$\sum_{\mathbf{n} \in \mathbf{N}} \mathbf{potlimit}_{n,tech}$ (GW)	Source
Onshore Wind	698	McKenna et al. (2015)
Offshore Wind	38	Gerhardt et al. (2015)
PV	235	Mainzer et al. (2014)

capacities are aggregated, no unit-specific end-of-life year is considered. Still, SecMOD includes aggregated shut-down curves for the initial renewable generation capacity. For each investment period y, the remaining infrastructure from the initial investment period is calculated and set as the existing infrastructure:

$$\mathrm{cex}^{\mathrm{initial}}_{n,y,yex,tech} = \mathrm{sdc}_{y,tech} \cdot \mathrm{cex}^{\mathrm{initial}}_{n,y=2016,yex,tech}. \tag{C.55}$$

Table C.16 shows the factors of the shut-down curve $\mathrm{sdc}_{y,tech}$. The factors are derived from Federal Network Agency (2018) by aggregating the capacities for onshore wind, offshore wind and PV according to their construction years. The share of the capacity still existing is calculated for each investment period considering the lifetime from Table C.6.

Table C.16: Factors of shut-down curve $\mathrm{sdc}_{y,tech}$ for onshore wind, offshore wind and PV: share of capacity from base year existing in investment period y based on Federal Network Agency (2018).

Technology	2016	2020	2025	2030	2035	2040	2045	2050
Onshore Wind	1.0000	0.9955	0.9426	0.6813	0.4791	0.1831	0	0
Offshore Wind	1.0000	1.0000	1.0000	0.9916	0.7589	0	0	0
PV	1.0000	0.9998	0.9992	0.9732	0.7416	0.0710	0	0

C.2.4 Life-cycle inventory data

The following tables show the LCI data used for the implementation of the power sector in SecMOD. Explanation for the tables and how to derive environmental impacts for the listed technologies can be found in Section C.1.5, as the table structure is the same for every sector in SecMOD.

Table C.17: LCI of power sector technologies, part 1. Processes without explicit source are based on ecoinvent 3.5 APOS (Wernet et al., 2016). The functional unit (FU) of infrastructural and operational processes is equal to 1 MW and 1 MWh, respectively.

$ecoinv_{tech,impact}$	Location	process quantity per FU	inv. share
Biomass*			
infrastructural processes			
heat and power co-generation unit construction, 160kW electrical, common components for heat+electricity	RER	6.25	
heat and power co-generation unit construction, 160kW electrical, components for electricity only	RER	6.25	
operational processes			
heat and power co-generation, biogas, gas engine	DE	4.27×10^2	4.87×10^{-5}
Combined cycle power plant (CCGT)			
infrastructural processes			
gas power plant construction, combined cycle, 400MW electrical	RER	2.50×10^{-3}	
operational processes			
electricity production, natural gas, combined cycle power plant	DE	1000	1.40×10^{-8}
Coal			
infrastructural processes			
hard coal power plant construction, 500MW	GLO	2×10^{-3}	
operational processes			
electricity production, hard coal	DE	1000	1.25×10^{-8}

* In SecMOD, biomass is not a combined heat and power producing technology. Therefore, the biomass LCI data is modified so that it only contains the components needed for power generation.

Table C.18: LCI of power sector technologies, part 2. Processes without explicit source are based on ecoinvent 3.5 APOS (Wernet et al., 2016). The functional unit (FU) of infrastructural and operational processes is equal to 1 MW and 1 MWh, respectively.

$ecoinv_{tech,impact}$	Location	process quantity per FU	inv. share
Gas			
infrastructural processes			
gas power plant construction, 100MW electrical	RER	0.01	
operational processes			
electricity production, natural gas, conventional power plant	DE	1000	6.96×10^{-8}
Geothermal			
infrastructural processes			
geothermal power plant construction 5.5MW	CH	1.82×10^{-1}	
operational processes			
electricity production, deep geothermal	DE	1	5.78×10^{-7}
H_2 electrolyzer and H_2 fuel-cell[A]			
infrastructural processes[B]			
balance of plant			
unreinforced concrete production, with cement CEM II/A	CH	$2.33\ m^3$ [C]	
electronic component production, passive, unspecified	GLO	1.10×10^{3} kg	
lubricating oil production	RER	2.00×10^{2} kg	
polyethylene production, high density, granulate	RER	3.00×10^{2} kg	
steel production, converter, low-alloyed	RER	4.80×10^{3} kg	
balance of plant and stack			
aluminum production, primary, ingot	IAI Area, EU27 & EFTA	1.90×10^{2} kg	
steel production, converter, chromium steel 18/8	RER	2.23×10^{3} kg	
copper production, primary	RER	1.15×10^{2} kg	

[A] H_2 fuel-cell process assumed as reverse H_2 electrolysis with equal costs and environmental impacts

Table C.19: LCI of power sector technologies, part 3. Processes without explicit source are based on ecoinvent 3.5 APOS (Wernet et al., 2016). The functional unit (FU) of infrastructural and operational processes is equal to 1 MW and 1 MWh, respectively.

$ecoinv_{tech,impact}$	Location	process quantity per FU
H_2 electrolyzer and H_2 fuel-cell[A] continued		
stack		
titanium production, primary	GLO	1.76×10^3 kg
market for platinum[D]	GLO	2.75 kg
tetrafluoroethylene production[E]	RER	5.33×10^1 kg
activated carbon production, granular from hard coal	RER	3.00×10^1 kg
balance of plant end-of-life		
treatment of waste concrete, inert material landfill	CH	5.60×10^3 kg
plastic (see Table C.3)	CH	1.10×10^3 kg
treatment of waste mineral oil, hazardous waste incineration	Europe w/o CH	2.00×10^2 kg
polyethylene treatment (see Table C.3)	CH	3.00×10^2 kg
balance of plant and stack end-of-life		
aluminum (see Table C.3)	CH	1.90×10^2 kg
steel (see Table C.3)	CH	7.03×10^3 kg
copper (see Table C.3)	RER	1.15×10^2 kg
stack end-of-life		
titanium (see Table C.3)	RER	1.76×10^3 kg
platinum (see Table C.3)	RER	2.50×10^{-1} kg
iridium (see Table C.3)	RER	2.50 kg
treatment of inert waste, inert material landfill[F]	CH	8.33×10^1 kg

[A] H_2 fuel-cell process assumed as reverse H_2 electrolysis with equal costs and environmental impacts
[B] Bareiß et al. (2019)
[C] concrete density = 2400 kg/m^3
[D] Platinum is assumed for Iridium and accounts for 2.5 kg of 2.75 kg.
[E] for naflon production
[F] for tetrafluoroethylene (naflon) and activated carbon

Table C.20: LCI of power sector technologies, part 4. Processes without explicit source are based on ecoinvent 3.5 APOS (Wernet et al., 2016). The functional unit (FU) of infrastructural and operational processes is equal to 1 MW and 1 MWh, respectively.

$ecoinv_{tech,impact}$	Location	process quantity per FU	inv. share
Lignite			
infrastructural processes			
lignite power plant construction	RER	2.60×10^{-3}	
operational processes			
electricity production, lignite	DE	1.08×10^{3}	1.06×10^{-8}
Lithium-ion battery			
infrastructural processes[A]			
lithium-hydroxide production[B]	GLO	5.13×10^{2} kg	
remaining inventory[C] derived from Ellingsen et al. (2014)			
end-of-life[D]			
steel (see Table C.3)		1.24×10^{3} kg	
aluminum (see Table C.3)		1.74×10^{3} kg	
copper (see Table C.3)		1.36×10^{3} kg	
tin (see Table C.3)		1.70 kg	
plastic (see Table C.3)		7.16×10^{2} kg	
pet (see Table C.3)		8.40 kg	
polypropylene (see Table C.3)		3.33×10^{2} kg	
polyvinylchloride (see Table C.3)		2.81×10^{-1} kg	
polyethylene (see Table C.3)		3.82×10^{-1} kg	
treatment of inert waste, inert material landfill	CH	1.68×10^{2} kg	
treatment of hazardous waste, hazardous waste incineration	Europe w/o CH	4.53×10^{3} kg	
treatment of wastewater, average, capacity	Europe w/o CH	1.69×10^{2} m3	
lithium-carbonate production, from concentrated brine[E]	GLO	-6.33×10^{2} kg	

[A] functional unit: 1 MWh
[B] LiOH production according to Majeau-Bettez et al. (2011)
[C] contains more than 250 entries
[D] recycling rates according to Zou et al. (2013): Ni, Mn, Co: 100 %
[E] according to Zou et al. (2013) 80 % of LiOH is recycled as Li_2CO_3

Table C.21: LCI of power sector technologies, part 5. Processes without explicit source are based on ecoinvent 3.5 APOS (Wernet et al., 2016). The functional unit (FU) of infrastructural and operational processes is equal to 1 MW and 1 MWh, respectively.

$ecoinv_{tech,impact}$	**Location**	**process quantity per FU**
Power line		100 km
infrastructural processes[A]		
foundation		
unreinforced concrete production, with cement CEM II/A	CH	7.06 m3B
cast iron production	RER	8.82×10^{2} kg
masts, conductors and insulators		
steel production, converter, low-alloyed	RER	3.63×10^{3} kg
masts		
primary zinc production from concentrate	RoW	9.41×10^{1} kg
insulators		
flat glass production, uncoated	RER	9.59×10^{1} kg
cement production, Portland	CH	5.12 kg
conductors		
aluminum production, primary, cast alloy slab from continuous casting	RoW	1.09×10^{3} kg
lubricating oil production	RER	3.50×10^{1} kg
foundation end-of-life		
treatment of waste concrete, inert material landfill	CH	1.69×10^{4} kg
masts, conductors and insulators end-of-life		
steel (see Table C.3)	CH	4.51×10^{3} kg
masts end-of-life		
zinc	CH	9.41×10^{1} kg
insulators end-of-life		
treatment of waste glass, inert material landfill	CH	9.59×10^{1} kg
treatment of waste cement, hydrated, residual material landfill	CH	5.12 kg
conductors end-of-life		
aluminum (see Table C.3)	CH	1.09×10^{3} kg
treatment of waste mineral oil, hazardous waste incineration	Europe w/o CH	3.50×10^{1} kg

[A] Jorge et al. (2012)
[B] concrete density = 2400 kg/m3

Table C.22: LCI of power sector technologies, part 6. Processes without explicit source are based on ecoinvent 3.5 APOS (Wernet et al., 2016). The functional unit (FU) of infrastructural and operational processes is equal to 1 MW and 1 MWh, respectively.

$ecoinv_{tech,impact}$	**Location**	**process quantity per FU**	**inv. share**
Nuclear			
infrastructural processes			
nuclear power plant construction, pressure water reactor 1000MW	DE	1×10^{-3}	
operational processes			
electricity production, nuclear, pressure water reactor	DE	1000	3.06×10^{-9}
Oil			
infrastructural processes			
oil power plant construction, 500MW	RER	2×10^{-3}	
operational processes			
electricity production, oil	DE	1000	1.19×10^{-8}
Other[A]			
infrastructural processes			
oil power plant construction, 500MW	RER	2×10^{-3}	
operational processes			
electricity production, oil	DE	1000	1.19×10^{-8}
Power-to-Methane			
infrastructural processes			
chemical factory construction, organics	RER	1.26×10^{-2}	
Pumped hydro			
infrastructural processes			
hydropower plant construction, reservoir, non-alpine regions	RER	1.10×10^{-4}	
operational processes			
lubricating oil production	RER	7.60×10^{-3} kg	
treatment of waste mineral oil, hazardous waste incineration	Europe w/o CH	7.60×10^{-3} kg	

[A] oil assumed as conservative choice

Table C.23: LCI of power sector technologies, part 7. Processes without explicit source are based on ecoinvent 3.5 APOS (Wernet et al., 2016). The functional unit (FU) of infrastructural and operational processes is equal to 1 MW and 1 MWh, respectively.

$ecoinv_{tech,impact}$	Location	process quantity per FU	inv. share
Photovoltaics			
infrastructural processes			
photovoltaic plant construction, 570kWp, multi-Si, on open ground	GLO	1.75	
operational processes			
electricity production, photovoltaic, 570kWp open ground installation, multi-Si	DE	1000	7.60×10^{-5}
Run of River			
infrastructural processes			
hydropower plant construction, run-of-river	Europe w/o CH	5.66×10^{-4}	
operational processes			
electricity production, hydro, run-of-river	DE	1000	8.07×10^{-10}
Switching of power level for power lines			
infrastructural processes[A]			
insulators and conductors			
steel production, converter, low-alloyed	RER	3.02×10^{2} kg	
insulators			
flat glass production, uncoated	RER	7.94×10^{1} kg	
cement production, Portland	CH	3.71 kg	
conductors			
aluminum production, primary, cast alloy slab from continuous casting	RoW	7.02×10^{2} kg	
lubricating oil production	RER	2.22×10^{1} kg	
insulators and conductors end-of-life			
steel (see Table C.3)	CH	3.02×10^{2} kg	
insulators end-of-life			
treatment of waste glass. inert material landfill	CH	7.94×10^{1} kg	
treatment of waste cement, hydrated, residual material landfill	CH	3.71 kg	
conductors end-of-life			
aluminum (see Table C.3)	CH	7.02×10^{2} kg	
treatment of waste mineral oil, hazardous waste incineration	Europe w/o CH	2.22×10^{1} kg	

[A] Jorge et al. (2012)

Table C.24: LCI of power sector technologies, part 8. Processes without explicit source are based on ecoinvent 3.5 APOS (Wernet et al., 2016). The functional unit (FU) of infrastructural and operational processes is equal to 1 MW and 1 MWh, respectively.

$ecoinv_{tech,impact}$	**Location**	**process quantity per FU**	**inv. share**
Waste			
infrastructural process			
municipal waste incineration facility construction	CH	$4.79 \times 10^{-2*}$	
operational processes			
electricity, from municipal waste incineration to generic market for electricity, medium voltage	DE	1000	-8.41724e-08[A]
Wind Onshore			
infrastructural processes			
wind turbine construction, 2MW, onshore	GLO	0.5	
wind turbine network connection construction, 2MW, onshore	RoW	0.5	
transport, freight, lorry 7.5-16 metric ton, EURO3	RER	9.84×10^{-5} tkm	
operational processes			
lubricating oil production	RER	4.91×10^{-2} kg	
treatment of waste mineral oil, hazardous waste incineration	Europe w/o CH	4.91×10^{-2} kg	
Wind Offshore			
infrastructural processes			
market for wind power plant, 2MW, offshore, fixed parts	GLO	0.5	
market for wind power plant, 2MW, offshore, moving parts	GLO	0.5	
operational processes			
electricity production, wind, 1-3MW turbine, offshore	DE	1000	-9.50×10^{-6}

[A] waste treatment capacity and power output interlinked by net energy output per kg waste according to ecoinvent, including allocation key according to ecoinvent

C.3 Heating sector

In this section, we describe the heating sector included in SecMOD. Heat is provided at different temperature levels, each representing an energy demand contained in the set of energy demands E: domestic heating (dom) and industrial heating at low temperature (lt), medium temperature (mt) or high temperature (ht). All temperature levels exist for both local (lh) and centralized (ch) heating and are differentiated in two subsets of E: LH for local and CH for centralized heating.

C.3.1 Technologies in the heating sector

Table C.25 lists the technologies in the heating sector, indicating which temperature level can be covered by the respective technology and if the technology is suitable for local and centralized heating. If a technology is not suitable for a specific energy demand e, the respective load variable $GENLOAD^{\mathrm{SO}}_{tech,n,t,y,e}$ is fixed to 0 at all nodes n, time steps t and investment periods y.

Table C.25: Suitability for technologies in the heating sector included in SecMOD depending on the temperature level.

Technology	**Local Heating (LH)**				**Centralized Heating (CH)**			
Thermal insulation	dom				dom			
Electrode boiler		lt	mt					
Gas boiler	dom	lt	mt	ht	dom	lt	mt	ht
Oil boiler	dom							
Ground heat pump	dom	lt						
Co-generation in conventional power plants					dom	lt	mt	ht

The generation portfolio in the domestic heating sector is based on data provided by ZIV (2017a) and ZIV (2017b). The sources include the number of existing gas and oil boilers in 5 different power classes. Further generation technologies are neglected due to relatively small capacities compared to oil and gas boilers. To prevent supply shortage of the initial energy system, the capacity of each boiler is estimated to be as high as the maximum capacity of the respective class. Thus, an existing capacity of 526 GW for oil boilers and 413 GW for gas boilers is estimated. The existing capacity is assigned to the nodes using the population density and the gross domestic product. The assignment is analogous to the procedure described in Section C.2.

Due to lack of data concerning existing capacities for industrial heat generation, the generation portfolio is estimated. We estimate the existing capacities for industrial heat generation based on an independent optimization of the base year 2016. The resulting capacities needed to meet the industrial heating demand are assumed as the existing capacities. The co-generation of heat in conventional power plants to meet centralized heat demand is taken from Umweltbundesamt (2018). The capacities are aggregated and matched to the closest node by ZIP code of the respective power plant, as described in the introduction of Section C.2.

The lifetime of the existing domestic and industrial heating capacity is set to 20 years according to Henning and Sauer (2015). The BDEW Bundesverband der Energie- und Wasserwirtschaft e.V. (2014) shows that a major part of the existing capacity is already older than 20 years in the year 2016. Despite their operability, a replacement is necessary due to high cost and emission intensity (BDEW Bundesverband der Energie- und Wasserwirtschaft e.V., 2014). Thus, SecMOD includes shut-down curves for the initial existing capacity using the start-up year from BDEW Bundesverband der Energie- und Wasserwirtschaft e.V. (2014) and a lifetime of 20 years. Table C.26 shows the remainder of the initial capacity of domestic and industrial heating in the respective investment period y. The shut-down curves are assumed to be valid for all oil and gas boilers.

Table C.26: Shut-down curve for oil and gas boilers: share of the initial existing capacity available in future investment periods. Calculated using data from BDEW Bundesverband der Energie- und Wasserwirtschaft e.V. (2014).

Investment period	**2016**	**2020**	**2025**	**2030**	**2035**	**2040**	**2050**
Share of initial capacity	1	0.62	0.45	0.29	0.14	0	0

The annualized investment costs $\text{invcost}_{tech,y}$ are calculated according to Eq. C.6. Maintenance costs are considered as a percentage of the investment costs and added to the investment costs. Table C.27 lists the lifetimes as well as investment and maintenance costs considered for the technologies in the heating sector.

The operational costs are determined according to the general Eq. C.7. As maintenance costs are considered as a percentage of the investment costs, the specific operational costs consist only of fuel costs and are calculated using Eq. C.24. Oil and gas boiler efficiencies are assumed to be 95 %. Analogous to ELMOD-DE, the fuel price is exogenously given as an hourly time series. Table C.7 lists the average fuel prices.

Table C.27: Investment and maintenance costs as well as lifetime ($lifetime_{tech}$) of the technologies in the heating sector.

Technology	Investment cost **2016** (k€/MW)	Investment cost **2050** (k€/MW)	Maintenance cost (% Invest)	Lifetime (a)	Source
Thermal insulation	2369.23	2369.23	1	50	A
Electrode boiler	237.5	237.5	2	20	B, C
Gas boiler	175	175	2	20	A
Oil boiler	175	140	2	20	A
Ground source heat pump	1540	1232	3.5	20	A

[A] Palzer (2016)

[B] Investment costs mean value from Brauner (2019).

[C] Maintenance costs and lifetime are assumed to be equal to gas and oil boilers.

C.3.2 Demand structure

At each node, the local and the centralized heating demand for all temperature levels is considered as a time series.

C.3.2.1 Industrial heating

The heating demand of different sub-industries is derived from Öko-Institut and ISI (2015). As an estimate for an industrial site's share of the heating demand in its sub-industry, the industrial site's share of CO_{2eq} emissions in its sub-industry is used. Data for fuel demand and emissions for the industrial sites is derived from the "European Pollutant Release and Transfer Register" (E-PRTR), which includes all industrial sites with demand over 100 MW. The sum of the industrial heat demand covered in the E-PRTR amounts to $507\,\mathrm{TW\,h\,a^{-1}}$. Thus, the heating demand of the sub-industry is split up to the industrial sites included in E-PRTR. The procedure is analogous to the electricity demand in Section C.2.2. The heating demand resulting from an industrial site is allocated to the closest node.

Industrial heating demand is implemented on three temperature levels according to Lutsch and Witterhold (2005):

- high-temperature heating (ht) at a temperature level greater than 400 °C
- medium-temperature heating (mt) at a temperature level between 100 °C and 400 °C
- low-temperature heating (lt) at a temperature level up to 100 °C.

The share of heating at each temperature level for every sub-industry is shown in Table C.28. According to their share, industrial heating demand at each temperature level is calculated for each industrial site.

C.3.2.2 Domestic heating

Data for the domestic heating demand per person depending on the household size are taken from Federal Statistical Office (2015) as shown in Table C.30. The population as well as the geographic coordinates of municipalities are provided by the database Open Geo DB (2017). For each municipality, the population is assigned to the closest node.

Table C.28: Share of temperature levels in industrial heating demand for each industry according to Lutsch and Witterhold (2005) as well as share of total industrial heat according to Öko-Institut and ISI (2015).

Industry	**lt** (%)	**mt** (%)	**ht** (%)	**Share of industrial heat (%)**
Mining and Quarrying	55.64	44.36	0.00	0.47
Food and Tobacco	55.47	44.53	0.00	8.20
Paper	26.17	57.16	16.74	8.67
Basic chemical processing	23.11	28.28	48.58	20.19
Other chemical industries	21.81	28.79	49.51	3.62
Rubber and Plastic	60.63	37.01	2.36	1.80
Glas and Ceramics	4.92	9.57	85.64	3.69
Stone and earth industry	6.70	3.92	82.81	8.68
Metal production	1.80	4.92	93.30	26.28
Non-ferrous metal, casting	17.89	12.42	69.89	4.70
Metalworking	45.35	34.09	20.55	3.32
Automobile industry	60.71	28.09	11.21	3.45
Other	48.13	34.55	17.24	6.92

Table C.30: Domestic heating demand depending household size according to Federal Statistical Office (2015) and Federal Statistical Office (2008).

Household size	**Annual domestic heating demand per person**	**Population**	**Share of total population**
1 person	10100 kWh	16.83 million	20.4 %
2 persons	7123 kWh	27.83 million	33.8 %
3 or more persons	4578 kWh	39.22 million	45.8 %

Assuming that the specific civil heating demand per person does neither depend on the population density nor on the state, the total annual civil heating demand at each node is calculated. The total household heating demand amounts to $540\,\mathrm{TW\,h\,a^{-1}}$. Of the $540\,\mathrm{TW\,h\,a^{-1}}$, $62.5\,\mathrm{TW\,h\,a^{-1}}$ are already supplied via electricity and already considered as electricity demand Arbeitsgemeinschaft Energiebilanzen (2017). Thus, to again avoid double accounting of this household heat demand, we only consider $477.5\,\mathrm{TW\,h\,a^{-1}}$ as household (HH) heating demand $\mathrm{Q}^{\mathrm{civ,HH}}$.
Further, $311.5\,\mathrm{TW\,h\,a^{-1}}$ results from the space heating demand of the commercial (GHD) sector as well as from the industrial heating demand from industries not included in E-PRTR (demand ≤ 100 MW) Arbeitsgemeinschaft Energiebilanzen (2017). Space heating in the commercial sector $\mathrm{Q}^{\mathrm{civ,GHD}}$ is incorporated as an additional civil heating demand. We assume that the heating demand in the commercial sector correlates with the population density as well. Following this assumption, the remaining heating demand is distributed among the nodes.

At each node n, the sum of household and commercial heating amounts to the total nodal domestic heating demand $\mathrm{Q}_n^{\mathrm{dom}}$:

$$\mathrm{Q}_n^{\mathrm{dom}} = \mathrm{Q}_n^{\mathrm{dom,HH}} + \mathrm{Q}_n^{\mathrm{dom,GHD}}. \tag{C.56}$$

The heating demand is temporally resolved based on weather data for each federal state. The hourly heating demand $\mathrm{Q}_{t,n}^{\mathrm{h}}$ is assumed to be proportional to the difference of the standardized indoor temperature $\mathrm{T^i} = 20\,^{\circ}\mathrm{C}$ (Schmidt, 2014) to the hourly outdoor temperature $\mathrm{T}_t^{\mathrm{o}}$ provided by DWD (2017). Thus, the domestic heating demand $\mathrm{Q}_{t,n}^{\mathrm{h,dom}}$ at a node n and in an hour t is calculated by:

$$\mathrm{Q}_{t,n}^{\mathrm{h,dom}} = \frac{\max\left\{0, (\mathrm{T^i} - \mathrm{T}_{t,n}^{\mathrm{o}})\right\} \cdot \mathrm{Q}_n^{\mathrm{dom}}}{\left(\sum_{t \in \mathrm{T}} (\mathrm{T^i} - \mathrm{T}_{t,n}^{\mathrm{o}})\right)}. \tag{C.57}$$

In accordance with VDI Guideline 2067/DIN4108T6, only degree days as days with average temperatures lower than $15\,^{\circ}\mathrm{C}$ are considered for heating demand to occur. Figure C.7 shows the result of the annual domestic heating demand qualitatively.

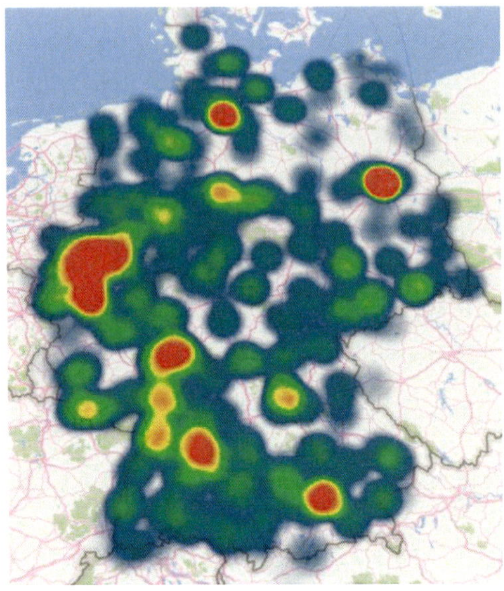

Figure C.7: Annual domestic heating demand in Germany shown qualitatively. Red indicates a high, blue indicates a low heating demand.

C.3.2.3 Centralized heating demand (CH)

The domestic heating demand is split into centralized and local heating demand. It is assumed that the share of centralized heating is constant until 2050. The share of the centralized heating demand at each node n is given by the Federal Cartel Office (2012).

The industrial heating demand is also split into centralized and local heating demand. We assume that co-generation in conventional power plants is used to meet the industrial heating demand if it is not already used for the centralized domestic heating demand. Thus, the centralized industrial heating demand is the difference between centralized thermal capacity at the node and centralized domestic heating demand. For each centralized energy demand ($e \in \mathrm{CH}$), the calculations result in a time series for the centralized heating load at each node n. The heating load $\mathrm{energyload}_{n,t,e}$ can be satisfied by the centralized heat generators from the set CHTECH (see Table C.25) as a subset of the set TECH. Thus, the energy balance results in:

$$\sum_{tech \in \text{CHTECH}} \sum_{\substack{yex \in \text{YEX} \\ yex \leq y}} GENLOAD^{\text{SO}}_{n,t,y,yex,tech,e} \cdot (1 - \text{dhloss}) \\ = \text{energyload}_{n,t,e} \quad \forall n \in \text{N}, t \in \text{T}, y \in \text{Y}, e \in \text{CH}. \tag{C.58}$$

The factor dhloss incorporates the distribution losses due to the district heating network. The loss factor is chosen as 10 %, which is in the range of literature data (9 % from Lutsch et al. (2004) and 12 % from Kempe (2014)).

C.3.2.4 Local heating demand (LH)

The remaining heat demand is met by local heat generation. For each local energy demand (e $\in$ LH), the calculations result in a time series for the local heating load at each node n. The energyload can be satisfied by the suitable local heat generators from the set LHTECH (see Table C.25) as a subset of the set TECH. Thus, the energy balance results in:

$$\sum_{tech \in \text{LHTECH}} \sum_{\substack{yex \in \text{YEX} \\ yex \leq y}} GENLOAD^{\text{SO}}_{n,t,y,yex,tech,e} \\ = \text{energyload}_{n,t,e} \quad \forall n \in \text{N}, t \in \text{T}, y \in \text{Y}, e \in \text{LH}. \tag{C.59}$$

C.3.2.5 Minimum installed capacity

To ensure the reliability of supply, the installed capacity must be able to meet the maximum heat load occurring over the year at each node. The necessary peak heating load is determined, according to DIN EN 12831. The minimum design temperature is −20 °C, following Jagnow and Wolff (2003). The peak heating load $\text{heat}_n^{\text{peak}}$ is calculated using Eq. C.57 with an outdoor temperature of −20 °C. For the local heating demand, the minimum capacity installed is constrained by

$$\sum_{tech \in \text{LHTECH}} \sum_{\substack{yex \in \text{YEX} \\ yex \leq y}} \left[\text{cex}_{n,y,yex,tech} + CNEW_{n,y,yex,tech}\right] \\ \geq \text{ybcap}_n + REPLACEOBOIL_n \quad \forall n \in \text{N}, \forall y \in \text{Y}. \tag{C.60}$$

The installed local capacity in the base year ybcap_n is the sum of all local heating technologies in the year 2016 at a node n. The variable $REPLACEOBOIL_n$ allows replacing oil boilers, so that alternative heating technologies can be used even if an oil boiler was installed in a house in the initial year. To serve as replacement, the newly built infrastructure must be built in addition to the old infrastructure. The possibility of oil boiler replacement is another measure to decarbonize the domestic heating sector.

For the centralized heating, the minimum capacity installed is constrained by

$$(1 - \mathrm{dhloss}) \cdot \sum_{tech \in \mathrm{CHTECH}} \sum_{\substack{yex \in \mathrm{YEX} \\ yex \leq y}} \left[\mathrm{cex}_{n,y,yex,tech} + CNEW_{n,y,yex,tech}\right] \geq \mathrm{heat}_n^{\mathrm{peak}} \cdot \mathrm{chshare}_n \quad \forall\, n \in \mathrm{N}, \forall\, y \in \mathrm{Y}. \tag{C.61}$$

The parameter $\mathrm{chshare}_n$ is the share of centralized heating at a node n. It is calculated by dividing the peak heat extraction by the peak heat demand at a node n derived from Federal Cartel Office (2012).

C.3.3 Description of specific technologies

In the following section, the specific technologies in the heating sector, including special equations and assumptions, are described.

C.3.3.1 Thermal insulation

Buildings with local heating as well as centralized heating can be renovated. The assumptions for modeling the building renovation follow Palzer (2016). Palzer (2016) names 3 renovation levels: "fully renovated+", the "fully renovated" and "not renovated". SecMOD considers the medium renovation level ("fully renovated") which is predicted by Palzer (2016) to have the highest importance in the future. This "full renovation" reduces the heating demand of a building by 50 %. The assumed costs correspond to the cost difference between an energy-efficiency increasing building renovation compared to a conventional building renovation. Thus, we only consider this cost difference in SecMOD. The assumed costs (assuming a lifetime of 50 years for the renovation) are offset by energy savings due to increased energy efficiency. According to Palzer (2016), the average renovation costs are 100 €/m^2 and the average house area is 462 m^2. This results in average total investments of 46,200 €. The average installed heating capacity of a building is 39 kW. From the saving of 50 % of the heating demand by the renovation, costs of $46.2\,\mathrm{k€} \cdot \frac{1}{0.5 \cdot 0.039\mathrm{MW}} = 2{,}369{,}231\,\mathrm{k€/MW}$ are calculated. This value is based on the assumption that only half of the heating capacity is needed to cover the heating demand, the other half of the heating demand is then provided by the renovation. The energetic renovation 'er' is modeled as heat generation without operational costs. As an additional constraint, not more than 50 % of the heating demand at each node can be met with energetic renovation:

$$GENLOAD^{\text{SO}}_{\text{'er'},n,t,y,e} \leq -0.5 \cdot \text{energyload}_{n,t,e} \qquad \forall n \in \text{N}, t \in \text{T}, y \in \text{Y}, e \in \text{CH} \tag{C.62}$$

C.3.3.2 Electrode boiler

Sector coupling between the electricity and the heating sector is achieved by electricity-based heat generators such as the electrode boiler 'eboil.' The electric power load $GENLOAD^{\text{SO}}_{\text{'eboil'},n,t,y,\text{'power'}}$ needed to cover the heating load of the boiler $GENLOAD^{\text{SO}}_{\text{'eboil'},n,t,y,e}$ is determined by

$$GENLOAD^{\text{SO}}_{\text{'eboil'},n,t,y,\text{'power'}} = -\sum_{e \in \text{DH} \cup \text{CH}} \frac{GENLOAD^{\text{SO}}_{\text{'eboil'},n,t,y,e}}{\eta_{\text{'eboil'}}} \quad \forall n \in \text{N}, t \in \text{T}, y \in \text{Y}. \tag{C.63}$$

In accordance with Gerhardt et al. (2015), the efficiency of electrode boilers $\eta_{\text{'eboil'}}$ is assumed to be 99 %.

C.3.3.3 Ground heat pump

The heat pump 'hpg' is implemented based on a ground heat pump with a size of 10 kW. For heat pumps of this type and size, we assume for the annualized coefficient of performance acop the upper value of the range given by Lucia et al. (2017) with a constant value of acop = 3.8. The sector coupling equation for the ground heat pump is

$$GENLOAD^{\text{SO}}_{\text{'hpg'},n,t,y,\text{'power'}} = -\sum_{e \in \text{DH} \cup \text{CH}} \frac{GENLOAD^{\text{SO}}_{\text{'hpg'},n,t,y,e}}{\text{acop}} \quad \forall n \in \text{N}, t \in \text{T}, y \in \text{Y}. \tag{C.64}$$

C.3.3.4 Oil boiler

Building new oil boilers (oboil) is not possible in SecMOD due to their emission intensity. Nevertheless, existing oil boilers must be used, because the local heating infrastructure makes it impossible to heat a house deployed with an oil boiler using a gas boiler. However, oil boilers can be replaced by different technologies. Thus, SecMOD includes a constraint that the existing oil boilers must be used proportionally for domestic heating ($\forall n \in \mathrm{N}, t \in \mathrm{T}, y \in \mathrm{Y}$):

$$GENLOAD^{\mathrm{SO}}_{\text{'oboil'},n,t,y,\text{'dom'}} \geq \frac{\mathrm{cex}_{n,y,\text{'oboil'}} - REPLACEOBOIL_n}{\mathrm{ybcap}_n} \cdot \mathrm{energyload}_{n,t,\text{'dom'}} \cdot \tag{C.65}$$

The variable $REPLACEOBOIL_n$ allows replacing oil boilers, so that they do not have to be used.

C.3.3.5 Co-generation in conventional power plants

Co-generation in conventional power plants meets the centralized heating demand. It is assumed that heat extraction is possible on each temperature level. After meeting the centralized domestic heating demand, the remaining thermal capacity at a node is used to meet the high-temperature heating demand first. If the heat extraction at the respective node is higher than the domestic and the high-temperature heating demand, the remaining heat extraction is used for low and medium temperature industrial heating. Co-generation in conventional power plants (subset CONV of TECH) imposes no additional costs and emissions. The generated heat of the conventional technologies with co-generation (lignite, coal, combined-cycle gas turbine, gas turbine, oil power plant, waste power plant) is expressed using the power-to-heat ratio in dependence of the produced electric power:

$$\sum_{e \in \mathrm{CH}} GENLOAD_{conv,n,t,y,e} \leq \sum_{conv \in \mathrm{TECH}} \mathrm{p2hratio}_n \cdot GENLOAD_{conv,n,t,y,\text{'power'}} \qquad \forall n \in \mathrm{N}, t \in \mathrm{T}, y \in \mathrm{Y}. \tag{C.66}$$

The power to heat ratio $\mathrm{p2hratio}_n$ is determined at each node as the ratio of maximum heat generation and maximum power generation of conventional power plants derived from Federal Network Agency (2018).

C.3.4 Life-cycle inventory data

Tables C.32, C.33 and C.34 show the LCI data used for the implementation of the heating sector in SecMOD. Explanation for the tables and how to derive environmental impacts for the listed technologies can be found in Section C.1.5, as the table structure is the same for every sector in SecMOD.

Table C.32: LCI of heating sector technologies part 1. Processes without explicit source are based on ecoinvent 3.5 APOS (Wernet et al., 2016). The functional unit (FU) of infrastructural and operational processes is equal to 1 MW and 1 MWh, respectively.

$\text{ecoinv}_{tech,impact}$	**Location**	**process quantity per FU**
Co-generation of heat in conventional power plants		
Environmental impacts are allocated to the power sector		
Energy-efficient building refurbishment		
infrastructural processes[A]		
glass fibre production[B]	RER	1.28×10^{4} kg
sodium silicate production, furnace liquor, product in 37 % solution state[B]	RER	3.05×10^{5} kg
cement mortar production[B]	CH	5.68×10^{5} kg
stone wool production, packed[C]	CH	4.09×10^{5} kg
Electrode boiler district heating[D,E]		
infrastructural processes		
boiler		
steel production, chromium steel 18/8. hot rolled	RER	3.12×10^{2} kg
insulation		
stone wool production, packed	CH	2.83×10^{1} kg
cover		
aluminum production, primary, ingot	IAI Area, EU27 & EFTA	2.33×10^{1} kg
boiler end-of-life		
steel (see Table C.3)		3.12×10^{2} kg
cover end-of-life		
aluminum (see Table C.3)		2.33×10^{1} kg
insulation end-of-life		
treatment of waste mineral wool, recycling	Europe without CH	2.83×10^{1} kg

[A] assuming $7.10 \times 10^{4}\ \frac{\text{m}^2_{\text{external wall area}}}{\text{MW}_{\text{heating demand}}}$

[B] according to thinkstep (2019)

[C] thickness of insulation based on a heat of transition coefficient of $0.28\frac{\text{W}}{\text{Km}^2}$ (according to Deutscher Bundestag (2013)), tightened by 25 percent (according to Palzer (2016))

[D] Product information: Parat (2019)

[E] Dimensions and masses for the boiler, cover and insulation are based on Parat (2019). The boilers thickness is derived with Barlow's formula ($p = 2 \cdot \sigma_t \cdot s/D_m$), with an assumed internal pressure of p=40 bars and steel as building material.

Table C.33: LCI of heating sector technologies part 2. Processes without explicit source are based on ecoinvent 3.5 APOS (Wernet et al., 2016). The functional unit (FU) of infrastructural processes equals 1 MW, and of operational processes 1 MWh , respectively.

$ecoinv_{tech,impact}$	**Location**	**process quantity per FU**	**inv. share**
Gas boiler 'mboil'			
infrastructural processes			
gas boiler production	RER	100	
operational processes			
heat production, natural gas, at boiler condensing modulating <100kW	Europe without CH	3.53×10^3	2.33×10^{-3}
Gas boiler 'hdboil'			
infrastructural processes			
industrial furnace production, natural gas	RER	1	
operational processes			
heat production, natural gas, at boiler condensing modulating <100kW	Europe without CH	3.60×10^3	9.90×10^{-6}
Oil boiler			
infrastructural processes			
oil boiler production, 10kW	CH	100	
operational processes			
heat production, light fuel oil, at boiler 10kW condensing, non-modulating	Europe without CH	3.60×10^3	-2.38×10^{-3}

Table C.34: LCI of heating sector technologies part 3. Processes without explicit source are based on ecoinvent 3.5 APOS (Wernet et al., 2016). The functional unit (FU) of infrastructural processes equals 1 MW, and of operational processes 1 MWh , respectively.

$\mathrm{ecoinv}_{tech,impact}$	**Location**	**process quantity per FU**
Heat pump ground (hpg)		
infrastructural processes		
borehole heat exchanger production, 150m	CH	1.00×10^{2} units
under floor heating system[A]		
polystyrene production, general purpose	RER	1.65×10^{4} kg
polyethylene production, low density, granulate	RER	2.53×10^{4} kg
aluminum production, primary, ingot	IAI Area, EU27 & EFTA	3.15×10^{4} kg
market for cement, unspecified	CH	2.25×10^{5} kg
silica sand production	DE	1.16×10^{6} kg
ground source[A]		
propylene production	RER	1.06×10^{3} kg
copper production, primary	RER	5.50×10^{3} kg
market for electricity, medium voltage	DE	3.50×10^{4} kWh
market for heat, district or industrial, natural gas	Europe without CH	3.33×10^{5} MJ
lubricating oil production	RER	4.25×10^{2} kg
polyvinylchloride production, bulk polymerisation	RER	2.50×10^{2} kg
reinforcing steel production	RER	1.88×10^{4} kg
steel production, low-alloyed, hot rolled	RER	5.00×10^{3} kg
tube insulation production, elastomere	DE	2.50×10^{3} kg
under floor heating system end-of-life[A]		
treatment of waste cement, hydrated, residual material landfill	CH	2.25×10^{5} kg
aluminum (see Table C.3)		3.15×10^{4} kg
polyethylene (see Table C.3)		2.53×10^{4} kg
polystyrene (see Table C.3)		1.65×10^{4} kg
ground source end-of-life[A]		
polyvinylchloride (see Table C.3)		2.50×10^{2} kg
steel (see Table C.3)	Europe without CH	2.38×10^{4} kg
copper (see Table C.3)		5.50×10^{3} kg
plastic (see Table C.3)		2.50×10^{3} kg

[A] Greening and Azapagic (2012)

C.4 Transportation sector

As part of the transportation sector, private transportation is modeled in SecMOD. Hence, with an amount of 104.8 Mt CO_{2eq} per year in 2016, approximately 60 % of the emissions of the entire German transportation sector are covered (Harthan and Herrmann, 2018). Apart from conventional technologies such as gasoline, diesel, and natural gas vehicles, the following technologies coupling the transportation and power sector are implemented: battery electric vehicles, plug-in hybrid electric vehicles, fuel-cell vehicles, and power-to-fuel technologies.

The demand structure is represented by a total annual mileage in vehicle kilometers (c.f. C.4.1) and an initial amount of vehicles (c.f. C.4.2), which are both exogenously given at each node. Furthermore, the vehicle technologies integrated into the model are described in C.4.3. The corresponding LCI data is stated in C.4.4.

C.4.1 Total annual mileage

The total annual mileage of all registered vehicles in Germany is measured in vehicle kilometers (vkm) and incorporates the demand of the transportation sector. The data is taken from Federal Motor Transport Authority (2017b). Until the year 2030, BVU Beratergruppe et al. (2014) predict a slight increase in the total annual mileage compared to 2016. Thus, the total annual mileage between 2016 and 2030 is linearly interpolated. Due to the lack of reliable data, the total annual mileage is assumed to be constant after 2030. The assumed total annual mileage for all investment periods is stated in Table C.35. BVU Beratergruppe et al. (2014) divides Germany into the assumed total annual mileage for all investment periods is stated in Table C.35.

Employing the data from BVU Beratergruppe et al. (2014), a node-to-node origin-destination matrix is determined. This origin-destination matrix shows that more than 80 % of all distances are covered within the respective registration district. Thus, the total annual mileage for all investment periods is allocated to the nodes based on vehicle density in the base year 2016, resulting in the nodal mileage demand $\text{nodalmileage}_{n,y}$. The heuristics to determine the vehicle density are explained in C.4.2. Furthermore, we assume that the nodal mileage demand increases equally at each node.

The fulfillment of the nodal mileage demand $\text{nodalmileage}_{n,y}$ in each investment period y is described by

$$\text{nodalmileage}_{n,y} = \sum_{car \in \text{CAR}} \sum_{\substack{yex \in YEX \\ yex \leq y}} (\text{mileage}_{car,yex} \cdot USEDCARS_{car,n,y,yex}) \; \forall n \in \text{N}, y \in \text{Y}. \tag{C.67}$$

The set CAR contains all vehicle technologies. The nodal mileage demand $\text{nodalmileage}_{n,y}$ is covered by the multiplication of the number of used vehicles $USEDCARS_{car,n,y,yex}$ of each vehicle technology car from the year yex, and the average annual mileage $\text{mileage}_{car,yex}$.

Table C.35: Assumed total annual mileage in billion vehicle kilometers (vkm) from 2016 to 2050.

Specification	unit	2016	2020	2030	2040	2050
total annual mileage	billion vkm	625.5	634.6	657.4	657.4	657.4

C.4.2 Allocation of the existing initial vehicle fleet

Federal Motor Transport Authority (2017a) publishes an overview of the German vehicle fleet divided both into the registration districts and the vehicle technologies. Conventional fuel technologies such as diesel and gasoline make up 98 % of the vehicle fleet. Thus, non-conventional modes of driving are neglected for the base year due to numerical instabilities. Based on the latest version from 2017 (Federal Motor Transport Authority, 2017a), the diesel and gasoline vehicle fleet is allocated to the nodes by the following steps:

1. Geographic coordinates are assigned to all registrations districts based on Federal Statistical Office (2018) which contains a list of all German municipalities and their corresponding geographic coordinates.
2. As there are fewer registration districts than nodes, each node is assigned to the closest registration district. Due to this allocation heuristics, several nodes are assigned to the same registration district whereas some registration districts remain unassigned to a node. Thus, each remaining unassigned registration district is assigned to the closest node.
3. After matching nodes and registration districts, the German diesel and gasoline vehicle fleet is allocated to the nodes. If several nodes are assigned to the same vehicle registration district, the allocation is carried out based on the nodal population.

In 2017, the amount of all registered vehicles in Germany sums up to slightly above 46 million vehicles (Federal Motor Transport Authority, 2017a). The nodal allocation of gasoline and diesel vehicles is shown qualitatively in Figure C.8 as a heat map.

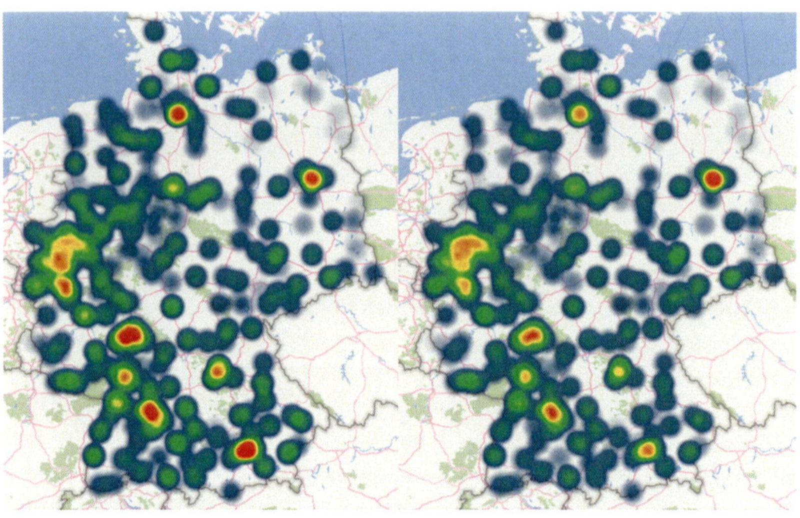

Figure C.8: Vehicle density of diesel (left) and gasoline vehicles (right) in Germany in 2017 based on Federal Motor Transport Authority (2017a). A red color implies a high vehicle density, a blue color implies a low vehicle density.

C.4.3 Vehicle technologies

Based on Deutsche Energie-Agentur (2018), Henning and Sauer (2015) and D. Kreyenberg et al. (2015), the following vehicle technologies are identified as currently relevant or relevant in the future and thus implemented in SecMOD:

1. conventional technologies: gasoline, diesel, and natural gas vehicles, see Section C.4.3.1
2. battery electric vehicles, see Section C.4.3.2
3. plug-in hybrid electric vehicles, see Section C.4.3.3
4. fuel-cell vehicles, see Section C.4.3.5

Based on Deutsche Energie-Agentur (2018), an average vehicle is assumed for each modeled vehicle technology. The modeling assumptions are briefly presented in the following subsections. The majority of the model equations of the transportation sector are structurally equal for each vehicle technology. For all vehicle technologies, the annualized investment costs of the average vehicle as well as the environmental impacts resulting from the construction and end-of-life of vehicles are calculated using Eq. C.5 and C.10 for synthesis optimization and Eq. C.15 and C.19 for operational optimization, respectively. Corresponding to Eq. C.8 in the synthesis optimization and Eq. C.18 in the operational optimization, vehicle usage is limited by the available vehicle capacity. All other relevant equations, especially those resulting from the coupling of the transportation and electricity sector, are presented in the respective subsections following. Information on the LCI data for all vehicle technologies is stated in Section C.4.4.

C.4.3.1 Conventional technologies

Gasoline, diesel, and natural gas vehicles are considered conventional vehicle technologies. Cost data for investment and maintenance for all investment periods up to 2050 are taken from Deutsche Energie-Agentur (2018). The average vehicle considered in this study is a Volkswagen Golf for diesel and gasoline and a mid-size vehicle for all other vehicle technologies.

Average annual mileages (in km per year) and average fuel consumption (in l per 100 km) are taken from Bundesministerium für Verkehr und digitale Infrastruktur (2018). The vehicle-specific annual mileage $\text{mileage}_{car,yex}$ is assumed to remain constant up to 2050. However, for the natural gas vehicles, no separate numbers have been published. Therefore, the average annual mileage of the natural gas vehicle is assumed to equal the average annual mileage of all vehicles taken from Bundesministerium für Verkehr und digitale Infrastruktur (2018).

The fuel consumption of gasoline and diesel vehicles is modeled using current values from Bundesministerium für Verkehr und digitale Infrastruktur (2018) for 2016 and relative efficiency gains between the years of 2020 to 2050 based on Deutsche Energie-Agentur (2018). According to Knörr et al. (2016), the average fuel consumption of a natural gas vehicle is about 4 % higher (based on the energy content) compared to gasoline vehicles. We assume that this gap remains constant.

According to Richter (2010), a total vehicle mileage of 200000 km is assumed for all vehicle technologies. The lifetime of a vehicle in years is calculated by dividing the total vehicle mileage by the average annual mileage. Due to the higher annual mileage the diesel vehicle has the shortest lifetime with 10 years. The lifetime of the gasoline vehicle is approximately 18 years, while the lifetime of the natural gas vehicle is 14 years. The conventional technologies can be operated with fossil gasoline, diesel or natural gas as well as with synthetic diesel or synthetic natural gas. The fossil fuel prices are set exogenously as net prices, i. e., excluding taxes. The average prices for petrol and diesel for the year 2015 are taken from ARAL (2017). The future fuel price development for gasoline and diesel until 2050 is taken from Federal Ministry for Economic Affairs and Energy (2017) not taking taxes into account as well. The natural gas price is assumed to be constant for all years and is stated in Table C.36. An overview of the data used for modeling conventional vehicle technologies is given in Table C.36.

For conventional vehicles, the operational costs $OPCOST_{car,y}$ caused by a vehicle car in investment period y are described by

$$
\begin{aligned}
OPCOST_{car,y} =& \text{fuelprice}_{y,car} \cdot \sum_{n \in \mathrm{N}} \sum_{\substack{yex \in \mathrm{YEX} \\ yex \leq y}} (USEDCARS_{car,n,y,yex} \\
& \cdot \text{fuelcon}_{car,yex} \cdot \text{mileage}_{car,yex}) \quad \forall\, car \in \mathrm{CAR}, y \in \mathrm{Y}.
\end{aligned} \tag{C.68}
$$

The variable $USEDCARS_{car,n,y,yex}$ indicates the number of vehicles of the technology car from the construction year yex used in investment period y at node n. The parameter $\text{fuelcon}_{car,yex}$ is the fuel consumption.

The environmental impacts during operation $OPEMIT^{\mathrm{SO}}_{car,impact,y}$ are calculated using

$$
\begin{aligned}
OPEMIT_{car,impact,y} =& \sum_{n \in \mathrm{N}} \sum_{\substack{yex \in \mathrm{YEX} \\ yex \leq y}} (USEDCARS_{car,n,y,yex} \\
& \cdot \text{opemit}_{car,yex,impact} \cdot \text{mileage}_{car,yex}) \\
& \forall\, car \in \mathrm{CAR}, impact \in \mathrm{IMPACT}, y \in \mathrm{Y}.
\end{aligned} \tag{C.69}
$$

The parameter $\text{opemit}_{car,yex,impact}$ indicates the corresponding specific environmental impacts per vehicle kilometer. The Eq. C.68 and C.69 are valid both for synthesis and operational optimization.Thus, for the sake of readability, the superscripts SO and OP at the variables are neglected.

Table C.36: Modeling assumptions for conventional vehicle technologies.

Specification	unit	2016	2030	2040	2050	Source
Total investment costs: $\text{invcost}^0_{car,yex}$						
diesel vehicle	€/car	28717	29499	29499	29499	A
gasoline vehicle	€/car	27382	28432	28432	28432	A
natural gas vehicle	€/car	28548	28322	28237	28151	A
Investment costs						
diesel vehicle	€/car	25592	26289	26289	26289	B
gasoline vehicle	€/car	23024	23907	23907	23907	B
natural gas vehicle	€/car	24631	24436	24363	24289	B
Maintenance costs						
diesel vehicle	%/(car · a)	1.6	1.6	1.6	1.6	B
gasoline vehicle	%/(car · a)	1.6	1.6	1.6	1.6	B
natural gas vehicle	%/(car · a)	1.6	1.6	1.6	1.6	B
Average annual mileage: $\text{mileage}_{car,yex}$						
diesel vehicle	km/a	20300	20300	20300	20300	C
gasoline vehicle	km/a	10900	10900	10900	10900	C
natural gas vehicle	km/a	14200	14200	14200	14200	C
Lifetime: lifetime_{car}						
diesel vehicle	a	9.85	9.85	9.85	9.85	C, D
gasoline vehicle	a	18.35	18.35	18.35	18.35	C, D
natural gas vehicle	a	14.08	14.08	14.08	14.08	C, D
Fuel consumption: $\text{fuelcon}_{car,yex}$						
diesel vehicle	l/100km	6.80	6.30	5.79	5.29	B, C
gasoline vehicle	l/100km	7.70	7.27	6.84	6.42	B, C
natural gas vehicle	kWh/100km	71.87	68.15	64.43	60.72	B, E
Fuel prices: $\text{fuelprice}_{y,car}$						
diesel price	€/l	0.65	0.89	1.00	1.13	F, G
gasoline price	€/l	0.61	0.65	0.75	0.87	F, G
natural gas price	€/MWh	72.05	72.05	72.05	72.05	H, I

[A] Own calculations based on Deutsche Energie-Agentur (2018) taking the lifetimes and maintenance costs of the different vehicles into account.
[B] Deutsche Energie-Agentur (2018)
[C] Bundesministerium für Verkehr und digitale Infrastruktur (2018)
[D] Richter (2010)
[E] Knörr et al. (2016)
[F] ARAL (2017)
[G] Federal Ministry for Economic Affairs and Energy (2017)
[H] ADAC (2019a)
[I] ADAC (2019b)

C.4.3.2 Battery electric vehicles (BEVs)

A battery electric vehicle (BEV) is driven by an electric engine using the necessary electrical energy from a battery. Cost data for acquisition and maintenance as well as average consumption per kilometer are taken from Deutsche Energie-Agentur (2018) for all investment periods. For the cost degression, Deutsche Energie-Agentur (2018) assumes that the average battery size will increase from 38 kWh in 2020 to 56 kWh in 2050.

In accordance with Anderson et al. (2016), the charging infrastructure consists of fast chargers (50 kW), public chargers (11 kW) and home chargers (3.7 kW). In SecMOD, the construction of a BEV results in the parallel construction of the vehicle as well as the charging infrastructure. Thus, the $\text{invcost}^{0}_{\text{'BEV'},yex}$ contains investment costs for the vehicle itself, prorated investment costs for the charging infrastructure on all three charging levels and maintenance costs for both the vehicle and the charging infrastructure. The individual lifetime of each cost component is considered and the total investment costs are adapted to the vehicle lifetime of 14 years.

Due to a lack of data for the average annual mileage of electric vehicles, the average mileage of an average vehicle of 14200 km per year is taken as the annual mileage of the BEV (Bundesministerium für Verkehr und digitale Infrastruktur, 2018). Similar to the natural gas vehicle, the lifetime of the BEV is assumed to be 14.08 years considering a vehicle lifetime of 200000 km.

Karlsruher Institute of Technology (2014) states that vehicles are bought based on their global use. Global use means that the daily range should also be sufficient for longer trips. Figure C.9 shows the percentage of the German passenger vehicle fleet that covers a maximum daily mileage in a specific time period. If this time period is 1 day, more than 90 % of all vehicle users cover less than 100 km. Given an average driving distance of 38.9 km per day for a passenger vehicle (Bundesministerium für Verkehr und digitale Infrastruktur, 2018), this result is in line with other studies (BMVBS, 2010b). However, if the observation period is extended to 1 year, less than 20 % of all vehicle users drive less than 100 km on at least one single day in this year. Assuming that the BEV can be charged once a day via fast charging, the daily range corresponds to the twofold range of a BEV. Using the data from Karlsruher Institute of Technology (2014), the maximum share of BEV's in the German vehicle fleet is

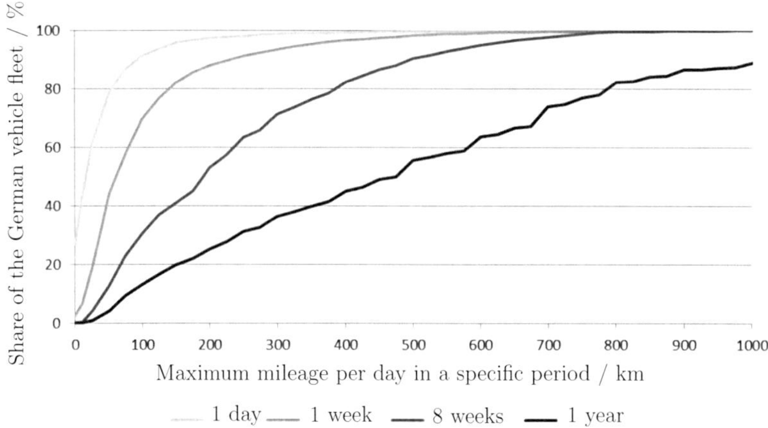

Figure C.9: Maximum share of the German vehicle fleet not exceeding a certain mileage within a given period (Karlsruher Institute of Technology, 2014).

determined given the assumed daily range. Hence, the maximum share of the nodal traffic demand covered by BEVs is limited by

$$\sum_{\substack{yex \in \mathrm{YEX} \\ yex \leq y}} (USEDCARS^{\mathrm{SO}}{}_{\text{'bev'},n,y,yex} \cdot \mathrm{mileage}_{\text{'bev'},yex}) \leq \\ \mathrm{nodalmileage}_{n,y} \cdot \mathrm{maxBEVshare}_{y} \; \forall n \in \mathrm{N} \; y \in \mathrm{Y}, \tag{C.70}$$

where $\mathrm{maxBEVshare}_y$ is the maximum share of BEVs in an investment period y. An overview of the data used for modeling the BEV is given in Table C.37.

Table C.37: Modeling assumptions regarding battery electric vehicles (BEVs) taking charging infrastructure with fast chargers (charging at 50 kW), public chargers (charging at 11 kW) and home charger (charging at 3.7 kW) into account.

Specification	Unit	2016	2030	2040	2050	Source
Total investment costs: $\text{invcost}^0_{\text{'BEV'},yex}$	€/car	56607	45479	42617	40841	A
Investment costs	€/car	41402	36604	33981	32354	A
Vehicle	€/car	37141	34093	31220	29437	B
Fast charger	€/charger	31578	19039	19039	19039	C
Public charger	€/charger	10242	6371	6371	6371	C
Home charger	€/charger	10539	5869	5869	5869	C
Maintenance costs	€/(car · a)	1530	893	869	854	A
Vehicle	%/(car · a)	0.9	0.9	0.9	0.9	B
Fast charger	€/(charger · a)	1500	750	750	750	C
Public charger	€/(charger · a)	1000	500	500	500	C
Home charger	€/(charger · a)	750	375	375	375	C
Lifetime						
Vehicle	a	14.08	14.08	14.08	14.08	D, E
Fast charger	a	12	12	12	12	B
Public charger	a	6	6	6	6	B
Home charger	a	15	15	15	15	B
Share of chargers						
Fast charger	charger/car	0.15	0.15	0.15	0.15	F
Public charger	charger/car	0.25	0.25	0.25	0.25	F
Home charger	charger/car	1	1	1	1	F
Average annual mileage						
$\text{mileage}_{car,yex}$	km/a	14200	14200	14200	14200	E
Fuel consumption						
$\text{fuelcon}_{car,yex}$	kWh_{el}/100km	19.0	17.3	15.7	14	E
Electric range	km/charge	200	254	319	400	B, E
Maximum electric-driven share						
maxBEVshare_y	%	45	56	66	82	G

[A] Own calculations based on Deutsche Energie-Agentur (2018) and Nationale Plattform Elektromobilität (2015) taking the share of chargers as well as the lifetimes of the different components into account.
[B] Deutsche Energie-Agentur (2018)
[C] Nationale Plattform Elektromobilität (2015)
[D] Richter (2010)
[E] Bundesministerium für Verkehr und digitale Infrastruktur (2018)
[F] Lucas et al. (2012)
[G] Own calculations based on Karlsruher Institute of Technology (2014)

C.4.3.3 Plug-in hybrid electric vehicles (PHEVs)

A plug-in hybrid vehicle (PHEV) is a so-called full hybrid that can be recharged by external electricity sources (Merker and Teichmann, 2018). The vehicle is equipped both with an internal combustion engine and an electric drive. Thus, the vehicle can be operated either purely electrical, purely conventional or with both engines (Merker and Teichmann, 2018). Due to the option of driving with the internal combustion engine, the batteries are significantly smaller than for the BEV. PHEVs offer an electric range of approximately 50 km (Anderson et al., 2016).
To model the PHEV, we assume an electric driving share. According to Knörr et al. (2016), the electric driving share is about 60 %. Thus, the PHEV is modeled as a mixture of a diesel (40 %) and an electric vehicle ($\text{phevelecshare}_{\text{yex}}$ of 60 %). Here, PHEVs are operated by a diesel engine and thus can be supplied by conventional diesel or synthetic diesel made by the Fischer-Tropsch process.

For the PHEV, the operational costs $OPCOST_{\text{'phev'},y}$ caused by a vehicle 'phev' in investment period y only take the conventional share of its operation $(1-\text{phevelecshare}_{yex})$ into account. The costs of the electrical operation are included in the electricity production. As a result the operational costs are described by

$$\begin{aligned} OPCOST_{\text{'phev'},y} =& \text{fuelprice}_{y,\text{'phev'}} \cdot \sum_{n \in \text{N}} \sum_{\substack{yex \in \text{YEX} \\ yex \leq y}} (USEDCARS_{\text{'phev'},n,y,yex} \\ & \cdot (1 - \text{phevelecshare}_{yex}) \cdot \text{fuelcon}_{\text{'phev'},yex} \cdot \text{mileage}_{\text{'phev'},yex}). \\ & \forall y \in \text{Y}. \end{aligned} \tag{C.71}$$

The variable $USEDCARS_{\text{'phev'},n,y,yex}$ indicates the number of PHEVs from the construction year yex being used in investment period y at node n. The parameter $\text{fuelcon}_{\text{'phev'},yex}$ is the fuel consumption. The parameter $\text{phevelecshare}_{yex}$ indicates the electric share of its operation. It is necessary to mention that the fuel price $\text{fuelprice}_{y,\text{'phev'}}$ is equivalent to the price of diesel.

PHEV cost data, including maintenance costs, are taken from Deutsche Energie-Agentur (2018). We assume that the annual mileage of the PHEV is the same as for an average vehicle of 14200 km. With a lifetime mileage of 200000 km, the lifetime of a PHEV is 14 a. The consumption data for the electrical driving share correspond to the BEV data, whereas the consumption data for the diesel driving share correspond to the diesel vehicle. The LCI data are calculated based on the assumptions for the

BEV. Only the battery size is adjusted to 10 kWh, so that the PHEV has a range of about 55 km in purely electric driving mode. Additionally, the conventional diesel drive-train, including a diesel engine, is considered in the LCI data of the PHEV. Table C.38 gives an overview of the modeling parameters.

Table C.38: Modeling assumptions for plug-in hybrid electric vehicles (PHEVs).

Specification	unit	2016	2030	2040	2050	Source
Total investment costs: $invcost^{0}_{'PHEV',yex}$	€/car	48653	41668	40733	39799	A
Investment costs	€/car	41402	36604	33981	32354	A
Vehicle	€/car	30254	29429	28603	27778	B
Fast charger	€/charger	31578	19039	19039	19039	C
Public charger	€/charger	10242	6371	6371	6371	C
Home charger	€/charger	10539	5869	5869	5869	C
Maintenance costs	€/(car · a)	1393	883	872	861	A
Vehicle	%/(car · a)	1.3	1.3	1.3	1.3	B
Fast charger	€/(charger · a)	1500	750	750	750	C
Public charger	€/(charger · a)	1000	500	500	500	C
Home charger	€/(charger · a)	750	375	375	375	C
Lifetime						
Vehicle	a	14.08	14.08	14.08	14.08	D, E
Fast charger	a	12	12	12	12	B
Public charger	a	6	6	6	6	B
Home charger	a	15	15	15	15	B
Share of chargers						
Fast charger	charger/car	0	0	0	0	F
Public charger	charger/car	0.25	0.25	0.25	0.25	G
Home charger	charger/car	1	1	1	1	G
Average annual mileage						
$mileage_{car,yex}$	km/a	14200	14200	14200	14200	E
Fuel consumption						
$fuelcon_{car,yex}$	l/100km	6.80	6.30	5.79	5.29	B, E
Electricity consumption	kWh_{el}/100km	19.0	17.3	15.7	14.0	B
Battery size	kWh	10	10	10	10	F

[A] Own calculations based on Deutsche Energie-Agentur (2018) and Nationale Plattform Elektromobilität (2015) taking the share of chargers as well as the lifetimes of the different components into account.
[B] Deutsche Energie-Agentur (2018)
[C] Nationale Plattform Elektromobilität (2015)
[D] Richter (2010)
[E] Bundesministerium für Verkehr und digitale Infrastruktur (2018)
[F] Battery size based on range from Anderson et al. (2016)
[G] Lucas et al. (2012)

C.4.3.4 Time series for electric vehicles

The use of electrically driven vehicles leads to direct coupling of the transportation sector with the electricity sector. Hence, it is important to model the charging behavior of electrically driven vehicles and resulting load curves for the electricity grid. The charging behavior of electrically driven vehicles is influenced by the mobility behavior. Data for the mobility behavior is, for example, investigated in BMVBS (2010b) and Karlsruher Institute of Technology (2014).

In SecMOD, we include load curves for the electricity grid based on Anderson et al. (2016) and BMVBS (2010b). Anderson et al. (2016) estimate the amount of charging infrastructure necessary for 1 million electric vehicles. Based on user data from BMVBS (2010b), weekly load curves are derived for all three charging levels. First, the weekly load curves of the reference scenario are aggregated over the given charging loads. Next, we scale the load curves by 2 % to match the energy consumption and the annual mileage per vehicle assumed in Table C.37 with the resulting overall load. We neglect seasonal deviations, as the average monthly mileage fluctuates by less than 10 % (BMVBS, 2010b). The resulting load curves are shown in Figure C.10. The load curves show large weekly fluctuations. Following Anderson et al. (2016), we

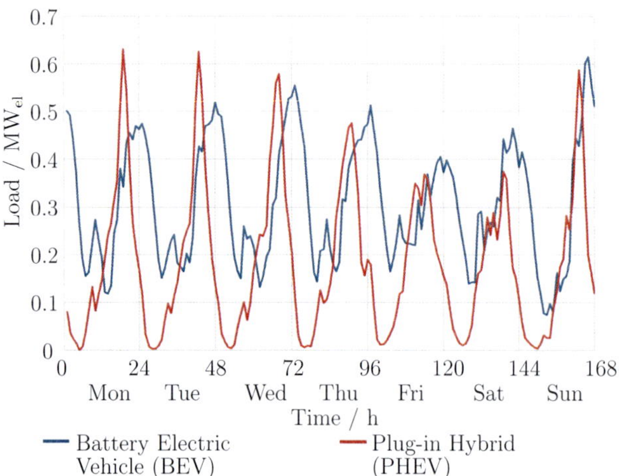

Figure C.10: Weekly load curves for charging 1000 BEVs and 1000 PHEVs based on Anderson et al. (2016) and BMVBS (2010a).

assume that electrically driven vehicles are not operated grid-friendly, but are charged on demand. Thus, the implemented load curves represent an extreme case.

The coupling equations with the electricity sector are formulated as following:

$$\begin{aligned} GENLOAD^{\mathrm{SO}}_{n,t,y,yex,\text{'bev'},\text{'power'}} = & -\mathrm{load}_{\text{'bev'},t} \\ & \cdot USEDCARS^{\mathrm{SO}}_{\text{'bev'},n,y,yex} \cdot \eta_{\text{'bev'},yex} \\ & \forall n \in \mathrm{N}, t \in \mathrm{T}, y \in \mathrm{Y}, yex \in \mathrm{YEX}, \end{aligned} \tag{C.72}$$

and

$$\begin{aligned} GENLOAD^{\mathrm{SO}}_{n,t,y,yex,\text{'phev'},\text{'power'}} = & -\mathrm{load}_{\text{'phev'},t} \\ & \cdot USEDCARS^{\mathrm{SO}}_{\text{'phev'},n,y,yex} \cdot \eta_{\text{'phev'},yex} \cdot \mathrm{phevelecshare}_{yex} \cdot \\ & \forall n \in \mathrm{N}, t \in \mathrm{T}, y \in \mathrm{Y}, yex \in \mathrm{YEX}. \end{aligned} \tag{C.73}$$

The load curves are represented by the parameter $\mathrm{load}_{tech,t}$, depending on the technology $tech$ and the time step t. Efficiency gains are incorporated in the efficiency $\eta_{tech,yex}$, depending on the technology $tech$ and the year of construction yex. For the PHEV, the share of electricity during operation is implemented using the parameter $\mathrm{phevelecshare}_{yex}$; thus, only the electric share of its operation is considered for the electricity load. The generated load is negative, as the technologies consume electricity.

C.4.3.5 Fuel-cell vehicles (FCVs)

In SecMOD, the construction of a fuel-cell vehicle (FCV) results in the parallel construction of its refueling infrastructure as well. The refueling infrastructure consists of decentralized on-site electrolysis at refueling stations and the refueling station itself. Cost data and technical assumptions are based on a pressure level of 700 bar. Thus, the total investment costs $\mathrm{invcost}^{0}_{\text{'FCV'},yex}$ contain investment costs for the vehicle and proportional shares of the investment costs for the hydrogen refueling stations and the PEM-electrolysis for hydrogen production. Additionally, maintenance costs both for the vehicle and the PEM-electrolyser are added to the total investment costs. The individual lifetime of each cost component is considered. Furthermore, the payments included in the total investment costs are adapted to the vehicle lifetime of 14 years. Again, vehicle lifetime is calculated by dividing the vehicle-specific total mileage from Richter (2010) by the annual mileage of an average vehicle.

To derive the refueling infrastructure as well as the electricity loads caused by refueling the FCV, the annual energy consumption $\mathrm{energycon}_{yex}$ from the construction year yex is considered. The annual energy consumption energycon_{yex} results from the production and the compression of hydrogen:

$$\text{energycon}_{yex} = \text{mileage}_{\text{'fcv'},yex} \cdot \text{fuelcon}_{\text{'fcv'},yex} \cdot \left(\frac{1}{\eta_{\text{'pem'},yex}} + \frac{\text{compressenergy}}{\text{LHV}^{\text{H2}}} \right), \tag{C.74}$$

with the efficiency of the PEM electrolysis $\eta_{\text{'pem'},yex}$ in the year of construction yex, the lower heating value of hydrogen LHV^{H2} of $33.33\,\text{kW h kg}^{-1}$ and the compression energy compressenergy. For further information regarding the modeled PEM electrolysis see Section C.2.3.2.

The compression energy compressenergy incorporates the energy requirements for hydrogen gas compression and liquefaction. We choose an average compression energy, according to the range given in Gardiner (2009). The annual fuel consumption in $\frac{\text{kWh}}{\text{a}}$ results from the multiplication of the specific hydrogen consumption per 100 km ($\text{fuelcon}_{\text{'fcv'},yex}$) with the average annual mileage $\text{mileage}_{\text{'fcv'},yex}$ of a FCV from construction year yex.
In order to generate the grid load demanded by the FCV, we assume a constant production of the hydrogen needed. Thus, we assume a constant hourly load curve $\text{h}_{t,yex,\text{'fcv'}}$ of the FCV,

$$\text{h}_{t,yex,\text{'fcv'}} = \frac{\text{energycon}_{yex}}{8760\,\text{h}}. \tag{C.75}$$

The share of the refueling station stationshare_y indicates how many stations are needed per vehicle. The share is calculated by dividing the annual hydrogen consumption of 1 FCV by the annual capacity of the hydrogen refueling station $\text{capacity}^{\text{station}}$ (365 times the daily capacity):

$$\text{stationshare}_{yex} = \frac{\text{mileage}_{\text{'fcv'},yex} \cdot \text{fuelcon}_{\text{'fcv'},yex}}{365\,\text{d/a} \cdot \text{capacity}^{\text{station}}} \cdot \frac{1}{\text{LHV}^{\text{H2}}}. \tag{C.76}$$

The necessary electrolysis power per refueling station pempower_y is calculated by

$$\text{pempower}_{yex} = \frac{\text{capacity}^{\text{station}} \cdot \text{LHV}^{\text{H2}}}{24 \cdot \eta_{\text{'pem'},yex} \cdot \text{utilizationrate}_{yex}} \tag{C.77}$$

where $\text{utilizationrate}_{yex}$ is the utilization rate of the refueling station. Table C.39 gives an overview of the modeling parameters.

Table C.39: Model data regarding fuel-cell vehicles (FCVs).

Specification	unit	2016	2030	2040	2050	Source
Total investment costs: $invcost^{0}_{'fcv',yex}$	€/car	65145	55755	46457	37226	A
Investment costs:	€/car	59405	50831	42391	33987	A
Vehicle	€/car	58045	49853	41662	33470	B
PEM-electrolysis	€/kW_{el}	800	633	467	300	B
Refueling station	€/station	2000000	2000000	2000000	2000000	C
Maintenance costs:	€/(car · a)	536	460	383	306	A
Vehicle	%/(car · a)	0.9	0.9	0.9	0.9	B
PEM-electrolysis	€/(kW_{el}·a)	14	11	8	5	B
Vehicle data						
Lifetime	a	14.08	14.08	14.08	14.08	D, E
Annual mileage ($mileage_{'fcv',y}$)	km/a	14200	14200	14200	14200	F
Hydrogen fuel consumption ($fuelcon_{yex}$)	kWh_{H_2}/100km	32	29.3	26.7	24	B
Energy consumption ($energycon_{yex}$)	kWh_{el}/a	5951	5414	4885	4364	Eq. C.74
Refueling station data						
Lifetime	a	30	30	30	30	F
Capacity ($capacity^{station}$)	kg/d	1500	1500	1500	1500	G
Utilization rate ($utilizationrate_y$)	%	80	80	80	80	H
Share ($stationshare_y$)	station/car	0.0002490	0.0002283	0.0002075	0.0001868	Eq. C.76
Compression energy (compressenergy)	kWh_{el}/kg_{H_2}	3	3	3	3	I
Electrolysis data						
Lifetime	a	11	12.7	14.3	16	A
Efficiency ($\eta_{'pem',yex}$)	%	82.0	82.7	83.3	84.0	A
Power per refueling station ($pempower_{yex}$)	kW_{el}/station	3175	3149	3124	3099	Eq. C.77

[A] Own calculations based on Deutsche Energie-Agentur (2018) and Nationale Plattform Elektromobilität (2015) taking the share of chargers as well as the lifetimes of the different components into account.
[B] Deutsche Energie-Agentur (2018)
[C] Schiebahn et al. (2015)
[D] Richter (2010)
[E] Bundesministerium für Verkehr und digitale Infrastruktur (2018)
[F] Reddi et al. (2017)
[G] Krieg (2012)
[H] McKinsey & Company (2010)
[I] Gardiner (2009)

C.4.4 Life-cycle inventory data

The following tables show the LCI data used for the implementation of the transportation sector in SecMOD. All LCI tables are structurally equal for every sector in SecMOD. Therefore, explanations for the tables and how to derive environmental impacts can be found in Section C.1.5. All environmental impacts are calculated for the base year 2016. Because certain technologies in the transport sector are subject to learning curves due to technological changes in the future, different amounts for 'process quantity' are built in every year. Those specific processes are marked with an L (learning). The learning curves are displayed in Table C.47. Note that the functional unit of the transportation sector for the infrastructural processes, including the end-of-life of the technologies, is 1000 vehicles and not 1 MW. The functional unit for the operational processes is 1×10^6 vehicle kilometers (vkm) instead of 1 MWh. For the technology 'charger', the functional unit is 1 charger.

Table C.40: LCI of transportation sector technologies part 1. Processes without explicit source are based on ecoinvent 3.5 APOS (Wernet et al., 2016). The functional unit (FU) of infrastructural and operational processes is 1000 vehicles and 10^6 vkm (vehicle kilometers).

$\text{ecoinv}_{tech,impact}$	**Location**	**process quantity per FU**
Battery electric vehicle		
infrastructural processes		
glider production, passenger car	GLO	9.12×10^5 kg
powertrain production, for electric passenger car	GLO	8.02×10^4 kg
lithium battery[A,L] (see p.199)		2x3.80×10^4 kWh
maintenance, passenger car, electric, without battery	GLO	1.00×10^3 unit
charger home (see p.236)		9.39×10^2 unit
charger normal (see p.236)		5.87×10^2 unit
charger fast (see p.235)		1.76×10^2 unit
end-of-life		
treatment of used glider, passenger car, shredding	GLO	9.12×10^5 kg
treatment of used powertrain for electric passenger car, manual dismantling	GLO	8.02×10^4 kg
manual dismantling of used electric passenger car	GLO	9.82×10^5 unit
operational processes[L]		1×10^6 km
road construction[L]	CH	5.45×10^2 m year
treatment of brake wear emissions, passenger car[L]	RER	1.18 kg
treatment of tire wear emissions, passenger car[L]	RER	7.57×10^1 kg
treatment of road wear emissions, passenger car[L]	RER	1.30×10^1 kg

$\text{ecoinv}_{tech,impact}$	**Location**	**process quantity per charger**
Charger fast		
infrastructural processes		
steel production, electric, chromium steel 18/8	RER	1.06×10^2 kg
copper production, primary	RER	1.78×10^2 kg
concrete block production	DE	2.82×10^3 kg
acrylonitrile-butadiene-styrene copolymer production	RER	4.46×10^2 kg
pig iron production	GLO	2.68×10^2 kg
end-of-life		
steel (see Table C.3)	CH	1.06×10^2 kg
treatment of waste concrete, not reinforced, recycling	Europe w/o CH	2.82×10^3 kg
copper (see Table C.3)		1.78×10^2 kg
plastic (see Table C.3)		4.46×10^2 kg

[L] process has a learning curve, see Table C.47

[A] The lithium battery process builds capacity equal to twice the base capacity of 1 battery electric vehicle, as the battery lifetime is exactly half the cars lifetime.

Table C.41: LCI of transportation sector technologies part 2. Processes without explicit source are based on ecoinvent 3.5 APOS (Wernet et al., 2016). The functional unit (FU) of infrastructural and operational processes is equal to 1000 vehicles and 10^6 vkm (vehicle kilometers), respectively.

$\mathrm{ecoinv}_{tech,impact}$	**Location**	**process quantity per charger**
Charger home		
infrastructural processes		
polyvinylidenchloride production, granulate	RER	7.04 kg
steel production. electric. chromium steel 18/8	RER	3.29 kg
end-of-life		
steel (see Table C.3)		3.29 kg
plastic (see Table C.3)		7.04 kg
Charger normal		
infrastructural processes		
concrete block production	DE	2816.04 kg
acrylonitrile-butadiene-styrene copolymer production	RER	1.23×10^2 kg
steel production, electric, chromium steel 18/8	RER	5.02 kg
end-of-life		
treatment of waste concrete, not reinforced, recycling	Europe w/o CH	2.82×10^3 kg
plastic (see Table C.3)		1.23×10^2 kg
steel (see Table C.3)		5.02 kg
$\mathrm{ecoinv}_{tech,impact}$	**Location**	**process quantity per FU**
Diesel vehicle		
infrastructural processes		
passenger car production, diesel	GLO	1.31×10^6 kg
maintenance, passenger car	RER	1.06×10^3 units
end-of-life		
treatment of used glider, passenger car, shredding	GLO	9.09×10^5 kg
treatment of used internal combustion engine, shredding	GLO	4.00×10^5 kg
operational processes		
non-exhaust emissions		
road construction	CH	7.56×10^2 m year
treatment of brake wear emissions, passenger car	RER	6.30 kg
treatment of tire wear emissions, passenger car	RER	8.06×10^1 kg
treatment of road wear emissions, passenger car	RER	1.38×10^1 kg
market group for diesel, low-sulfur[L]	RER	5.66×10^4 kg
exhaust emissions, only direct emissions are accounted		
transport, passenger car, medium size, diesel, EURO 5[L]	RER	1.02×10^6 km

[L] process has a learning curve, see Table C.47

Table C.42: LCI of transportation sector technologies part 3. Processes without explicit source are based on ecoinvent 3.5 APOS (Wernet et al., 2016). The functional unit (FU) of infrastructural and operational processes is 1000 vehicles and 10^6 vkm (vehicle kilometers).

$\text{ecoinv}_{tech,impact}$	**Location**	**process quantity per FU**
Fuel-cell car		
infrastructural processes		
car		
glider production, passenger car	GLO	8.38×10^5 kg
electric motor production, vehicle (electric powertrain)	GLO	5.99×10^4 kg
cable production, three-conductor cable	GLO	3.39×10^3 m
power distribution unit production, for electric passenger car	GLO	4.41×10^3 kg
inverter production, for electric passenger car	GLO	1.07×10^4 kg
converter production, for electric passenger car	GLO	5.09×10^3 kg
Lithium battery (see p.199)		3.20×10^3 kWh
H_2 fuel-cell (see p.197)		8.03×10^1 MW
maintenance, passenger car, electric, without battery	GLO	1.00×10^3 unit
H_2 station		
natural gas service station construction[A]	CH	2.09×10^{-1} unit
H_2 electrolyzer (see p.197)		7.14×10^{-1} MW
hydrogen tank		
polyethylene production, high density, granulate	RER	3.72×10^4 kg
glass fibre production	RER	3.02×10^3 kg
epoxy resin production, liquid	RER	3.72×10^4 kg
steel production, chromium steel 18/8, hot rolled	RER	4.27×10^2 kg
carbon fibres for hydrogen tank		
Sohio process	RER	5.87×10^4 kg
vinyl acetate production	RER	5.22×10^3 kg
natural gas production	DE	2.01×10^5 kg
market for electricity, high voltage	DE	5.76×10^5 kWh

[A] assumption due to the lack of data

Table C.43: LCI of transportation sector technologies part 4. Processes without explicit source are based on ecoinvent 3.5 APOS (Wernet et al., 2016). The functional unit (FU) of infrastructural and operational processes is 1000 vehicles and 10^6 vkm (vehicle kilometers).

$ecoinv_{tech,impact}$	Location	process quantity per FU
Fuel-cell car continued		
vehicle end-of-life		
treatment of used glider, passenger car, shredding	GLO	8.38×10^5 kg
treatment of used powertrain for electric passenger car, manual dismantling	GLO	8.35×10^4 kg
hydrogen tank end-of-life		
steel (see Table C.3)		4.27×10^2 kg
polyethylene (see Table C.3)		3.72×10^4 kg
plastic (see Table C.3)		3.72×10^4 kg
treatment of inert waste, inert material landfill[A]	CH	3.02×10^3 kg
carbon fibers for hydrogen tank end-of-life		
treatment of inert waste, inert material landfill	CH	6.39×10^4 kg
operational processes		
vehicle		
road construction	CH	6.19×10^2 m year
treatment of brake wear emissions, passenger car	RER	5.12 kg
treatment of tire wear emissions, passenger car	RER	6.60×10^1 kg
treatment of road wear emissions, passenger car	RER	1.13×10^1 kg
Gasoline vehicle		
infrastructural processes		
passenger car production, petrol/natural gas	GLO	1.23×10^6 kg
maintenance, passenger car	RER	9.92×10^2 units
end-of-life		
treatment of used glider, passenger car, shredding	GLO	9.09×10^5 kg
treatment of used internal combustion engine, shredding	GLO	3.20×10^5 kg
operational processes		
non-exhaust emissions		
road construction	CH	7.13×10^2 m year
treatment of brake wear emissions, passenger car	RER	5.90 kg
treatment of tire wear emissions, passenger car	RER	7.60×10^1 kg
treatment of road wear emissions, passenger car	RER	1.30×10^1 kg
market for petrol, low-sulfur[L]	CH	1.56×10^3 kg
market for petrol, low-sulfur[L]	Europe w/o CH	5.55×10^4 kg
exhaust emissions, only direct emissions are accounted		
transport, passenger car, medium size, petrol, EURO 5[L]	RER	9.20×10^5 km

[L] process has a learning curve, see Table C.47
[A] for glass fibers

Table C.44: LCI of transportation sector technologies part 5. Processes without explicit source are based on ecoinvent 3.5 APOS (Wernet et al., 2016). The functional unit (FU) of infrastructural and operational processes is 1000 vehicles and 10^6 vkm (vehicle kilometers).

$\text{ecoinv}_{tech,impact}$	**Location**	**process quantity per FU**
natural gas vehicle		
infrastructural processes		
passenger car production, petrol/natural gas	GLO	1.23×10^6 kg
maintenance, passenger car	RER	9.92×10^2 units
end-of-life		
treatment of used glider, passenger car, shredding	GLO	9.09×10^5 kg
treatment of used internal combustion engine, shredding	GLO	3.20×10^5 kg
operational processes		
non-exhaust emissions		
road construction	CH	7.13×10^2 m year
treatment of brake wear emissions, passenger car	RER	5.90 kg
treatment of tire wear emissions, passenger car	RER	7.60×10^1 kg
treatment of road wear emissions, passenger car	RER	1.30×10^1 kg
market group for natural gas, high pressure^L	Europe without CH	7.27×10^4 kg
market for natural gas, high pressure^L	CH	4.70×10^2 kg
exhaust emissions, only direct emissions are accounted		
transport, passenger car, medium size, natural gas, EURO 5^L	RER	9.55×10^5 km
Plug-in hybrid electric vehicle		
infrastructural processes		
glider production, passenger car	GLO	9.13×10^5 kg
internal combustion engine production, passenger car	GLO	3.21×10^5 kg
Lithium battery (see p.199)		2.00×10^4 kWh
converter production, for electric passenger car	GLO	2.70×10^3 kg
inverter production, for electric passenger car	GLO	5.70×10^3 kg
charger production, for electric passenger car	GLO	3.72×10^3 kg
electric motor production, vehicle (electric powertrain)	GLO	3.18×10^4 kg
maintenance, passenger car	RER	1.00×10^3 kg
charger home (see p.236)		9.39×10^2 unit
charger normal (see p.236)		5.87×10^2 unit
end-of-life		
manual dismantling of used passenger car with internal combustion engine	GLO	1.23×10^6 unit
treatment of used glider, passenger car, shredding	GLO	9.13×10^5 kg
treatment of used internal combustion engine, shredding	GLO	3.21×10^5 kg
treatment of used powertrain for electric passenger car, manual dismantling	GLO	3.18×10^4 kg

^{L} process has a learning curve, see Table C.47

Table C.45: LCI of transportation sector technologies part 6. Processes without explicit source are based on ecoinvent 3.5 APOS (Wernet et al., 2016). The functional unit (FU) of infrastructural and operational processes is 1000 vehicles and 10^6 vkm (vehicle kilometers).

$ecoinv_{tech,impact}$	**Location**	**process quantity per FU**
Plug-in hybrid electric vehicle continued		
end-of-life		
treatment of used glider, passenger car, shredding	GLO	9.13×10^5 kg
treatment of used internal combustion engine, shredding	GLO	3.21×10^5 kg
treatment of used powertrain for electric passenger car, manual dismantling	GLO	3.18×10^4 kg
operational processes		
consists of 60 % battery electric vehicle (see p.235) operational processes and 40 % diesel vehicle operational[L] processes (see p.236)		
Power-to-Diesel (Fischer-Tropsch process)		
infrastructural processes		
building and foundation		
steel production, low-alloyed, hot rolled	RER	1.47×10^6 kg
oriented strand board production	RER	4.07×10^1 m3
bitumen seal production	RER	1.58×10^4 kg
titanium zinc plate production, without pre-weathering	DE	7.67×10^2 kg
glass wool mat production	CH	3.94×10^3 kg
stone wool production	CH	1.07×10^5 kg
aluminum production, primary, cast alloy slab from continuous casting	RoW	1.53×10^3 kg
concrete production 35MPa, RNA only	RoW	2.09×10^3 m3
polyethylene production, low density, granulate	RER	1.41×10^3 kg
bitumen adhesive compound production, hot	RER	1.02×10^2 kg
reinforcing steel production	RER	4.82×10^5 kg
autoclaved aerated concrete block production	CH	1.09×10^5 kg
reactors, pumps and heat exchangers		
steel production, converter, unalloyed	RER	1.50×10^5 kg
steel production, converter, low-alloyed	RER	2.68×10^5 kg
steel production, chromium steel 18/8, hot rolled	RER	7.75×10^4 kg
aluminum oxide production	IAI Area, EU27 & EFTA	1.76×10^3 kg
cobalt production	GLO	4.40×10^2 kg
nickel mine operation, sulfidic ore	GLO	1.02×10^2 kg
frit production, for ceramic tile	GLO	3.41×10^2 kg
activated silica production	GLO	4.60×10^2 kg
zirconium oxide production	RoW	2.45×10^3 kg
silica sand production	DE	2.45×10^3 kg
H_2 electrolysis (see p.197)		1.24×10^2 unit

[L] process has a learning curve, see Table C.47

Table C.46: LCI of transportation sector technologies part 7. Processes without explicit source are based on ecoinvent 3.5 APOS (Wernet et al., 2016). The functional unit (FU) of infrastructural and operational processes is 1000 vehicles and 10^6 vkm (vehicle kilometers).

$\text{ecoinv}_{tech,impact}$	**Location**	**process quantity per FU**
Power to Diesel (Fischer-Tropsch process) continued		
infrastructural processes		
pipes		
drawing of pipe, steel	RER	4.67×10^4 kg
building and foundation end-of-life		
aluminum (see Table C.3)		1.53×10^3 kg
steel (see Table C.3)		1.96×10^6 kg
treatment of waste concrete, not reinforced, recycling	Europe w/o CH	5.11×10^6 kg
treatment of waste wood, post-consumer, sorting and shredding	CH	2.47×10^4 kg
treatment of waste mineral wool, recycling	Europe w/o CH	1.10×10^5 kg
treatment of waste bitumen, sanitary landfill	Europe w/o CH	1.59×10^4 kg
polyethylene (see Table C.3)		1.41×10^3 kg
zinc (see Table C.3)		7.67×10^2 kg
reactors, pumps and heat exchangers end-of-life		
steel (see Table C.3)		4.95×10^5 kg
treatment of inert waste, inert material landfill	CH	7.47×10^3 kg
nickel (see Table C.3)		1.02×10^2 kg
cobalt (see Table C.3)		4.40×10^2 kg
pipes end-of-life		
steel (see Table C.3)		4.67×10^4 kg
operational processes		
market group for diesel, low-sulfur	RER	-8.32×10^2 kg[A]

[L] process has a learning curve, see Table C.47
[A] negative value accounted as negative emissions in the overall emissions balance

As already mentioned in the beginning of Section C.4.4, all environmental impacts are calculated for the base year 2016. Because certain technologies in the transport sector are subject to learning curves due to technological changes in the future, different amounts for 'process quantity' are build in every year for the same process $ecoinv_{tech,impact}$. These learning curves are displayed in Table C.47. At the end of every row, the page is mentioned where each learning curve is used, for easier document navigation.

Table C.47: learning curves of transportation sector technologies

$ecoinv_{tech,impact}$	**2016**	**2020**	**2025**	**2030**
battery electric vehicle operational processes	1	1	1.0203	1.0405
transport, passenger car, medium size, diesel, EURO 5	1017550	1017550	979860	942180
transport, passenger car, medium size, petrol, EURO 5	920032	920032	894473	868915
transport, passenger car, medium size, natural gas, EURO 5	954519	954519	929835	905151
lithium battery (battery electric vehicle)	76000	76000	82000	88000
market group for diesel, low-sulfur	5662.58	56624.58	54527.21	52430.4
market for petrol. low-sulfur [CH]	1557.12	1557.12	1513.86	1470.6
market for petrol. low-sulfur [EU w/o CH]	55546.33	55546.33	54003.25	52460.18
market for natural gas, high pressure [CH]	469.62	469.62	457.48	445.33
market group for natural gas, high pressure [EU w/o CH]	72735.48	72735.48	70854.54	68973.6

$ecoinv_{tech,impact}$	**2035**	**2040**	**2045**	**2050**
battery electric vehicle operational processes	1.0608	1.0811	1.1013	1.1216
transport, passenger car, medium size, diesel, EURO 5	904490	866800	829109	791430
transport, passenger car, medium size, petrol, EURO 5	843366	817807	792249	766690
transport, passenger car, medium size, natural gas, EURO 5	880458	855774	831090	806406
lithium battery (battery electric vehicle)	94000	100000	106000	112000
market group for diesel, low-sulfur	50333.03	48235.65	46138.28	44041.47
market for petrol. low-sulfur [CH]	1427.36	1384.1	1340.85	1297.59
market for petrol. low-sulfur [EU w/o CH]	50917.66	49374.58	47831.5	46288.42
market for natural gas, high pressure [CH]	433.18	421.04	408.9	396.75
market group for natural gas, high pressure [EU w/o CH]	67091.93	65210.99	63330.05	61449.11

Appendix D

Additional notes on DeLoop

In DeLoop, critical time steps can be added to each subproblem to improve performance, step(ii)(4). Critical time steps are case-study specific and often unknown beforehand. For the network-connection fee and the peak-power price, the peak-power demand represents a critical time step. Thus, in this version of DeLoop, we add the time step with the expected peak-power demand as a critical time step. Here, we employ a heuristic based on time series analysis and component characteristics to identify the time step with the expected peak-power demand. The heuristic neglects all storage units because storage levels are unknown beforehand. The goal of the heuristic is to approximate the minimum peak-power demand from the electricity grid while satisfying the electricity and thermal energy demand.
The approximation of the minimum power demand $P_t^{el,calc}$ is calculated for all time steps t (Eq. (D.1)):

$$
\begin{aligned}
P_t^{el,calc} &= \dot{E}_t^{el} - \sum_{n\in\mathcal{PV}} \left(\dot{V}_{n,t}^{el,calc}\right) + \dot{V}_t^{cool} - \dot{V}_t^{heating} \\
\dot{V}_t^{cool} &= \frac{\dot{V}_t^{cool-CC,calc}}{\max_{n\in\mathcal{CC}}(\eta_n^N)} = \frac{\dot{E}_t^{cool} - \dot{V}_t^{cool-AbC,calc}}{\max_{n\in\mathcal{CC}}(\eta_n^N)}, \\
\dot{V}_t^{heating} &= \dot{V}_t^{heating,calc} \cdot \max_{n\in\mathcal{CHP}} \left(\frac{\eta_n^{N,el}}{\eta_n^{N,th}}\right) \\
&= \min\left\{ \dot{E}_t^{heat,lt} + \frac{\dot{V}_t^{cool-AbC,calc}}{\max_{n\in\mathcal{A}\mathrm{b}\mathcal{C}}(\eta_n^N)}, \sum_{n\in\mathcal{CHP}} \dot{V}_n^N \right\} \cdot \max_{n\in\mathcal{CHP}} \left(\frac{\eta_n^{N,el}}{\eta_n^{N,th}}\right), \\
\dot{V}_t^{cool-AbC,calc} &= \min\left\{ \dot{E}_t^{cool}, \sum_{n\in\mathcal{A}\mathrm{b}\mathcal{C}} \dot{V}_n^N \right\}.
\end{aligned}
\tag{D.1}
$$

In Eq. (D.1), we use subsets to describe all absorption chillers by $\mathcal{A}\mathrm{b}\mathcal{C}$, all compression by $\mathcal{CC}$, all combined heat and power engines by $\mathcal{CHP}$, and the subset of all photovoltaics by $\mathcal{PV}$. To minimize the peak-power demand, power-generating units

($\mathcal{CHP}$ and $\mathcal{PV}$) are assumed to operate at the highest possible output level. At the same time, the use of power-consuming compression chillers is reduced to a minimum. The time step with the highest minimum electricity demand $P_t^{el,calc}$ is selected as the critical time step representative. The maximal possible output power of the PV units $\sum\limits_{n \in PV} \left(\dot{V}_{n,t}^{el,calc} \right)$ is a time-dependent parameter and thus known prior to the optimization. The CHP power output depends on the amount of low-temperature heat supplied by the CHP $\dot{V}_t^{heating,calc}$. The power generated by the combined heat and power engine is approximated using the maximal ratio of the nominal thermal and electrical efficiency $\max_{n \in \mathcal{CHP}} \left(\frac{\eta_n^{N,el}}{\eta_n^{N,th}} \right)$. Additional demand for low-temperature heat can be generated by using absorption chillers to supply the cooling demand $\dot{V}_t^{cool-AbC,calc}$. The absorption chillers reduce the amount of cooling energy provided by the compression chillers $\dot{V}_t^{cool-CC,calc}$ and thus the electricity demand of the compression chillers. The absorption chillers are hence operated at the highest output rate possible.

We calculate the electricity demand of the compression chillers for providing the remaining cooling energy $\dot{V}_t^{cool-CC,calc}$ with the maximal nominal efficiency $\max_{n \in \mathcal{CC}}(\eta_n^N)$. Then, the time step with the highest approximated minimum peak-power demand $P_t^{el,calc}$ is added to the subproblems as critical time step in step(ii)(4).

Appendix E

Additional notes on RiSES

E.1 Heuristic for the adaptation of the time resolution in RiSES

In the master branch (4.2.1), we increase the time resolution of either the number of periods N_p or the number of segments N_s until the optimality gap $\varepsilon_{\text{RiSES}}$ is satisfied:

$$\varepsilon_{\text{RiSES}} \geq \varepsilon = \frac{TAC^{\text{UB}} - TAC^{\text{LB}}}{TAC^{\text{UB}}}. \tag{E.1}$$

We employ a heuristic based on finite backward differences to choose whether to increase the number of periods N_p or the number of segments N_s (Bahl et al., 2018b). We calculate the finite backward differences $\delta(\varepsilon)$ of the optimality gap ε for the number of periods N_p by

$$\delta(\varepsilon)^p = \frac{\varepsilon(N_p - 1, N_s) - \varepsilon(N_p, N_s)}{N_p \times N_s - ((N_p - 1) \times N_s)}, \tag{E.2}$$

and the number of segments N_s by

$$\delta(\varepsilon)^s = \frac{\varepsilon(N_p, N_s - 1) - \varepsilon(N_p, N_s)}{N_p \times N_s - (N_p \times (N_s - 1))}. \tag{E.3}$$

In the numerator, the difference of the gap of the previous gap $\varepsilon(N_p - 1, N_s)$ to the current iteration loop $\varepsilon(N_p, N_s)$ is calculated. This difference is divided by the difference of the current number of time steps $N_p \times N_s$ to the previously considered time steps $(N_p - 1) \times N_s$. If $\delta(\varepsilon)^p \geq \delta(\varepsilon)^s$, the number of periods N_p is increased, else the number of segments N_s is increased, as the method aims for small optimality gaps ε. For the **R&A** branch (4.2.1), we employ $\delta(TAC^{\text{LB}})$ instead of $\delta(\varepsilon)$. Here, if $\delta(TAC^{\text{LB}})^p \leq \delta(TAC^{\text{LB}})^s$ the number of periods N_p is increased, else the number of segments N_s in increased, as the method aims for high TAC^{LB} as lower bounds.

To calculate the backward difference, we need to initialize the method. To initialize RiSES[3], we calculate the solutions for ($N_p = 1$, $N_s = 2$), ($N_p = 2$, $N_s = 1$) and ($N_p = 2$, $N_s = 2$). If a feasible solution for the initialization of the heuristic is missing, the number of segments N_s or the number of periods N_p is increased until the finite backward differences $\delta(\varepsilon)$ can be calculated. After each iteration, the solution from the previous run is also used in the next iteration to calculate $\delta(\varepsilon)$. For example, if in the beginning the number of periods is increased from $N_p = 2$ to $N_p = 3$, the solution $\delta(\varepsilon(N_p = 3, N_s = 1))$ is unknown. To calculate $\delta(\varepsilon)^s$, the solution of the previous run $\delta(\varepsilon(N_p = 2, N_s = 1))$ is used instead of the unknown $\delta(\varepsilon(N_p = 3, N_s = 1))$. For further information, see Bahl et al. (2018b).

For unchronological time steps, the increase of the time resolution is trivial, as no segments exist. In this case, the number of typical periods N_p, each with a length of 1 time step, is increased in every iteration and no initialization is necessary. Employing this method, the time resolution of the time-series aggregation in the master branch (4.2.1) and the **R&A** branch (4.2.3) is increased.

E.2 Examples for the aggregated time series used in RiSES

As example for the aggregated time series of case study ***ES***, the highest time resolution in the master branch with 246 aggregated time steps is shown in Fig. E.1. In general, the demands and electricity prices are well fitted. In particular, the electricity and low-temperature heating demand is fitted more accurately compared to the aggregated time-series consisting of only 13 time steps, compare Fig. E.1 and Fig. 4.9.

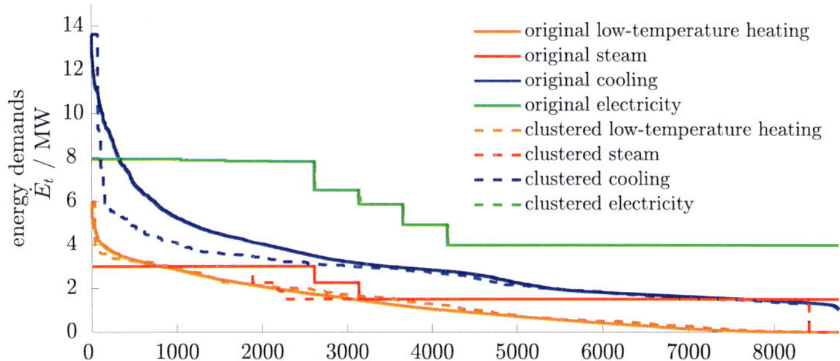

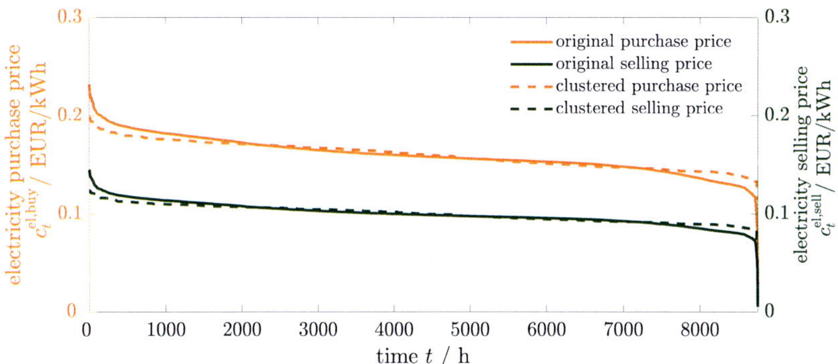

Figure E.1: Sorted time series of case study ***ES*** for low temperature heat, steam, cooling, and electricity demands for 1 year with a time step length of $\Delta t_t = 2\,\text{h}$. Additionally, the used aggregated time series with 246 time steps (P=246) for the best found design of RiSES[3] within the time limit is shown. Time series for wind speed, solar radiation, and temperature are omitted for simplicity.

As example for the aggregated time series of case study ***PS***, the highest time resolution in the master branch with 39 aggregated time steps is shown in Fig. E.2. The volume flow is fitted very accurately, compare Fig. E.2 and Figure Fig. 4.14. However, the resulting design is equivalent to the first design employing the aggregated time series with 4 aggregated time steps.

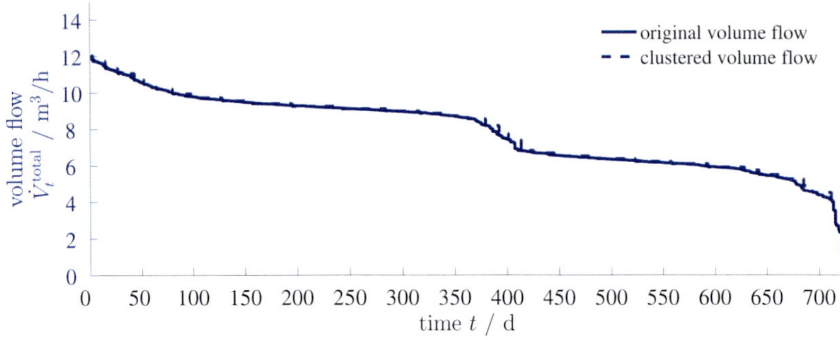

Figure E.2: Sorted time series of the volume flow rate $\dot{V}_t^{total}$ for case study ***PS***. The time series consists of two years with a time step length of 1 day. Additionally, the used aggregated time series with 39 time steps (P=39) for the best found design of RiSES[3] within the time limit is shown.

Appendix F

Additional notes on Chapter 5

This appendix consists of 3 parts presenting further details on Chapter 5:

- F.1: Data from ecoinvent
- F.2: Data from ENTSO-E
- F.3: Logarithmic efficiency regression

In F.1, for the calculation of the time-dependent emission factors, we need emission data for each type of electricity generator present in the investigated electricity system. We present the assumptions and the assignment of the emission data to the generator types. We use emission data from ecoinvent version 3.3 Wernet et al. (2016).
In F.2, to calculate the time-dependent emission factors, we additionally use time-resolved data of the electricity production per generator type and electricity imports and exports per neighboring country. We show how we handle data gaps in the original data from ENTSO-E (2017).
In F.3, for the calculation of the time-dependent marginal emissions factor we require a merit order. To determine the merit order, power plant-specific costs and emissions are needed which both depend on the power plant's efficiency. Here, we present the employed size-dependent regression for the power plants efficiency from Heuck et al. (2013).

F.1 Data from ecoinvent

For the calculation of the time-dependent emission factors, Section 3.1 and Section 3.2, data from ecoinvent 3.3 Wernet et al. (2016) is matched to the reported energy production from ENTSO-E (2017). Here, the cradle-to-grave emission $\mathrm{EF}_{\mathrm{el},k}$ data from eco-invent (Wernet et al., 2016) for the energy carriers k is assigned to the net energy production $P_{\mathrm{net},k,t}$ per energy carrier k from the ENTSO-E database as shown in Table F.1. We use the 'Allocation At Point Of Substitution' method to obtain the global warming potential for 100 years GWP_{100} from the ecoinvent version 3.3.

Table F.1: Assignment of the cradle-to-grave emission $EF_{el,k}$ data from ecoinvent (Wernet et al., 2016) to the energy carriers k from the net energy production $P_{net,k,t}$.

energy carriers k from ENTSO-E	**assignment/ assumptions**
biomass	DE, electricity production, heat and power co-generation, wood chips, 6667 kW, state-of-the-art 2014
fossil coal-derived gas	DE, treatment of coal gas, in power plant
fossil hard coal	DE, electricity production, hard coal
fossil gas	The emission factor is based on the weighted sum for following generation types based on the installed capacity for every utility $\geq 100\,MW_{el}$ (Umweltbundesamt, 2018). DE, heat and power co-generation, natural gas, combined cycle power plant, 400MW electrical DE, heat and power co-generation, natural gas, conventional power plant, 100MW electrical DE, natural gas, burned in gas turbine, for compressor station DE, electricity production, natural gas, combined cycle power plant DE, electricity production, natural gas, conventional power plant.
fossil lignite	DE, electricity production, lignite
fossil oil	DE, electricity production, oil
geothermal	DE, electricity production, deep geothermal
hydro pumped storage	DE, electricity production, hydro, pumped storage
hydro run-of-river and poundage	DE, electricity production, hydro, run-of-river
hydro water reservoir	DE, electricity production, hydro, reservoir, non-alpine region
nuclear	DE, electricity production, pressure water reactor 77.06 % and *Electricity, high voltage, DE, electricity production, boiling water reactor* 22.94 % - based on shares of the electricity production in 2016 (DAtF, 2017).

Table F.1: Assignment of the cradle-to-grave emission $EF_{el,k}$ data from ecoinvent (Wernet et al., 2016) to the energy carriers k from the net energy production $P_{net,k,t}$.

energy carriers k from ENTSO-E	**assignment/ assumptions**
other	The average non-renewable technology mix is used for the emission factor.
other renewable	The average renewable technology mix is used for the emission factor.
solar	DE, electricity production, photovoltaic, 3kWp slanted-roof installation, single-Si, panel, mounted
waste	DE, electricity, from municipal waste incineration to generic market for electricity, medium voltage
wind offshore	DE, electricity production, wind, 1-3MW turbine, off-shore
wind onshore	DE, electricity production, wind, 1-3MW turbine, on-shore

Employing the assignment from Table F.1, the emissions of the energy production from ENTSO-E (2017) can be calculated.

F.2 Data from ENTSO-E

For the calculation of time-dependent emission factors, Section 3.1 and Section 3.2, data from ENTSO-E (2017) is used. The data has some gaps, we informed ENTSO-E (2017) about every gap in the original data. In most cases, we gratefully received the missing data sets from the administrators of ENTSO-E (2017). For few sets, no correction was possible.
We used the following data sets for Germany with assumptions and corrections as listed below :

1. Export and import electricity in hourly resolution for 2016 between Germany and neighboring countries.
   For the electricity import and export with Luxembourg, no data is available. Thus, the electricity exchange with Luxembourg is neglected. Additionally,

Table F.2 states the periods for which no data was available and how the gaps are filled for the emission calculation.

Table F.2: Periods in which no export and import data was available from ENTSO-E (2017) and how the gap is filled for the emission calculation

country	**period in 2016**	**data used for the emission calculation**
all countries	27.03, 02 h - 03 h	Interpolation between existing data points
Czech Republic	10.08, 21 h - 22 h; 08.09, 11 h - 12 h; 20.09, 12 h - 13 h ; 02.10, 08 h - 09 h; 05.10, 15 h - 16 h; 06.10, 23 h - 24 h; 07.10, 13 h - 14 h; 28.10, 06 h - 07 h; 28.10, 13 h - 14 h; 23.11, 02 h - 03 h; 23.11, 03 h - 04 h	interpolation between existing data points
Denmark	26.09, 19 h - 20 h; 04.10, 13 h - 14 h; 05.10, 15 h - 16 h; 06.10, 14 h - 15 h; 06.10, 23 h - 24 h; 15.10, 21 h - 22 h, 26.10 10 h - 11 h; 03.11, 16 h - 17 h	interpolation between existing data points
Sweden	27.11 0 h - 24 h	data of the previous week is used instead

2. Net electricity production in 15 minute resolution for 2016 per energy carrier k. In ENTSO-E (2017), the aggregated German national value for production from the energy carrier 'fossil oil' is missing. Instead, we used the sum of the reported aggregated production from the energy carrier 'fossil oil' of all transmission system operators (TSOs) for the German national area. A comparison with the other energy carriers shows that ENTSO-E (2017) internally uses the data of the TSOs in order to aggregate national values as well. Additionally, Table F.3 states the periods in which no electricity production data was available and how the gaps are filled in the emission calculation.

Table F.3: Periods in which no data was available from ENTSO-E (2017) for the net electricity production per aggregated energy carrier

period in 2016	**data used for the emission calculation**
30.10, 0 h - 1 h; 30.10, 23 h - 24 h	interpolation between existing data points
27.03, 0 h - 24 h	data of the previous week is used instead
10.03, $16^{\underline{45}}$ h - 14.03, $17^{\underline{15}}$ h; 06.10, $21^{\underline{15}}$ h - 07.10, $11^{\underline{45}}$ h	data of previous week is used instead

With these corrections in the data, the time-dependent emission factors in Section 3.1 and Section 3.2 are calculated.

F.3 Logarithmic efficiency regression

To determine the merit order in Section 3.2, the marginal costs c_p^{m} and the power plant-specific emissions $\mathrm{EF}_{\mathrm{el},p}$ are needed. Both, the marginal costs $c_{m,p}$ and the power plant-specific emissions $\mathrm{EF}_{\mathrm{el},p}$, depend on the efficiency $\eta_{\mathrm{el},p}$, see Eq. (17) and Eq. (18). In the power plant list of the Umweltbundesamt (2018) only the capacity is given. Thus, we use the mean function of a logarithmic size-dependent regression of existing power plants for the efficiency based on Heuck et al. (2013). For nuclear power plants, we use a size-independent efficiency as in Heuck et al. (2013) with a value of 32 % (Wernet et al., 2016).

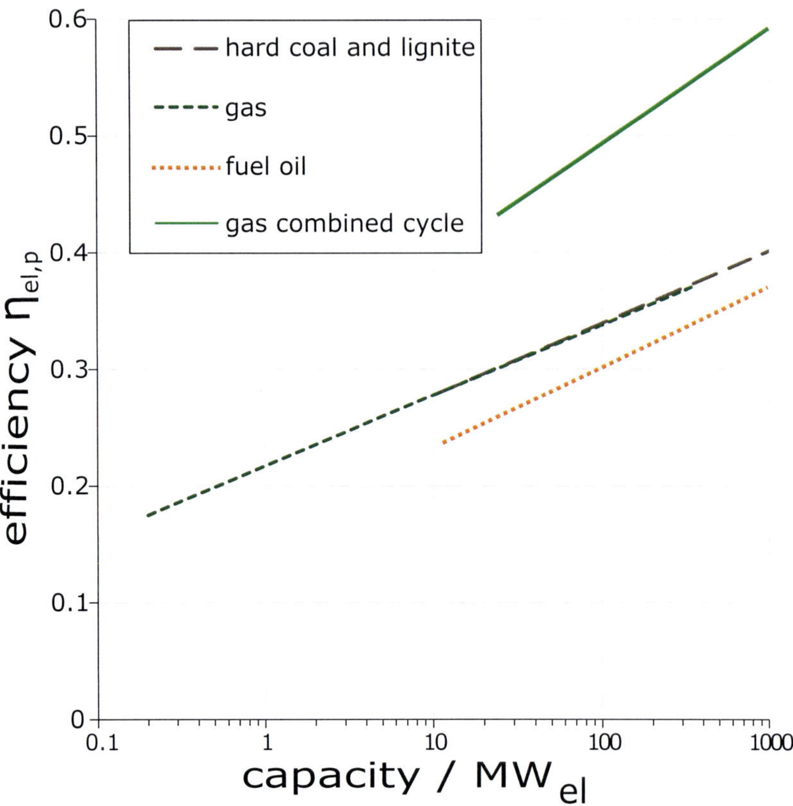

Figure F.1: Logarithmic size-dependent regression for the efficiency of existing power plants. For the conventional technologies based on the energy carriers lignite, hard coal, gas (combined cycle), gas and fuel oil, from Heuck et al. (2013)

Appendix G

Additional LCA results of SecMOD

In Chapter 6, we compute an optimal transition pathway for the German energy system towards a low-carbon future. Further we extend the model by a life-cycle assessment (LCA). In this section, we provide additional LCA results in the following tables.

Table G.1: Environmental impacts of the electricity sector per generated MWh.

environmental impacts	2016	2020	2030	2040	2050	unit
climate change	4.20×10^{-1}	2.81×10^{-1}	2.17×10^{-1}	1.53×10^{-1}	1.18×10^{-1}	t CO_{2eq}
ozone depletion	4.13×10^{-5}	4.23×10^{-5}	2.68×10^{-5}	2.10×10^{-5}	1.63×10^{-5}	kg CFC-11eq
particulate matter	8.47×10^{-6}	9.01×10^{-6}	6.70×10^{-6}	6.81×10^{-6}	6.03×10^{-6}	Disease incidences
acidication, freshwater and terrestrial	1.50	1.40	1.02	8.70×10^{-1}	7.03×10^{-1}	Mole of H+ eq
eutrophication, freshwater	5.59×10^{-1}	2.21×10^{-1}	1.69×10^{-1}	8.22×10^{-2}	6.46×10^{-2}	kg P eq
eutrophication, marine	3.10×10^{-1}	2.11×10^{-1}	1.61×10^{-1}	1.52×10^{-1}	1.72×10^{-1}	kg N eq
eutrophication, terrestrial	4.98	5.05	3.49	2.68	1.84	Mole of N eq
ionizing radiation	7.78×10^{1}	8.68×10^{1}	3.60	5.47	6.33	kg ^{235}U eq
photochemical ozone formation	5.09×10^{-1}	4.04×10^{-1}	3.38×10^{-1}	3.42×10^{-1}	3.27×10^{-1}	kg NMVOC eq
human toxicity, cancer	3.23×10^{-6}	2.69×10^{-6}	2.89×10^{-6}	4.24×10^{-6}	4.71×10^{-6}	CTUh
human toxicity, non-cancer	2.20×10^{-5}	1.68×10^{-5}	1.66×10^{-5}	2.31×10^{-5}	2.68×10^{-5}	CTUh
ecotoxicity, freshwater	9.66×10^{1}	8.01×10^{1}	7.66×10^{1}	1.07×10^{2}	1.24×10^{2}	CTUe
land use	1.64×10^{3}	1.68×10^{3}	2.06×10^{3}	3.48×10^{3}	3.00×10^{3}	Pt
water scarcity	7.23×10^{1}	6.48×10^{1}	4.45×10^{1}	6.28×10^{1}	6.46×10^{1}	m^3
resource depletion, energy	7.39×10^{3}	5.86×10^{3}	3.23×10^{3}	2.31×10^{3}	1.76×10^{3}	MJ
resource depletion, mineral and metals	4.68×10^{-4}	5.25×10^{-4}	7.63×10^{-4}	1.94×10^{-3}	2.90×10^{-3}	kg Sb eq

Table G.2: Share in % of the infrastructural environmental impacts in the electricity sector.

environmental impacts	2016	2020	2030	2040	2050
climate change	3.04	4.97	10.23	33.00	55.28
ozone depletion	2.94	3.14	7.99	23.82	40.62
particulate matter	11.59	11.75	24.27	50.46	70.02
acidication, freshwater and terrestrial	5.60	6.58	14.03	37.27	60.08
eutrophication, freshwater	1.47	4.19	8.72	44.32	82.12
eutrophication, marine	5.21	8.96	18.19	54.78	80.41
eutrophication, terrestrial	3.21	3.44	7.72	22.20	41.37
ionizing radiation	1.44	1.41	54.57	82.86	91.50
photochemical ozone formation	10.59	14.48	27.10	58.63	76.89
human toxicity, cancer	33.43	44.60	65.00	86.79	93.84
human toxicity, non-cancer	20.60	30.16	49.57	80.19	91.20
ecotoxicity, freshwater	22.78	30.74	51.41	80.07	90.86
land use	49.70	50.02	71.38	88.29	91.66
water scarcity	15.55	18.56	46.20	77.88	88.63
resource depletion, energy	2.40	3.31	9.73	30.97	51.85
resource depletion, mineral and metals	74.61	76.65	88.48	96.71	98.48

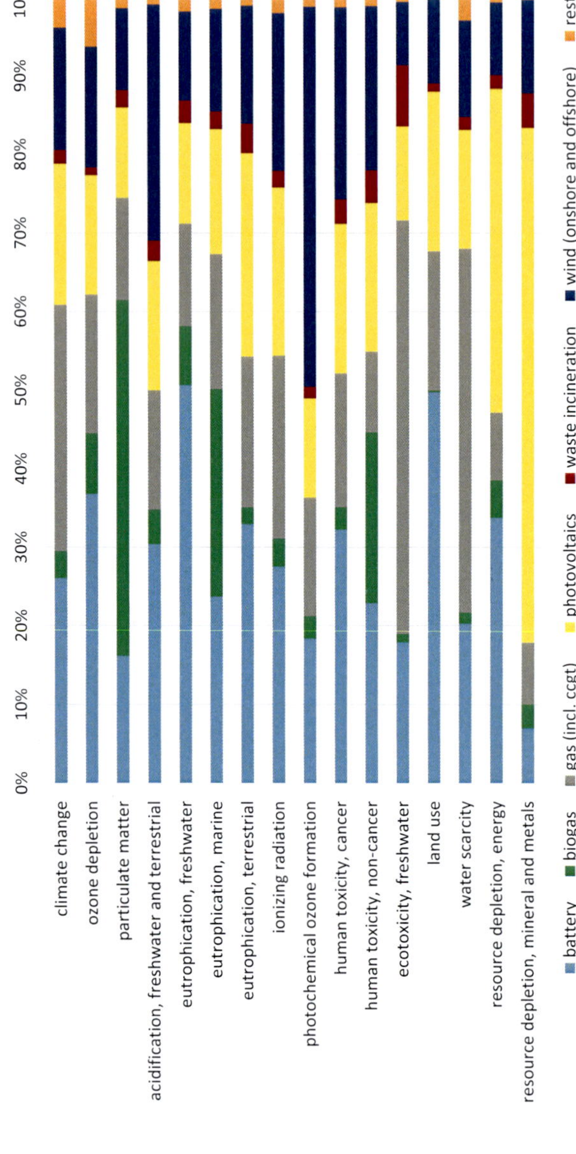

Figure G.1: Contribution of the environmental impacts of the electricity sector for the year 2050

Table G.3: Environmental impacts of the heating sector per generated MWh.

environmental impacts	2016	2020	2030	2040	2050	unit
climate change	2.55×10^{-1}	2.49×10^{-1}	1.53×10^{-1}	1.06×10^{-1}	8.45×10^{-2}	t CO_{2eq}
ozone depletion	3.35×10^{-5}	3.16×10^{-5}	1.79×10^{-5}	1.14×10^{-5}	1.08×10^{-5}	kg CFC-11 eq
particulate matter	1.84×10^{-6}	1.69×10^{-6}	1.63×10^{-6}	1.62×10^{-6}	2.08×10^{-6}	Disease incidences
acidication, freshwater and terrestrial	3.61×10^{-1}	3.36×10^{-1}	2.77×10^{-1}	2.29×10^{-1}	2.83×10^{-1}	Mole of H+ eq
eutrophication, freshwater	8.75×10^{-3}	8.22×10^{-3}	2.02×10^{-2}	1.48×10^{-2}	2.49×10^{-2}	kg P eq
eutrophication, marine	5.34×10^{-2}	5.05×10^{-2}	4.28×10^{-2}	3.98×10^{-2}	5.41×10^{-2}	kg N eq
eutrophication, terrestrial	5.67×10^{-1}	5.36×10^{-1}	6.15×10^{-1}	5.59×10^{-1}	5.76×10^{-1}	Mole of N eq
ionizing radiation	7.28	6.43	3.38	2.09	3.39	kg ^{235}U eq
photochemical ozone formation	2.49×10^{-1}	2.39×10^{-1}	1.65×10^{-1}	1.36×10^{-1}	1.59×10^{-1}	kg NMVOC eq
human toxicity, cancer	9.97×10^{-7}	9.25×10^{-7}	8.62×10^{-7}	9.86×10^{-7}	1.73×10^{-6}	CTUh
human toxicity, non-cancer	5.74×10^{-6}	5.35×10^{-6}	4.92×10^{-6}	5.41×10^{-6}	1.05×10^{-5}	CTUh
ecotoxicity, freshwater	6.24×10^{1}	6.14×10^{1}	4.30×10^{1}	3.85×10^{1}	5.73×10^{1}	CTUe
land use	1.23×10^{2}	1.17×10^{2}	2.83×10^{2}	5.19×10^{2}	7.31×10^{2}	Pt
water scarcity	5.64	5.28	7.75	1.09×10^{1}	1.50×10^{1}	
resource depletion, energy	3.68×10^{3}	3.61×10^{3}	2.22×10^{3}	1.55×10^{3}	1.46×10^{3}	MJ
resource depletion, mineral and metals	1.60×10^{-4}	1.41×10^{-4}	1.87×10^{-4}	3.43×10^{-4}	1.04×10^{-3}	kg Sb eq

Table G.4: Share in % of the infrastructural environmental impacts in the heating sector.

environmental impacts	2016	2020	2030	2040	2050
climate change	0.80	0.78	3.38	5.94	14.48
ozone depletion	0.56	0.55	2.22	4.15	10.19
particulate matter	9.11	9.27	22.56	28.33	43.81
acidication, freshwater and terrestrial	5.58	5.58	15.05	20.07	40.19
eutrophication, freshwater	34.45	32.49	15.97	20.65	56.00
eutrophication, marine	5.13	5.03	13.70	20.25	31.92
eutrophication, terrestrial	4.79	4.83	11.92	16.39	32.08
ionizing radiation	3.26	3.27	14.17	24.42	49.33
photochemical ozone formation	3.49	3.41	12.15	18.80	32.45
human toxicity, cancer	38.24	35.91	37.14	29.68	51.56
human toxicity, non-cancer	31.03	29.17	33.40	28.08	53.91
ecotoxicity, freshwater	10.50	9.38	15.63	17.12	40.54
land use	15.72	16.23	22.26	12.55	36.62
water scarcity	14.42	14.19	23.78	18.77	30.06
resource depletion, energy	0.76	0.73	2.99	5.21	11.88
resource depletion, mineral and metals	74.69	73.86	56.86	28.30	58.22

Table G.5: Share in % of the environmental impacts due to used electricity in the heating sector.

environmental impacts	2016	2020	2030	2040	2050
climate change	0.00	0.00	11.90	17.81	9.75
ozone depletion	0.00	0.00	12.54	22.67	27.90
particulate matter	0.00	0.00	34.37	51.69	64.11
acidication, freshwater and terrestrial	0.00	0.00	30.77	46.84	60.51
eutrophication, freshwater	0.00	0.00	70.05	68.23	82.04
eutrophication, marine	0.00	0.00	31.43	46.97	63.68
eutrophication, terrestrial	0.00	0.00	47.54	59.04	62.96
ionizing radiation	0.00	0.00	8.92	32.25	61.68
photochemical ozone formation	0.00	0.00	17.15	31.02	46.48
human toxicity, cancer	0.00	0.00	28.14	52.97	74.78
human toxicity, non-cancer	0.00	0.00	12.78	56.76	77.89
ecotoxicity, freshwater	0.00	0.00	28.20	52.53	77.07
land use	0.00	0.00	61.08	82.71	88.20
water scarcity	0.00	0.00	48.11	70.79	79.67
resource depletion, energy	0.00	0.00	12.21	18.35	24.01
resource depletion, mineral and metals	0.00	0.00	34.27	69.59	90.11

Table G.6: Environmental impacts of the private transportation sector per vkm.

environmental impacts	**2016**	**2020**	**2030**	**2040**	**2050**	**unit**
climate change	2.80×10^{-4}	2.75×10^{-4}	2.58×10^{-4}	2.75×10^{-4}	1.76×10^{-4}	t CO_{2eq}
ozone depletion	5.73×10^{-8}	5.69×10^{-8}	5.35×10^{-8}	9.89×10^{-8}	8.09×10^{-8}	kg CFC-11 eq
particulate matter	1.04×10^{-8}	1.05×10^{-8}	1.05×10^{-8}	1.28×10^{-8}	1.36×10^{-8}	Disease incidences
acidication, freshwater and terrestrial	1.07×10^{-3}	1.17×10^{-3}	1.22×10^{-3}	2.73×10^{-3}	2.83×10^{-3}	Mole of H+ eq
eutrophication, freshwater	4.22×10^{-5}	4.15×10^{-5}	4.16×10^{-5}	2.04×10^{-4}	2.50×10^{-4}	kg P eq
eutrophication, marine	2.55×10^{-4}	3.18×10^{-4}	3.63×10^{-4}	2.41×10^{-4}	2.56×10^{-4}	kg N eq
eutrophication, terrestrial	2.75×10^{-3}	3.44×10^{-3}	3.92×10^{-3}	2.80×10^{-3}	2.21×10^{-3}	Mole of N eq
ionizing radiation	2.24×10^{-2}	2.22×10^{-2}	2.12×10^{-2}	1.49×10^{-2}	1.41×10^{-2}	kg ^{235}U eq
photochemical ozone formation	9.28×10^{-4}	1.06×10^{-3}	1.15×10^{-3}	9.57×10^{-4}	1.14×10^{-3}	kg NMVOC eq
human toxicity, cancer	3.96×10^{-9}	3.94×10^{-9}	3.98×10^{-9}	6.52×10^{-9}	7.88×10^{-9}	CTUh
human toxicity, non-cancer	2.72×10^{-8}	2.75×10^{-8}	2.78×10^{-8}	6.11×10^{-8}	7.06×10^{-8}	CTUh
ecotoxicity, freshwater	5.86×10^{-1}	5.85×10^{-1}	5.94×10^{-1}	9.71×10^{-1}	1.04	CTUe
land use	2.13	2.15	2.16	3.09	2.80	Pt
water scarcity	2.65×10^{-2}	2.62×10^{-2}	2.61×10^{-2}	4.83×10^{-2}	6.13×10^{-2}	m^3
resource depletion, energy	4.09	4.04	3.82	4.86	2.69	MJ
resource depletion, mineral and metals	2.78×10^{-6}	2.78×10^{-6}	2.84×10^{-6}	1.57×10^{-5}	1.91×10^{-5}	kg Sb eq

Table G.7: Share in % of the infrastructural environmental impacts in the transportation sector.

environmental impacts	**2016**	**2020**	**2030**	**2040**	**2050**
climate change	19.37	19.62	21.32	36.3	78.82
ozone depletion	7.34	7.39	8.04	65.65	91.09
particulate matter	35.25	34.72	35.48	57.02	72.44
acidication, freshwater and terrestrial	30.65	28.02	27.48	83.57	94.94
eutrophication, freshwater	88.93	89.44	90.62	91.36	96.66
eutrophication, marine	21.67	17.29	15.47	56.04	85.35
eutrophication, terrestrial	19.87	15.81	14.14	47.61	80.85
ionizing radiation	25.40	25.67	27.51	61.26	82.00
photochemical ozone formation	23.38	20.25	19.12	58.62	88.53
human toxicity, cancer	87.45	87.14	87.59	82.34	91.04
human toxicity, non-cancer	64.01	63.49	64.2	81.17	89.22
ecotoxicity, freshwater	75.66	75.54	76.09	85.79	92.31
land use	17.05	16.92	17.2	26.47	38.03
water scarcity	74.59	75.11	76.71	66.15	83.53
resource depletion, energy	19.10	19.25	20.75	29.86	74.13
resource depletion, mineral and metals	98.51	98.63	98.82	97.51	98.16

Table G.8: Share in % of the environmental impacts due to used electricity in the transportation sector.

environmental impacts	**2016**	**2020**	**2030**	**2040**	**2050**
climate change	0.00	0.00	0.00	10.52	7.79
ozone depletion	0.00	0.00	0.00	4.03	2.34
particulate matter	0.00	0.00	0.00	10.09	5.17
acidication, freshwater and terrestrial	0.00	0.00	0.00	6.04	2.89
eutrophication, freshwater	0.00	0.00	0.00	7.63	3.00
eutrophication, marine	0.00	0.00	0.00	11.93	7.83
eutrophication, terrestrial	0.00	0.00	0.00	18.12	9.68
ionizing radiation	0.00	0.00	0.00	6.96	5.20
photochemical ozone formation	0.00	0.00	0.00	6.77	3.35
human toxicity, cancer	0.00	0.00	0.00	12.33	6.94
human toxicity, non-cancer	0.00	0.00	0.00	7.16	4.41
ecotoxicity, freshwater	0.00	0.00	0.00	2.09	1.38
land use	0.00	0.00	0.00	21.39	12.45
water scarcity	0.00	0.00	0.00	24.63	12.25
resource depletion, energy	0.00	0.00	0.00	9.00	7.58
resource depletion, mineral and metals	0.00	0.00	0.00	2.34	1.76

Appendix H

Publications and student theses

List of publications

Journal papers

Baumgärtner, N., Shu, D., Bahl, B., Hennen, M., Hollermann, D., and Bardow, A. (2020). DeLoop: Decomposition-based Long-term Operational Optimization of energy systems with time-coupling constraints. *Energy*, 198, 117272.

Baumgärtner, N., Bahl, B., Hennen, M., and Bardow, A. (2019). RiSES3: Rigorous Synthesis of Energy Supply and Storage Systems via time-series relaxation and aggregation. *Computers & Chemical Engineering*, 127, 127-139.

Baumgärtner, N., Delorme, R., Hennen, M., and Bardow, A. (2019). Design of low-carbon utility systems: Exploiting time-dependent grid emissions for climate-friendly demand-side management. *Applied Energy*, 247, 755-765.

Baumgärtner, N., Deutz, S., Reinert, C., Nolzen, N., Küpper, E., Hennen, M., Hollermann, D., and Bardow, A. (2020). Life-cycle assessment of sector-coupled national energy systems: Environmental impacts of future electricity, heat & transportation in Germany. *submitted*.

Peer-reviewed conference papers

Baumgärtner, N., Temme, F., Bahl, B., Hennen, M., and Bardow, A. (2019). RiSES4: Rigorous Synthesis of Energy Supply Systems with Seasonal Storage by relaxation and time-series aggregation to typical periods. *Proceedings of ECOS 2019 - the 32ND International conference on Efficiency, Cost, Optimization, Simulation and environmental impact of energy systems*, June 23-28, 2019, Wroclaw, Poland

Baumgärtner, N., Shu, D., Bahl, B., Hennen, M., and Bardow, A. (2019). From peak power prices to seasonal storage: Long-term operational optimization of energy systems by time-series decomposition. In Kiss, A.A., Zondervan, E., Lakerveld, R., Özkan, L., editors, *Proceedings of the 29th European Symposium on Computer-Aided Process Engineering (ESCAPE-29)*, volume 46 of *Computer Aided Chemical Engineering*, pages 703-708.

Baumgärtner, N., Leisin, M., Bahl, B., Hennen, M., and Bardow, A. (2018). Rigorous synthesis of energy systems by relaxation and time-series aggregation to typical periods. In Eden, M., Ierapetritou M., Towler G., editors, *Proceedings of the 13th International Symposium on Process Systems Engineering (PSE 2018)*, volume 44 of *Computer Aided Chemical Engineering*, pages 793-798.

Student theses supervised during this work

Delorme, R. (2017). Multikriterielle Optimierung dezentraler Energieversorgungssysteme unter Berücksichtigung zeitabhängiger Emissionsfaktoren der Elektrizitätsversorgung (in German). Bacherlor's thesis, RWTH Aachen University.

Küpper, E. (2017). Infrastrukturoptimierung von Energiesystemen mit Sektorenkopplung (in German). Bacherlor's thesis, RWTH Aachen University.

Leisin, M. (2017). Multikriterielle Strukturoptimierung von Energieversorgungssystemen mittels Aggregation gekoppelter Zeitreihen (in German). Master's thesis, RWTH Aachen University.

Jörres, S. (2018). Technologiebewertung zur Dekarbonisierung des Energiesystems eines Chemiestandorts mittels Strukturoptimierung (in German). Master's thesis, RWTH Aachen University.

Shu, D. (2018). A Method for Handling Time Coupling Constraints and Variables in Operational Optimization of Renewable Energy Systems. Master's thesis, RWTH Aachen University.

Reinert, C. (2018). Zielkonflikte der Energiewende - Life Cycle Assessment der Dekarbonisierung Deutschlands durch sektorenübergreifende Infrastrukturoptimierung (in German). Master's thesis, RWTH Aachen University.

Granacher, J. (2018). New energy pathways for the Swiss Industry: A techno-economic analysis of selected industrial energy applications in the light of the Swiss Energy Strategy 2050. Master's thesis, RWTH Aachen University.

Temme, F. (2018). Rigerous Synthesis of Energy Systems with Seasonal Storage by Means of Two Time Grids. Bacherlor's thesis, RWTH Aachen University.

Nolzen, N. (2018). Sektorübergreifende Infrastrukturoptimierung und Life Cycle Assessment eines Transitionspfades des deutschen Energiesystems (in German). Master's thesis, RWTH Aachen University.

Lindenmeyer, C. (2018). Impact of extrem value representation and clustering methods on two-stage design and operation optimization problems. Master's thesis, RWTH Aachen University.

Schröer, S. (2019). Multikriterielle Infrastrukturoptimierung des deutschen Energiesystems auf Basis ökonomischer und ökologischer Indikatoren (in German). Master's thesis, RWTH Aachen University.

Söhler, T. (2019). Räumliche Aggregations- und Rückaggregationsmethodik zur Infrastrukturoptimierung sektorübergreifender Energiesysteme (in German). Master's thesis, RWTH Aachen University.

Sauer, C. (2019). Infrastrukturoptimierung unter Berücksichtigung lokaler Emissionsgrenzwerte durch Life Cycle Assessment für die deutsche Energiewende (in German). Master's thesis, RWTH Aachen University.

Küpper, E. (2019). Investigation of time-series-aggregation methods for infrastructure optimization of low emissions energy systems. Master's thesis, RWTH Aachen University.

Bibliography

50Hertz, Amprion, Tennet, and Transnet BW (2012). Bericht der deutschen Übertragungsnetzbetreiber zur Leistungsbilanz nach EnWG § 12 Abs. 4 und 5: (German).

ADAC (2019a). ADAC Kostenvergleich: Erd- und Autogas gegen Benziner und Diesel: (German). Online available at `https://www.adac.de/_mmm/pdf/g-b-d-vgl_47097.pdf`.

ADAC (2019b). Erdgas - ein Antrieb mit Zukunft? (German). Online available at `https://www.adac.de/verkehr/tanken-kraftstoff-antrieb/alternative-antriebe/erdgas`.

Adhau, S. P., Moharil, R. M., and Adhau, P. G. (2014). K-Means clustering technique applied to availability of micro hydro power. *Sustainable Energy Technologies and Assessments*, 8:191–201.

Agora Verkehrswende, Agora Energiewende, and Frontier Economics (2018). The Future Cost of Electricity-Based Synthetic Fuels.

Al-Agtash, S. and Su, R. (1998). Augmented Lagrangian approach to hydro-thermal scheduling. *IEEE Transactions on Power Systems*, 13(4):1392–1400.

Alarcon-Rodriguez, A., Ault, G., and Galloway, S. (2010). Multi-objective planning of distributed energy resources: A review of the state-of-the-art. *Renewable and Sustainable Energy Reviews*, 14(5):1353–1366.

Algunaibet, I. and Guillén-Gosálbez, G. (2019). Life cycle burden-shifting in energy systems designed to minimize greenhouse gas emissions: Novel analytical method and application to the United States. *Journal of Cleaner Production*, 229:886–901.

Anderson, J., Bergfeld, M., Hoffmann, N., Kuhnimhof, T., Lenz, B., Steck, F., Friedrich, H., Kleiner, F., Klötzke, M., Schmid, S., Chlond, B., Soylu, T., and Weiß, C. (2016). LADEN2020 Schlussbericht: Konzept zum Aufbau einer bedarfsgerechten Ladeinfrastruktur in Deutschland von heute bis 2020: (German).

Andiappan, V. (2017). State-Of-The-Art Review of Mathematical Optimisation Approaches for Synthesis of Energy Systems. *Process Integration and Optimization for Sustainability*, 1(3):165–188.

ARAL (2017). Kraftstoffpreise - Monatsübersicht 2015: (German). Online available at `https://www.aral.de/content/dam/aral/PDFs/Kraftstoffpreis-Archiv/Kraftstoffpreise_Monatsuebersicht_12.pdf`.

Arbeitsgemeinschaft Energiebilanzen (2017). Anwendungsbilanzen für die Endenergiesektoren in Deutschland in den Jahren 2013 bis 2016: (German). Berlin.

Asrari, A., Lotfifard, S., and Payam, M. S. (2016). Pareto Dominance-Based Multiobjective Optimization Method for Distribution Network Reconfiguration. *IEEE Transactions on Smart Grid*, 7(3):1401–1410.

Atabay, D. (2017). An open-source model for optimal design and operation of industrial energy systems. *Energy*, 121:803–821.

Bahl, B., Kümpel, A., Seele, H., Lampe, M., and Bardow, A. (2017a). Time-series aggregation for synthesis problems by bounding error in the objective function. *Energy*, 135:900–912.

Bahl, B., Lampe, M., Voll, P., and Bardow, A. (2017b). Optimization-based identification and quantification of demand-side management potential for distributed energy supply systems. *Energy*, 135:889–899.

Bahl, B., Lützow, J., Shu, D., Hollermann, D. E., Lampe, M., Hennen, M., and Bardow, A. (2018a). Rigorous synthesis of energy systems by decomposition via time-series aggregation. *Computers & Chemical Engineering*, 112:70–81.

Bahl, B., Söhler, T., Hennen, M., and Bardow, A. (2018b). Typical Periods for Two-Stage Synthesis by Time-Series Aggregation with Bounded Error in Objective Function. *Frontiers in Energy Research*, 5:35.

Bareiß, K., de La Rua, C., Möckl, M., and Hamacher, T. (2019). Life cycle assessment of hydrogen from proton exchange membrane water electrolysis in future energy systems. *Applied Energy*, 237:862–872.

Baumgärtner, N., Bahl, B., Hennen, M., and Bardow, A. (2019a). RiSES3: Rigorous Synthesis of Energy Supply and Storage Systems via time-series relaxation and aggregation. *Computers & Chemical Engineering*, 127:127–139.

Baumgärtner, N., Delorme, R., Hennen, M., and Bardow, A. (2019b). Design of low-carbon utility systems: Exploiting time-dependent grid emissions for climate-friendly demand-side management. *Applied Energy*, 247:755–765.

Baumgärtner, N., Deutz, D., Reinert, C., Nolzen, N., Küpper, L. E., Hennen, M., Hollermann, D. E., and Bardow, A. (2020). Life-cycle assessment of sector-coupled national energy systems: Environmental impacts of electricity, heat & transportation in Germany till 2050. *In preperation.*

Baumgärtner, N., Leisin, M., Bahl, B., Hennen, M., and Bardow, A. (2018). Rigorous synthesis of energy systems by relaxation and time-series aggregation to typical periods. In Eden, M. R., Ierapetritou, M. G., and Towler, G. P., editors, *13th International Symposium on Process Systems Engineering (PSE 2018)*, volume 44 of *Computer-aided chemical engineering*, pages 793–798. Elsevier, Amsterdam and Boston and Heidelberg.

Baumgärtner, N., Temme, F., Bahl, B., Hennen, M., Hollermann, D., and Bardow, A. (2019c). RiSES4 Rigorous Synthesis of Energy Supply Systems with Seasonal Storage by relaxation and time- series aggregation to typical periods. In *Proceedings of ECOS 2019 - The 32nd international conference on efficiency, cost, optimization, simulation and environmental impact of energy systems. June 23-28, 2019 Wroclaw, Poland.*

Bazzanella, A. M. and Ausfelder, F. (2017). Low carbon energy and feedstock for the European chemical industry: Report of DECHEMA: Gesellschaft für Chemische Technik und Biotechnologie e.V.

BDEW Bundesverband der Energie- und Wasserwirtschaft e.V. (2014). Wie heizt Deutschland? BDEW-Studie zum Heizungsmarkt: (German). Berlin.

Bergner, M., Caprara, A., Ceselli, A., Furini, F., Lübbecke, M. E., Malaguti, E., and Traversi, E. (2015). Automatic Dantzig–Wolfe reformulation of mixed integer programs. *Mathematical Programming*, 149:391–424.

Berrill, P., Arvesen, A., Scholz, Y., Gils, H., and Hertwich, E. (2016). Environmental impacts of high penetration renewable energy scenarios for Europe. *Environmental Research Letters*, 11(1):014012.

Bettle, R., Pout, C. H., and Hitchin, E. R. (2006). Interactions between electricity-saving measures and carbon emissions from power generation in England and Wales. *Energy Policy*, 34(18):3434–3446.

Bischi, A., Taccari, L., Martelli, E., Amaldi, E., Manzolini, G., Silva, P., Campanari, S., and Macchi, E. (2014). A detailed MILP optimization model for combined cooling, heat and power system operation planning. *Energy*, 74:12–26.

Bischi, A., Taccari, L., Martelli, E., Amaldi, E., Manzolini, G., Silva, P., Campanari, S., and Macchi, E. (2019). A rolling-horizon optimization algorithm for the long term operational scheduling of cogeneration systems. *Energy*, 184:73–90.

Blanco, H., Codina, V., Laurent, A., Nijs, W., Maréchal, F., and Faaij, A. (2020). Life cycle assessment integration into energy system models: An application for Power-to-Methane in the EU. *Applied Energy*, 259:114160.

Blarke, M. B. and Dotzauer, E. (2011). Intermittency-friendly and high-efficiency cogeneration: Operational optimisation of cogeneration with compression heat pump, flue gas heat recovery, and intermediate cold storage. *Energy*, 36(12):6867–6878.

BMVBS (2010a). Mobilität in Deutschland 2008: Ergebnisbericht: Struktur - Aufkommen - Emissionen - Trends (German).

BMVBS (2010b). Mobilität in Deutschland 2008: Tabellenband: (German).

Bongartz, D., Doré, L., Eichler, K., Grube, T., Heuser, B., Hombach, L. E., Robinius, M., Pischinger, S., Stolten, D., Walther, G., and Mitsos, A. (2018). Comparison of light-duty transportation fuels produced from renewable hydrogen and green carbon dioxide. *Applied Energy*, 231:757–767.

Box, G. E. P. and Jenkins, G. M. (2015). *Time series analysis: Forecasting and control*. Holden-Day, San Francisco, FL, USA.

Bracco, S., Dentici, G., and Siri, S. (2013). Economic and environmental optimization model for the design and the operation of a combined heat and power distributed generation system in an urban area. *Energy*, 55:1014–1024.

Bracco, S., Dentici, G., and Siri, S. (2016). DESOD: a mathematical programming tool to optimally design a distributed energy system. *Energy*, 100:298–309.

Bradley, S. P., Hax, A. C., and Magnanti, T. L. (1977). *Applied mathematical programming*. Addison-Wesley, Reading, Mass.

Brauner, G. (2019). *Systemeffizienz bei regenerativer Stromerzeugung: (German)*. Springer Fachmedien Wiesbaden, Wiesbaden.

Breyer, C., Bogdanov, D., Aghahosseini, A., Gulagi, A., and Fasihi, M. (2020). On the Techno-economic Benefits of a Global Energy Interconnection. *Economics of Energy & Environmental Policy*, 9(1).

Breyer, C., Fasihi, M., and Aghahosseini, A. (2018). CO2 Direct Air Capture for effective Climate Change Mitigation: A new Type of Energy System Sector Coupling. In *International Conference on Negative CO2 Emissions*.

Bruche, S. and Tsatsaronis, G. (2019). A Multi-Stage Optimization Approach for Energy Supply Systems With Discrete Design Decisions. In *Volume 6: Energy*. IMECE.

Bruno, J. C., Fernandez, F., Castells, F., and Grossmann, I. E. (1998). A Rigorous MINLP Model for the Optimal Synthesis and Operation of Utility Plants. *Chemical Engineering Research and Design*, 76(3):246–258.

Buchholz, S., Gamst, M., and Pisinger, D. (2019). A comparative study of time aggregation techniques in relation to power capacity expansion modeling. *TOP*, 27(3):353–405.

Bundesministerium für Verkehr und digitale Infrastruktur (2018). *Verkehr in Zahlen 2017/2018: (German)*, volume 2018. DVV Media Group, Hamburg, 45. aktualisierte neuauflage, revidierte ausgabe edition.

Buoro, D., Casisi, M., de Nardi, A., Pinamonti, P., and Reini, M. (2013). Multicriteria optimization of a distributed energy supply system for an industrial area. *Energy*, 58:128–137.

Burger, B. (2017). Stromerzeugung in Deutschland im Jahr 2016: (German). Fraunhofer-Institut für Solare Energiesysteme ISE.

Burre, J., Bongartz, D., Brée, L., Roh, K., and Mitsos, A. (2020). Power–to–X: Between Electricity Storage, e–Production, and Demand Side Management. *Chemie Ingenieur Technik*, 92(1-2):74–84.

BVU Beratergruppe, Intraplan Consult GmbH, IVV GmbH & Co. KG, and Planco Consulting GmbH (2014). Verkehrsverflechtungsprognose 2030 - Abschlussbericht: im Auftrag des Bundesministeriums für Verkehr und digitale Infrastruktur (German).

Cebulla, F., Haas, J., Eichman, J., Nowak, W., and Mancarella, P. (2018). How much electrical energy storage do we need? A synthesis for the U.S., Europe, and Germany. *Journal of Cleaner Production*, 181:449–459.

Cebulla, F., Naegler, T., and Pohl, M. (2017). Electrical energy storage in highly renewable European energy systems: Capacity requirements, spatial distribution, and storage dispatch. *Journal of Energy Storage*, 14:211–223.

Cedillos Alvarado, D., Acha, S., Shah, N., and Markides, C. N. (2016). A Technology Selection and Operation (TSO) optimisation model for distributed energy systems: Mathematical formulation and case study. *Applied Energy*, 180:491–503.

Chatzimouratidis, A. and Pilavachi, P. (2009). Technological, economic and sustainability evaluation of power plants using the Analytic Hierarchy Process. *Energy Policy*, 37(3):778–787.

Child, M., Bogdanov, D., and Breyer, C. (2018). The role of storage technologies for the transition to a 100% renewable energy system in Europe. *Energy Procedia*, 155:44–60.

Chiu, W.-Y., Sun, H., and Vincent Poor, H. (2015). A Multiobjective Approach to Multimicrogrid System Design. *IEEE Transactions on Smart Grid*, 6(5):2263–2272.

Christidis, A., Koch, C., Pottel, L., and Tsatsaronis, G. (2012). The contribution of heat storage to the profitable operation of combined heat and power plants in liberalized electricity markets. *Energy*, 41(1):75–82.

Christoffer, J. and Ulbricht-Eissing, M. (1989). *Die bodennahen Windverhältnisse in der Bundesrepublik Deutschland: (mit 15 Tab. im Anh.): (German)*, volume 147 of *Berichte des Deutschen Wetterdienstes*. Deutscher Wetterdienst Zentralamt, Offenbach am Main, 2., vollst. neu bearb. edition.

Conejo, A. J., Castillo, E., Mínguez, R., and García-Bertrand, R. (2006). *Decomposition Techniques in Mathematical Programming*. Springer-Verlag, Berlin/Heidelberg.

Cox, B., Mutel, C. L., Bauer, C., Mendoza Beltran, A., and van Vuuren, D. P. (2018). Uncertain Environmental Footprint of Current and Future Battery Electric Vehicles. *Environmental science & technology*, 52(8):4989–4995.

Cronin, J., Anandarajah, G., and Dessens, O. (2018). Climate change impacts on the energy system: a review of trends and gaps. *Climatic change*, 151(2):79–93.

D. Kreyenberg, A. Lischke, F. Bergk, F. Duennebeil, C. Heidt, W. Knörr, T. Raksha, P. Schmidt, W. Weindorf, K. Naumann, S. Majer, and F. Müller-Langer (2015). Erneuerbare Energien im Verkehr: Potenziale und Entwicklungsperspektiven verschiedener erneuerbarer Energieträger und Energieverbrauch der Verkehrsträger: (German). Berlin.

DAtF (2017). Betriebsergebnisse Kernkraftwerke 2016: (German). deutsches atomform e.v. Berlin.

DeCarolis, J., Daly, H., Dodds, P., Keppo, I., Li. F., McDowall, F., Pye, S., Strachan, N., Trutnevyte, E., Usher, W., Winning, M., Yeh, S., and Zeyringer, M. (2017). Formalizing best practice for energy system optimization modelling. *Applied Energy*, 194:184–198.

DEHSt (2006). Emissionsfaktoren und Kohlenstoffgehalte (Stoffliste 2006): Report of Deutsche Emissionshandelsstelle (DEHSt) im Umweltbundesamt: (German).

DEHSt (2017). Treibhausgasemissionen 2016 - Emissionshandelspflichtige stationäre Anlagen und Luftverkehr in Deutschland: Report of Deutsche Emissionshandelsstelle (DEHSt) im Umweltbundesamt: (German).

Demirhan, C. D., Tso, W. W., Ogumerem, G. S., and Pistikopoulos, E. N. (2019). Energy systems engineering - a guided tour. *BMC Chemical Engineering*, 1(1):131.

Dena (2010). dena Grid Study II: Integration of Renewable Energy Sources in the German Power Supply System from 2015 - 2020 with an Outlook to 2025. Berlin.

Destatis (2019). Daten zur Energiepreisentwicklung: Lange Reihen von Januar 2005 bis September 2019: Report of Statistische Bundesamt: (German).

Deutsche Energie-Agentur (2010). Kurzanalyse der Kraftwerksplanung in Deutschland bis 2020: Annahmen, Ergebnisse und Schlussfolgerungen: (German). Berlin.

Deutsche Energie-Agentur (2018). dena-Leitstudie Integrierte Energiewende: Impulse für die Gestaltung des Energiesystems 2050: (German). Berlin.

Deutscher Bundestag (2013). Verordnung über energiesparenden Wärmeschutz und energiesparende Anlagentechnik bei Gebäuden: EnEV (German).

Deutz, S. and Bardow, A. (2020). How (carbon) negative is direct air capture? Life cycle assessment of an industrial temperature-vacuum swing adsorption process. *Submittted*.

Deutz, S., Bongartz, D., Heuser, B., Kätelhön, A., Schulze Langenhorst, L., Omari, A., Walters, M., Klankermayer, J., Leitner, W., Mitsos, A., Pischinger, S., and Bardow, A. (2018). Cleaner production of cleaner fuels: wind-to-wheel – environmental assessment of CO 2 -based oxymethylene ether as a drop-in fuel. *Energy & Environmental Science*, 11(2):331–343.

Di Somma, M., Yan, B., Bianco, N., Graditi, G., Luh, P. B., Mongibello, L., and Naso, V. (2015). Operation optimization of a distributed energy system considering energy costs and exergy efficiency. *Energy Conversion and Management*, 103:739–751.

DWD (2017). Wetterdaten und -statistiken Express: (German). Online available at `https://www.dwd.de/DE/klimaumwelt/cdc/klinfo_systeme/weste/weste_node.html`.

Egerer, J. (2016). Open Source Electricity Model for Germany (ELMOD-DE). Berlin.

Egerer, J., Gerbaulet, C., Ihlenburg, R., and Kunz, F. (2014). Electricity Sector Data for Policy-Relevant Modeling: Data Documentation and Applications to the German and European Electricity Markets. Berlin.

Eicker, U. (2005). *Solar technologies for buildings*. Wiley, Chichester.

Ekvall, T. and Weidema, B. P. (2004). System boundaries and input data in consequential life cycle inventory analysis. *The International Journal of Life Cycle Assessment*, 9(3):161–171.

Elia, J. A. and Floudas, C. A. (2014). Energy supply chain optimization of hybrid feedstock processes: a review. *Annual review of chemical and biomolecular engineering*, 5:147–179.

Eller, D. (2015). *Integration erneuerbarer Energien mit Power-to-Heat in Deutschland: (German)*. Springer Fachmedien Wiesbaden, Wiesbaden.

Ellingsen, L. A.-W., Majeau-Bettez, G., Singh, B., Srivastava, A. K., Valøen, L. O., and Strømman, A. H. (2014). Life Cycle Assessment of a Lithium-Ion Battery Vehicle Pack. *Journal of Industrial Ecology*, 18(1):113–124.

Elsido, C., Bischi, A., Silva, P., and Martelli, E. (2017). Two-stage MINLP algorithm for the optimal synthesis and design of networks of CHP units. *Energy*, 121:403–426.

ENTSO-E (2017). ENTSO-E Transparency Platform: Central collection and publication of electricity generation, transportation and consumption data and information for the pan-European market. online available at `https://transparency.entsoe.eu/`.

EPA (2016). Greenhouse Gas Inventory Guidance: Report of U.S. Environmental Protection Agency.

epexspot (2017). EPEX-SPOT prices: (German). Online available at https://www.epexspot.com/de/marktdaten/dayaheadauktion/auction-table/.

European Commission (2018). Product Environmental Footprint Category Rules Guidance: Version 6.3.

European Environment Agency (2019). E-PRTR. Online available at https://prtr.eea.europa.eu/#/home.

Eurostat (2018). Greenhouse gas emission statistics. Online available at https://ec.europa.eu/eurostat/statistics-explained/index.php/Greenhouse_gas_emission_statistics.

Facci, A. L., Andreassi, L., Ubertini, S., and Sciubba, E. (2014). Analysis of the Influence of Thermal Energy Storage on the Optimal Management of a Trigeneration Plant. *Energy Procedia*, 45:1295–1304.

Falke, T., Krengel, S., Meinerzhagen, A.-K., and Schnettler, A. (2016). Multi-objective optimization and simulation model for the design of distributed energy systems. *Applied Energy*, 184:1508–1516.

Fasihi, M. and Breyer, C. (2020). Baseload electricity and hydrogen supply based on hybrid PV-wind power plants. *Journal of Cleaner Production*, 243:118466.

Fazlollahi, S., Becker, G., and Maréchal, F. (2014). Multi-objectives, multi-period optimization of district energy systems: II—Daily thermal storage. *Computers & Chemical Engineering*, 71:648–662.

Fazlollahi, S., Mandel, P., Becker, G., and Maréchal, F. (2012). Methods for multi-objective investment and operating optimization of complex energy systems. *Energy*, 45(1):12–22.

Federal Cartel Office (2012). Sektoruntersuchung Fernwärme: Abschlussbericht: (German).

Federal Ministry for Economic Affairs and Energy (2017). Langfristszenarien für die Transformation des Energiesystems in Deutschland: Modul 3: Referenzszenario und Basisszenario: (German).

Federal Ministry for the Environment, Nature Conservation and Nuclear Safety (2018). Klimaschutz in Zahlen (2018) – Fakten, Trends und Impulse deutscher Klimapolitik: Ausgabe 2018: (German). Berlin.

Federal Ministry of Justice and Consumer Protection and Federal Office of Justice (1985). Atomgesetz in der Fassung der Bekanntmachung vom 15. Juli 1985 (BGBl. I S. 1565), zuletzt geändert durch Artikel 2 des Gesetzes vom 12. Dezember 2019 (BGBl. I S. 2510) (German).

Federal Motor Transport Authority (2017a). Bestand am 1. Januar 2017 nach Zulassungsbezirken und Gemeinden: (German). Online available at `https://www.kba.de/DE/Statistik/Fahrzeuge/Bestand/ZulassungsbezirkeGemeinden/2017/2017_zulassungsbezirke_node.html`.

Federal Motor Transport Authority (2017b). Verkehr in Kilometern der deutschen Kraftfahrzeuge im Jahr 2016: Kurzbericht: (German).

Federal Network Agency (2018). Kraftwerksliste Bundesnetzagentur (bundesweit; alle Netz- und Umspannebenen) Stand 02.02.2018: (German).

Federal Statistical Office (2008). Energieverbrauch der privaten Haushalte: (German).

Federal Statistical Office (2015). Bevölkerung und Erwerbstätigkeit: Haushalte und Familien - Ergebnisse des Mikrozensus: (German).

Federal Statistical Office (2018). Gemeindeverzeichnis-Informationssystem GV-ISys: Regionalgliederung: Quartalsausgabe 30.06.2018: (German). Online available at `https://www.destatis.de/DE/ZahlenFakten/LaenderRegionen/Regionales/Gemeindeverzeichnis/Gemeindeverzeichnis.html`.

Finenko, A. and Cheah, L. (2016). Temporal CO2 emissions associated with electricity generation: Case study of Singapore. *Energy Policy*, 93:70–79.

Finnveden, G., Hauschild, M. Z., Ekvall, T., Guinée, J., Heijungs, R., Hellweg, S., Koehler, A., Pennington, D., and Suh, S. (2009). Recent developments in Life Cycle Assessment. *Journal of environmental management*, 91(1):1–21.

Floudas, C. A. (1995). *Nonlinear and mixed-integer optimization: Fundamentals and applications*. Topics in chemical engineering. Oxford Univ. Press, New York, NY.

Floudas, C. A., Niziolek, A. M., Onel, O., and Matthews, L. R. (2016). Multi-scale systems engineering for energy and the environment: Challenges and opportunities. *AIChE Journal*, 62(3):602–623.

Frangopoulos, C. A. (2018). Recent developments and trends in optimization of energy systems. *Energy*, 164:1011–1020.

Fraunhofer ISE (2018). Energy Charts - Jährliche Stromerzeugung in Deutschland in 2012: (German). Online available at www.energy-charts.de.

Frew, B. A. and Jacobson, M. Z. (2016). Temporal and spatial tradeoffs in power system modeling with assumptions about storage: An application of the POWER model. *Energy*, 117:198–213.

Gabrielli, P., Gazzani, M., Martelli, E., and Mazzotti, M. (2018). Optimal design of multi-energy systems with seasonal storage. *Applied Energy*, 219:408–424.

Gamrath, G. and Lübbecke, M. E. (2010). Experiments with a Generic Dantzig-Wolfe Decomposition for Integer Programs. In Festa, P., editor, *Experimental Algorithms*, LNCS sublibrary, pages 239–252. Springer, Berlin and New York.

GAMS Development Corporation (2019). General Algebraic Modeling System (GAMS) Release 27.3. Fairfax, VA, USA. Available for download at https://www.gams.com/download.

García-Gusano, D., Iribarren, D., Martín-Gamboa, M., Dufour, J., Espegren, K., and Lind, A. (2016). Integration of life-cycle indicators into energy optimisation models: the case study of power generation in Norway. *Journal of Cleaner Production*, 112:2693–2696.

Gardiner, M. (2009). DOE Hydrogen and Fuel Cells Program Record: Energy requirements for hydrogen gas compression and liquefaction as related to vehicle storage needs.

Geißler, B. (2011). *Towards Globally Optimal Solutions For MINLPs by Discretization Techniuqes with Applications in Gas Network Optimization*. PhD thesis, Naturwissenschaftlichen Fakultät der Friedrich-Alexander-Universität, Erlangen-Nürnberg.

Geißler, B., Martin, A., Morsi, A., and Schewe, L. (2012). Using Piecewise Linear Functions for Solving MINLPs. In Lee, J. and Leyffer, S., editors, *Mixed Integer Nonlinear Programming*, volume 154 of *The IMA Volumes in Mathematics and its Applications*, pages 287–314. Springer Science+Business Media, LLC, New York, NY.

Gençer, E., Torkamani, S., Miller, I., Wu, T. W., and O'Sullivan, F. (2020). Sustainable energy system analysis modeling environment: Analyzing life cycle emissions of the energy transition. *Applied Energy*, 277:115550.

Gerhardt, N., Sandau, F., Scholz, A., Hahn, H., Schumacher, P., Sager, C., Bergk, F., Kämper, C., Knörr, W., Kräck, J., Lambrecht, U., Antoni, O., Hilpert, J., Merkel, K., and Müller, T. (2015). Interaktion EE-Strom, Wärme und Verkehr: Analyse der Interaktion zwischen den Sektoren Strom, Wärme/Kälte und Verkehr in Deutschland in Hinblick auf steigende Anteile fluktuierender Erneuerbarer Energien im Strombereich unter Berücksichtigung der europäischen Entwicklung: (German).

GHG Protocol Initiative (2004). The Greenhouse Gas Protocol: A Corporate Accounting and Reporting Standard: Revised Edition.

Gils, H. C., Scholz, Y., Pregger, T., Luca de Tena, D., and Heide, D. (2017). Integrated modelling of variable renewable energy-based power supply in Europe. *Energy*, 123:173–188.

Gleixner, A., Bastubbe, M., Eifler, L., Gally, T., Gamrath, G., Gottwald, R. L., and et al. (2018). The SCIP Optimization Suite 6.0, Available at http://www.optimization-online.org/DB_HTML/2018/07/6692.html and as Report at https://opus4.kobv.de/opus4-zib/frontdoor/index/index/docId/6936. Takustr. 7, 14195 Berlin.

Glover, F. (1975). Improved Linear Integer Programming Formulations of Nonlinear Integer Problems. *Management Science*, 22(4):455–460.

Goderbauer, S., Bahl, B., Voll, P., Lübbecke, M. E., Bardow, A., and Koster, A. M. (2016). An adaptive discretization MINLP algorithm for optimal synthesis of decentralized energy supply systems. *Computers & Chemical Engineering*, 95:38–48.

Goderbauer, S., Comis, M., and Willamowski, F. (2019). The synthesis problem of decentralized energy systems is strongly NP-hard. *Computers & Chemical Engineering*, 124:343–349.

Gong, J. and You, F. (2015). Sustainable design and synthesis of energy systems. *Current Opinion in Chemical Engineering*, 10:77–86.

Graff Zivin, J. S., Kotchen, M. J., and Mansur, E. T. (2014). Spatial and temporal heterogeneity of marginal emissions: Implications for electric cars and other electricity-shifting policies. *Journal of Economic Behavior & Organization*, 107:248–268.

Greening, B. and Azapagic, A. (2012). Domestic heat pumps: Life cycle environmental impacts and potential implications for the UK. *Energy*, 39(1):205–217.

Grossmann, I. E. (2012). Advances in mathematical programming models for enterprise-wide optimization. *Computers & Chemical Engineering*, 47:2–18.

Guillén-Gosálbez, G. (2011). A novel MILP-based objective reduction method for multi-objective optimization: Application to environmental problems. *Computers & Chemical Engineering*, 35(8):1469–1477.

Haas, J., Cebulla, F., Cao, K., Nowak, W., Palma-Behnke, R., Rahmann, C., and Mancarella, P. (2017). Challenges and trends of energy storage expansion planning for flexibility provision in low-carbon power systems – a review. *Renewable and Sustainable Energy Reviews*, 80:603–619.

Haikarainen, C., Pettersson, F., and Saxén, H. (2014). A model for structural and operational optimization of distributed energy systems. *Applied Thermal Engineering*, 70(1):211–218.

Haller, M., Ludig, S., and Bauer, N. (2012). Decarbonization scenarios for the EU and MENA power system: Considering spatial distribution and short term dynamics of renewable generation. *Energy Policy*, 47(C):282–290.

Hank, C., Sternberg, A., Köppel, N., Holst, M., Smolinka, T., Schaadt, A., Hebling, C., and Henning, H.-M. (2020). Energy efficiency and economic assessment of imported energy carriers based on renewable electricity. *Sustainable Energy Fuels*, 4(5):2256–2273.

Hanne, T. and Dornberger, R. (2017). *Computational Intelligence in Logistics and Supply Chain Management*, volume 244. Springer International Publishing, Cham.

Harb, H., Reinhardt, J., Streblow, R., and Müller, D. (2016). MIP approach for designing heating systems in residential buildings and neighbourhoods. *Journal of Building Performance Simulation*, 9(3):316–330.

Harthan, R. O. and Herrmann, H. (2018). Sektorale Abgrenzung der deutschen Treibhausgasemissionen mit einem Schwerpunkt auf die verbrennungsbedingten CO2-Emissionen: Arbeitspapier: (German). Berlin.

Hawkes, A. D. (2010). Estimating marginal CO2 emissions rates for national electricity systems. *Energy Policy*, 38(10):5977–5987.

He, G., Lin, J., Sifuentes, F., Liu, X., Abhyankar, N., and Phadke, A. (2020). Rapid cost decrease of renewables and storage accelerates the decarbonization of China's power system. *Nature communications*, 11(1):2486.

Hennen, M., Lampe, M., Voll, P., and Bardow, A. (2017). SPREAD – Exploring the decision space in energy systems synthesis. *Computers and Chemical Engineering*, 106:297–308.

Henning, H.-M. and Sauer, D. U. (2015). Demand-Side-Management im Wärmemarkt: Technologiesteckbrief zur Analyse „Flexibilitätskonzepte für die Stromversorgung 2050“: (German). Freiburg.

Heuberger, C., Bains, P., and Dowell, N. (2020). The EV-olution of the power system: A spatio-temporal optimisation model to investigate the impact of electric vehicle deployment. *Applied Energy*, 257:113715.

Heuberger, C., Staffell, I., Shah, N., and Mac Dowell, N. (2018). Impact of myopic decision-making and disruptive events in power systems planning. *Nature Energy*, 3:634–640.

Heuck, K., Dettmann, K.-D., and Schulz, D. (2013). *Elektrische Energieversorgung: Erzeugung, Übertragung und Verteilung elektrischer Energie für Studium und Praxis (German)*. Springer Fachmedien Wiesbaden, Wiesbaden and s.l., 9. edition.

Hoffmann, M., Kotzur, L., Stolten, D., and Robinius, M. (2020). A Review on Time Series Aggregation Methods for Energy System Models. *Energies*, 13(3):641.

Howells, M., Rogner, H., Strachan, N., Heaps, C., Huntington, H., Kypreos, S., Hughes, A., Silveira, S., DeCarolis, J., Bazillian, M., and Roehrl, A. (2011). OSeMOSYS: The Open Source Energy Modeling System. *Energy Policy*, 39(10):5850–5870.

IBM (2016). IBM ILOG and CPLEX Optimization and Studio and CPLEX User's Manual, Release 12.7. Incline Village, NV, USA. Available for download at `https://www.ibm.com/products/ilog-cplex-optimization-studio`.

Icha, P. and Kuhs, G. (2018). Entwicklung der spezifischen Kohlendioxid-Emissionen des deutschen Strommix in den Jahren 1990-2017: Report of Umweltbundesamt: (German).

IEA-ETSAP and IRENA (2015). Solar Heat for Industrial Processes Technology Brief.

ifeu, Fraunhofer IEE, Consentec (2019). Building sector Efficiency: A crucial Component of the Energy Transition: A study commissioned by Agora Energiewende.

ILCD (2010). ILCD Handbook International Reference Life Cycle Data System: General guide on LCA - Detailed guidance.

International Organization for Standardization (2006a). ISO 14040: Environmental management - Life cycle assessment - Principles and framework.

International Organization for Standardization (2006b). ISO 14044: Environmental management - Life cycle assessment - Requirements and guidelines.

IPCC (2013). Climate Change 2013: The Physical Science Basis: Contribution of Working Group I to the Fifth Assessment Report of the Intergovernmental Panel on Climate Change.

IPCC (2018). Global Warming of 1.5°C. An IPCC Special Report on the impacts of global warming of 1.5°C above pre-industrial levels and related global greenhouse gas emission pathways, in the context of strengthening the global response to the threat of climate change, sustainable development, and efforts to eradicate poverty.

Jagnow, K. and Wolff, D. (2003). Bemessung von Wärmeerzeugern, Speichern und Heizflächen nach EN 12831 und DIN 4708: (German).

Jorge, R., Hawkins, T., and Hertwich, E. (2012). Life cycle assessment of electricity transmission and distribution—part 1: Power lines and cables. *The International Journal of Life Cycle Assessment*, 17(1):9–15.

JRC (2010). *International Reference Life Cycle Data System (ILCD) Handbook - General guide for Life Cycle Assessment - Detailed guidance*, volume 24708 of *EUR (Luxembourg)*. Publications Office.

JRC (2011). *European Commission - Joint Research Centre - Institute of Environment and Sustainability. International Reference Life Cycle Data System (ILCD) Handbook: Recommendations for Life Cycle Impact Assessment in the European context from the European Commission.* Publications Office of the European Union.

JRC (2018). Product Environmental Footprint Category Rules Guidance: Version 6.3 – May 2018.

Kannengießer, T., Hoffmann, M., Kotzur, L., Stenzel, P., Schuetz, F., Peters, K., Nykamp, S., Stolten, D., and Robinius, M. (2019). Reducing Computational Load for Mixed Integer Linear Programming: An Example for a District and an Island Energy System. *Energies*, 12(14):2825.

Karlsruher Institute of Technology (2014). Deutsches Mobilitätspanel (MOP) - Wissenschaftliche Begleitung und Auswertungen: Bericht 2012/2013: Alltagsmobilität und Fahrleistungen: im Auftrag des Bundesministeriums für Verkehr und digitale Infrastruktur (German). Karlsruhe.

Kavvadias, K. C. and Maroulis, Z. B. (2010). Multi-objective optimization of a trigeneration plant. *Energy Policy*, 38:945–954.

Kazarlis, S. A., Bakirtzis, A. G., and Petridis, V. (1996). A Genetic Algorithm Solution to the Unit Commitment Problem. *IEEE Transactions on Power Systems*, 11(1):83–92.

Keck, F., Lenzen, M., Vassallo, A., and Li, M. (2019). The impact of battery energy storage for renewable energy power grids in Australia. *Energy*, 173:647–657.

Kelley, M. T., Baldick, R., and Baldea, M. (2019). Demand Response Operation of Electricity-Intensive Chemical Processes for Reduced Greenhouse Gas Emissions: Application to an Air Separation Unit. *ACS Sustainable Chemistry & Engineering*, 7(2):1909–1922.

Kempe, S. (2014). *Räumlich detaillierte Potenzialanalyse der Fernwärmeversorgung in Deutschland mit einem hoch aufgelösten Energiesystemmodell: (German)*. PhD thesis, Universität Stuttgart, Stuttgart.

Khalili, Rantanen, Bogdanov, and Breyer (2019). Global Transportation Demand Development with Impacts on the Energy Demand and Greenhouse Gas Emissions in a Climate-Constrained World. *Energies*, 12(20):3870.

Kießling, F., Nefzger, P., and Kaintzyk, U. (2001). *Freileitungen: Planung, Berechnung, Ausführung: (German)*. Springer Berlin Heidelberg, Berlin, Heidelberg.

Kim, S. H., Yoon, S.-G., Chae, S. H., and Park, S. (2010). Economic and environmental optimization of a multi-site utility network for an industrial complex. *Journal of environmental management*, 91(3):690–705.

Kirschbaum, S., Voll, P., Scheffler, R., Pleßow, M., and Bardow, A. (2013). Simultaneous optimisation of structure and infrastructure for distributed energy supply systems. In ECOS Guilin, editor, *Proceedings of the 26th International Conference on Effciency, Cost, Optimization, Simulation and Environmental Impact of Energy Systems*.

Knörr, W., Heidt, C., Gores, S., and BErgk, F. (2016). Aktualisierung Daten- und Rechenmodell: Energieverbrauch und Schadstoffemissionen des motorisierten Verkehrs in Deutschland 1960-2035 (TREMOD) für die Emissionsberichterstattung 2016: (Berichtsperiode 1990-2014): (German).

König, D. H., Baucks, N., Dietrich, R.-U., and Wörner, A. (2015). Simulation and evaluation of a process concept for the generation of synthetic fuel from 5CO26 and 5H26. *Energy*, 91:833–841.

Kono, J., Ostermeyer, Y., and Wallbaum, H. (2017). The trends of hourly carbon emission factors in Germany and investigation on relevant consumption patterns for its application. *The International Journal of Life Cycle Assessment*, 22:1493–1501.

Konstantin, P. (2017). *Praxisbuch Energiewirtschaft: Energieumwandlung, -transport und -beschaffung, Übertragungsnetzausbau und Kernenergieausstieg (German)*. VDI-Buch. Springer Vieweg, Berlin, 4., aktualisierte edition.

Kopsakangas-Savolainen, M., Mattinen, M. K., Manninen, K., and Nissinen, A. (2017). Hourly-based greenhouse gas emissions of electricity – cases demonstrating possibilities for households and companies to decrease their emissions. *Journal of Cleaner Production*, 153:384–396.

Kotzur, L., Markewitz, P., Robinius, M., and Stolten, D. (2018). Time series aggregation for energy system design: Modeling seasonal storage. *Applied Energy*, 213:123–135.

Kouloumpis, V., Stamford, L., and Azapagic, A. (2015). Decarbonising electricity supply: Is climate change mitigation going to be carried out at the expense of other environmental impacts? *Sustainable Production and Consumption*, 1:1–21.

Krämer, S. and Engell, S., editors (2018). *Resource efficiency of processing plants: Monitoring and improvement*. Wiley-VCH, Weinheim, Germany.

Krieg, D. (2012). *Konzept und Kosten eines Pipelinesystems zur Versorgung des deutschen Straßenverkehrs mit Wasserstoff: Zugl.: Aachen, Techn. Hochsch., PhD thesis, 2012 (German)*, volume 144 of *Schriften des Forschungszentrums Jülich Reihe Energie & Umwelt*. Forschungszentrum Jülich, Jülich.

Leenders, L., Bahl, B., Lampe, M., Hennen, M., and Bardow, A. (2019). Optimal design of integrated batch production and utility systems. *Computers & Chemical Engineering*, 128:496–511.

Lenzen, M. (2008). Double-Counting in Life Cycle Calculations. *Journal of Industrial Ecology*, 12(4):583–599.

Li, L., Mu, H., Li, N., and Li, M. (2016). Economic and environmental optimization for distributed energy resource systems coupled with district energy networks. *Energy*, 109:947–960.

Li, X., Damartzis, T., Stadler, Z., Moret, S., Meier, B., Friedl, M., and Maréchal, F. (2020). Decarbonization in complex energy systems: a study on the feasibility of carbon neutrality for Switzerland in 2050. *Frontiers in Energy Research.*

Lin, F., Leyffer, S., and Munson, T. (2016). A two-level approach to large mixed-integer programs with application to cogeneration in energy-efficient buildings. *Computational Optimization and Applications*, 65(1):1–46.

Lindner, C., Schmitt, J., Hiestermann, L., Fischer, E., Beylage, H., and Hein, J. (2017). Post-consumer Plastic Waste Management in European Countries 2016.

Lopion, P., Markewitz, P., Robinius, M., and Stolten, D. (2018). A review of current challenges and trends in energy systems modeling. *Renewable and Sustainable Energy Reviews*, 96(C):156–166.

Louis, J.-N., Allard, S., Kotrotsou, F., and Debusschere, V. (2020). A multi-objective approach to the prospective development of the European power system by 2050. *Energy*, 191:116539.

Loulou, R., Lehtilä, A., Kanudia, A., Remme, U., and Goldstein, G. (2016). Documentation for the TIMES Model Part II: Reference Manual.

Lucas, A., Alexandra Silva, C., and Costa Neto, R. (2012). Life cycle analysis of energy supply infrastructure for conventional and electric vehicles. *Energy Policy*, 41:537–547.

Lucia, U., Simonetti, M., Chiesa, G., and Grisolia, G. (2017). Ground-source pump system for heating and cooling: Review and thermodynamic approach. *Renewable and Sustainable Energy Reviews*, 70:867–874.

Luderer, G., Pehl, M., Arvesen, A., Gibon, T., Bodirsky, B. L., de Boer, H. S., Fricko, O., Hejazi, M., Humpenöder, F., Iyer, G., Mima, S., Mouratiadou, I., Pietzcker, R. C., Popp, A., van den Berg, M., van Vuuren, D., and Hertwich, E. G. (2019). Environmental co-benefits and adverse side-effects of alternative power sector decarbonization strategies. *Nature communications*, 10(1):5229.

Lund, H., Mathiesen, B. V., Christensen, P., and Schmidt, J. H. (2010). Energy system analysis of marginal electricity supply in consequential LCA. *The International Journal of Life Cycle Assessment*, 15(3):260–271.

Luo, X., Zhang, B., Chen, Y., and Mo, S. (2012). Operational planning optimization of multiple interconnected steam power plants considering environmental costs. *Energy*, 37(1):549–561.

Luo, X., Zhu, X., and Lim, E. G. (2019). A parametric bootstrap algorithm for cluster number determination of load pattern categorization. *Energy*, 180:50–60.

Lutsch, W., Neuffer, H., and Witterhold, F.-G. (2004). Strategien und Technologien einer pluralistischen Fern- und Nahwärmeversorgung in einem liberalisierten Energiemarkt unter besonderer Berücksichtigung der Kraft-Wärme-Kopplung und regenerativer Energien: (German).

Lutsch, W. and Witterhold, F.-G. (2005). Perspektiven der Fernwärme und der Kraft-Wärme-Kopplung: Ergebnisse und Schlussfolgerungen aus der AGFW-Studie Pluralistische Wärmeversorgung: (German).

Mainzer, K., Fath, K., McKenna, R., Stengel, J., Fichtner, W., and Schultmann, F. (2014). A high-resolution determination of the technical potential for residential-roof-mounted photovoltaic systems in Germany. *Solar Energy*, 105:715–731.

Majeau-Bettez, G., Hawkins, T. R., and Strømman, A. H. (2011). Life Cycle Environmental Assessment of Lithium-Ion and Nickel Metal Hydride Batteries for Plug-In Hybrid and Battery Electric Vehicles. *Environmental science & technology*, 45(10).

Mallapragada, D., Papageorgiou, D., Venkatesh, A., Lara, C., and Grossmann I. (2018). Impact of model resolution on scenario outcomes for electricity sector system expansion. *Energy*, 163:1231–1244.

Mancarella, P. (2014). MES (multi-energy systems): An overview of concepts and evaluation models. *Energy*, 65:1–17.

Maravelias, C. T. and Sung, C. (2009). Integration of production planning and scheduling: Overview, challenges and opportunities. *Computers & Chemical Engineering*, 33(12):1919–1930.

Markewitz, P. and Stein, G., editors (2003). *Das IKARUS-Projekt: Energietechnische Perspektiven für Deutschland: Abschlussbericht des Projektes IKARUS: (German)*, volume 39 of *Schriften des Forschungszentrums Jülich Reihe Umwelt*. Forschungszentrum Jülich, Jülich.

Marnay, C., Fisher, D., Murtishaw, S., Phadke, A., Price, L., and Sathaye, J. (2002). Estimating carbon dioxide emissions factors for the California electric power sector: In the Proceedings of the 2002 ACEEE's Summer Study on Energy Efficiency.

Maroufmashat, A., Sattari, S., Roshandel, R., Fowler, M., and Elkamel, A. (2016). Multi-objective Optimization for Design and Operation of Distributed Energy Systems through the Multi-energy Hub Network Approach. *Industrial & Engineering Chemistry Research*, 55(33):8950–8966.

Martelli, E., Freschini, M., and Zatti, M. (2020). Optimization of renewable energy subsidy and carbon tax for multi energy systems using bilevel programming. *Applied Energy*, 267:115089.

MathWorks, I. (2017). MATLAB and Statistics Toolbox Release 2017b. Natick, MA, USA. Available for download at `https://mathworks.com/downloads`.

Mavrotas, G. (2009). Effective implementation of the epsilon-constraint method in Multi-Objective Mathematical Programming problems. *Applied Mathematics and Computation*, 213(2):455–465.

McDowall, W., Rodriguez, B., Usubiaga, A., and Fernández, J. (2018). Is the optimal decarbonization pathway influenced by indirect emissions? Incorporating indirect life-cycle carbon dioxide emissions into a European TIMES model. *Journal of Cleaner Production*, 170:260–268.

Mckay, M. D., Beckman, R. J., and Conover, W. J. (2000). A Comparison of Three Methods for Selecting Values of Input Variables in the Analysis of Output From a Computer Code. *Technometrics*, 42(1):55–61.

McKenna, R., Hollnaicher, S., Ostman v. d. Leye, P., and Fichtner, W. (2015). Cost-potentials for large onshore wind turbines in Europe. *Energy*, 83:217–229.

McKinsey & Company (2010). A portfolio of power-trains for Europe: a fact-based analysis: The role of Batter Electric Vehicles, Plug-in Hybrids and Fuel Cell Electric Vehicles.

Mencarelli, L., Chen, Q., Pagot, A., and Grossmann, I. E. (2020). A review on superstructure optimization approaches in process system engineering. *Computers & Chemical Engineering*, page 106808.

Merker, G. P. and Teichmann, R. (2018). *Grundlagen Verbrennungsmotoren: (German)*. Springer Fachmedien Wiesbaden, Wiesbaden.

Mirakyan, A. and de Guio, R. (2015). Modelling and uncertainties in integrated energy planning. *Renewable and Sustainable Energy Reviews*, 46:62–69.

Misener, R., Gounaris, C. E., and Floudas, C. A. (2009). Global Optimization of Gas Lifting Operations: A Comparative Study of Piecewise Linear Formulations. *Industrial & Engineering Chemistry Research*, 48(13):6098–6104.

Mitra, S., Sun, L., and Grossmann, I. E. (2013). Optimal scheduling of industrial combined heat and power plants under time-sensitive electricity prices. *Energy*, 54:194–211.

Morvaj, B., Evins, R., and Carmeliet, J. (2016). Optimization framework for distributed energy systems with integrated electrical grid constraints. *Applied Energy*, 171:296–313.

Morvaj, B., Evins, R., and Carmeliet, J. (2017). Decarbonizing the electricity grid: The impact on urban energy systems, distribution grids and district heating potential. *Applied Energy*, 191:125–140.

Nahmmacher, P., Schmid, E., Hirth, L., and Knopf, B. (2016). Carpe diem: A novel approach to select representative days for long-term power system modeling. *Energy*, 112:430–442.

Nasrolahpour, E., Kazempour, S. J., Zareipour, H., and Rosehart, W. D. (2016). Strategic Sizing of Energy Storage Facilities in Electricity Markets. *IEEE Transactions on Sustainable Energy*, 7(4):1462–1472.

Nationale Plattform Elektromobilität (2015). Ladeinfrastruktur für Elektrofahrzeuge in Deutschland: Statusbericht und Handlungsempfehlungen 2015: AG 3 – Ladeinfrastruktur und NetzintegrationAG 3 – Ladeinfrastruktur und Netzintegration (German).

Netztransparenz.de (2018). EEG-Anlagenstammdaten zur Jahresabrechnung 2017: (German). Online available at `https://www.netztransparenz.de/EEG/Anlagenstammdaten`.

Noack, C., Burggraf, F., Hosseiny, S. S., Lettenmeier, P., Kolb, S., Belz, S., Kallo, J., Friedrich, K. A., Pregger, T., Cao, K.-K., Heide, D., Naegler, T., Borggrefe, F., Bünger, U., Michalski, J., Raksha, T., Voglstätter, C., Smolinka, T., Crotogino, F., Donadei, S., Horvath, P.-L., and Schneider, G.-S. (2015). Studie über die Planung einer Demonstrationsanlage zur Wasserstoff-Kraftstoffgewinnung durch Elektrolyse mit Zwischenspeicherung in Salzkavernen unter Druck: (german). Stuttgart.

Öko-Institut and ISI, F. (2015). Klimaschutzszenario 2050 – 2. Endbericht: (German).

Olivier, J., Schure, K., and Peters, J. (2017). Trends in global CO2 and total greenhouse gas emissions: 2017.

Olkkonen, V. and Syri, S. (2016). Spatial and temporal variations of marginal electricity generation: the case of the Finnish, Nordic, and European energy systems up to 2030. *Journal of Cleaner Production*, 126:515–525.

Onishi, V. C., Ravagnani, M. A., Jiménez, L., and Caballero, J. A. (2017). Multi-objective synthesis of work and heat exchange networks: Optimal balance between economic and environmental performance. *Energy Conversion and Management*, 140:192–202.

Open Geo DB (2017). Datensatz DE.tab: (German). Online available at `http://www.fa-technik.adfc.de/code/opengeodb`.

Ortiga, J., Bruno, J. C., and Coronas, A. (2011). Selection of typical days for the characterisation of energy demand in cogeneration and trigeneration optimisation models for buildings. *Energy Conversion and Management*, 52(4):1934–1942.

Overbye, T. J., Cheng, X., and Sun, Y. (2004). A comparison of the AC and DC power flow models for LMP calculations. In *Proceedings of the 37th Hawaii International Conference on System Sciences*, page 9 pp.

Palmintier, B. S. and Webster, M. D. (2014). Heterogeneous Unit Clustering for Efficient Operational Flexibility Modeling. *IEEE Transactions on Power Systems*, 29(3):1089–1098.

Palzer, A. (2016). *Sektorübergreifende Modellierung und Optimierung eines zukünftigen deutschen Energiesystems unter Berücksichtigung von Energieeffizienzmaßnahmen im Gebäudesektor: (German)*. PhD thesis, Fraunhofer-Institut für Solare Energiesysteme and Fraunhofer IRB-Verlag.

Panuschka, S. and Hofmann, R. (2019). Impact of thermal storage capacity, electricity and emission certificate costs on the optimal operation of an industrial energy system. *Energy Conversion and Management*, 185:622–635.

Parat (2019). Hochspannungs Elektrodenkessel: Von ENERGIE zu Wärme für Dampf- und Heißwasser: (German). Flekkefjord, Norwegen.

Park, J.-B., Park, Y.-M., Won, J.-R., and Lee, K. Y. (2000). An Improved Genetic Algorithm for Generation Expansion Planning. *IEEE Transactions on Power Systems*, 15(3):916–922.

Perez, R., Seals, R., Ineichen, P., Stewart, R., and Menicucci, D. (1987). A new simplified version of the perez diffuse irradiance model for tilted surfaces. *Solar Energy*, 39(3):221–231.

Petersen, C. (1971). *A note on transforming the product of variables to linear form in linear programs*. PhD thesis, Purdue Univerity.

Pfenninger, S., Hawkes, A., and Keirstead, J. (2014). Energy systems modeling for twenty-first century energy challenges. *Renewable and Sustainable Energy Reviews*, 33:74–86.

Piacentino, A. and Cardona, F. (2008). EABOT – Energetic analysis as a basis for robust optimization of trigeneration systems by linear programming. *Energy Conversion and Management*, 49(11):3006–3016.

Pina, E. A., Lozano, M. A., and Serra, L. M. (2020). A multiperiod multiobjective framework for the synthesis of trigeneration systems in tertiary sector buildings. *International Journal of Energy Research*, 44(2):1140–1166.

Pintarič, Z. N. and Kravanja, Z. (2015). The importance of proper economic criteria and process modeling for single- and multi-objective optimizations. *Computers & Chemical Engineering*, 83:35–47.

Plevin, R. J., Delucchi, M. A., and Creutzig, F. (2014). Using Attributional Life Cycle Assessment to Estimate Climate-Change Mitigation Benefits Misleads Policy Makers. *Journal of Industrial Ecology*, 18(1):73–83.

Pöstges, A. and Weber, C. (2019). Time series aggregation – A new methodological approach using the "peak-load-pricing" model. *Utilities Policy*, 59:100917.

Pregger, T., Nitsch, J., and Naegler, T. (2013). Long-term scenarios and strategies for the deployment of renewable energies in Germany. *Energy Policy*, 59:350–360.

Psomopoulos, C. S., Skoula, I., Karras, C., Chatzimpiros, A., and Chionidis, M. (2010). Electricity savings and CO2 emissions reduction in buildings sector: How important the network losses are in the calculation? *Energy*, 35(1):485–490.

Quoilin, S., van den Broek, M., Declaye, S., Dewallef, P., and Lemort, V. (2013). Techno-economic survey of Organic Rankine Cycle (ORC) systems. *Renewable and Sustainable Energy Reviews*, 22:168–186.

Rager, J. M. F. (2015). *Urban Energy System Design from the Heat Perspective using mathematical Programming including thermal Storage*. PhD thesis, École polytechnique fédérale de Lausanne.

Rahmaniani, R., Gabriel, T., Gendreau, M., and Rei, W. (2017). The Benders decomposition algorithm: A literature review. *European Journal of Operational Research*, 259(3):801–817.

Rauner, S. and Budzinski, M. (2017). Holistic energy system modeling combining multi-objective optimization and life cycle assessment. *Environmental Research Letters*, 12(12):124005.

Reddi, K., Elgowainy, A., Rustagi, N., and Gupta, E. (2017). Impact of hydrogen refueling configurations and market parameters on the refueling cost of hydrogen. *International Journal of Hydrogen Energy*, 42(34):21855–21865.

Regett, A., Baing, F., and Conrad, J. (2018). Emission Assessment of Electricity: Mix vs. Marginal Power Plant Method. In *15th International Conference on the European Energy Market*, pages 1–5. IEEE.

Reinert, C., Baumgärtner, N., Söhler, T., and Bardow, A. (2020a). Optimization of Regionally Resolved Energy Systems by Spatial Aggregation and Disaagregation. In *16th Symposium Energieinnovation Graz*. ENINNOV.

Reinert, C., Deutz, S., Minten, H., Dörpinghaus, L., von Pfingsten, S., Baumgärtner, N., and Bardow, A. (2020b). Dynamic Life Cycle Assessment in National Energy Systems – A Game Changer for Long-Term Assessments? *In preperation*.

Ren, H., Zhou, W., Nakagami, K., Gao, W., and Wu, Q. (2010). Multi-objective optimization for the operation of distributed energy systems considering economic and environmental aspects. *Applied Energy*, 87(12):3642–3651.

Renaldi, R. and Friedrich, D. (2017). Multiple time grids in operational optimisation of energy systems with short- and long-term thermal energy storage. *Energy*, 133:784–795.

Reuter, M., Oyj, O., Hudson, C., van Schaik, A., Heiskanen, K., Meskers, C., and Hagelüken, C. (2013). *Metal recycling: Opportunities, limits, infrastructure : this is report 2b of the Global Metal Flows Working Group of the International Resource Panel of UNEP*. United Nations Environment Programme, Nairobi, Kenya.

Richter, J. (2010). Potenziale der Elektromobilität bis 2050 – Eine szenarienbasierte Analyse der Wirtschaftlichkeit, Umweltauswirkungen und Systemintegration: Endbericht, Juni 2010: (German).

Ringkjøb, H., Haugan, P., and Solbrekke, I. (2018). A review of modelling tools for energy and electricity systems with large shares of variable renewables. *Renewable and Sustainable Energy Reviews*, 96:440–459.

Rippel, K., Wiede, T., Meinecke, M., and König, R. (2017). Netzentwicklungsplan Strom 2030: Erster Entwurf der Übertragungsnetzbetreiber: (German).

Rong, A., Lahdelma, R., and Luh, P. B. (2008). Lagrangian relaxation based algorithm for trigeneration planning with storages. *European Journal of Operational Research*, 188(1):240–257.

Roustaei, M., Niknam, T., Salari, S., Chabok, H., Sheikh, M., Kavousi-Fard, A., and Aghaei, J. (2020). A scenario-based approach for the design of Smart Energy and Water Hub. *Energy*, 195:116931.

Roux, C., Schalbart, P., and Peuportier, B. (2016). Accounting for temporal variation of electricity production and consumption in the LCA of an energy-efficient house. *Journal of Cleaner Production*, 113:532–540.

Ruhnau, O., Bannik, S., Otten, S., Praktiknjo, A., and Robinius, M. (2019). Direct or indirect electrification? A review of heat generation and road transport decarbonisation scenarios for Germany 2050. *Energy*, 166:989–999.

Ryan, N. A., Johnson, J. X., and Keoleian, G. A. (2016). Comparative Assessment of Models and Methods To Calculate Grid Electricity Emissions. *Environmental science & technology*, 50(17):8937–8953.

Salb, C., Gül, S., Cuntz, C., Monschauer, Y., and Weishäupl, J. (2018). Climate Action in Figures: Facts, Trends and Incentives for German Climate Policy.

Samsatli, S., Staffell, I., and Samsatli, N. J. (2016). Optimal design and operation of integrated wind-hydrogen-electricity networks for decarbonising the domestic transport sector in Great Britain. *International Journal of Hydrogen Energy*, 41(1):447–475.

Schiebahn, S., Grube, T., Robinius, M., Tietze, V., Kumar, B., and Stolten, D. (2015). Power to gas: Technological overview, systems analysis and economic assessment for a case study in Germany. *International Journal of Hydrogen Energy*, 40(12):4285–4294.

Schmidt, M. (2014). Heizlastberechnung nach DIN EN 12831: (German).

Schneider, C. and Ketzler, G. (2014). Klimamessstation Aachen-Hörn - Monatsbericht 2014: Report of RWTH Aachen University, Department of Geography, Physical Geography and Climatology (German).

Schütz, T., Schraven, M. H., Fuchs, M., Remmen, P., and Müller, D. (2018). Comparison of clustering algorithms for the selection of typical demand days for energy system synthesis. *Renewable Energy*, 129:570–582.

Shamsi, S. and Omidkhah, M. R. (2012). Optimization of Steam Pressure Levels in a Total Site Using a Thermoeconomic Method. *Energies*, 5(3):702–717.

Siler-Evans, K., Azevedo, I. L., and Morgan, M. G. (2012). Marginal emissions factors for the U.S. electricity system. *Environmental science & technology*, 46:4742–4748.

Simoes, S., Nijs, W., Ruiz, P., Sgobbi, A., Radu, D., Bolat, P., Thiel, C., and Peteves, S. (2013). *The JRC-EU-TIMES model: Assessing the long-term role of the SET plan energy technologies*, volume 26292 of *EUR, Scientific and technical research series*. Publications Office, Luxembourg.

Smith, R. (2005). *Chemical process design and integration*. Wiley, Chichester, West Sussex, United Kingdom.

Spittler, Gladkykh, Diemer, and Davidsdottir (2019). Understanding the Current Energy Paradigm and Energy System Models for More Sustainable Energy System Development. *Energies*, 12(8):1584.

StromNEV (2005). Verordnung über die Entgelte für den Zugang zu Elektrizitätsversorgungsnetzen (Stromnetzentgeltverordnung - StromNEV) at http://www.gesetze-im-internet.de/stromnev/StromNEV.pdf (German).

Teichgräber, H. and Brandt, A. R. (2019). Clustering methods to find representative periods for the optimization of energy systems: An initial framework and comparison. *Applied Energy*, 239:1283–1293.

Teichgraeber, H., Lindenmeyer, C. P., Baumgärtner, N., Kotzur, L., Stolten, D., Robinius, M., Bardow, A., and Brandt, A. R. (2020). Extreme events in time series aggregation: A case study for optimal residential energy supply systems. *Applied Energy, accepted.*

The Royal Society (2019). Sustainable synthetic carbon based fuels for transport: Policy briefing.

The World Bank (2014). Electric power transmission and distribution losses - Germany. Online available at https://data.worldbank.org/indicator/EG.ELC.LOSS.ZS?locations=DE.

Theo, W. L., Lim, J. S., Ho, W. S., Hashim, H., and Lee, C. T. (2017). Review of distributed generation (DG) system planning and optimisation techniques: Comparison of numerical and mathematical modelling methods. *Renewable and Sustainable Energy Reviews*, 67:531–573.

Thind, M. P. S., Wilson, E. J., Azevedo, I. L., and Marshall, J. D. (2017). Marginal Emissions Factors for Electricity Generation in the Midcontinent ISO. *Environmental science & technology*, 51(24):14445–14452.

thinkstep (2019). GaBi Datasets: Process data set: External wall insulation system (Silicate dispersion plaster). Online available at http://gabi-documentation-2019.gabi-software.com/xml-data/processes/2f094bd0-4ecf-4b2e-a54c-10d42b21a788.xml.

Timmerman, J., Hennen, M., Bardow, A., Lodewijks, P., Vandevelde, L., and van Eetvelde, G. (2017). Towards low carbon business park energy systems: A holistic techno-economic optimisation model. *Energy*, 125:747–770.

Turconi, R., Tonini, D., Nielsen, C., Simonsen, C., and Astrup, T. (2014). Environmental impacts of future low-carbon electricity systems: Detailed life cycle assessment of a Danish case study. *Applied Energy*, 132:66–73.

Umweltbundesamt (2018). Kraftwerke in Deutschland (ab 100 Megawatt elektrischer Leistung): (German).

van der Heijde, B., Vandermeulen, A., Salenbien, R., and Helsen, L. (2019). Representative days selection for district energy system optimisation: a solar district heating system with seasonal storage. *Applied Energy*, 248:79–94.

Varbanov, P., Perry, S., Klemeš, J., and Smith, R. (2005). Synthesis of industrial utility systems: cost-effective de-carbonisation. *Applied Thermal Engineering*, 25(7):985–1001.

Varbanov, P. S., Doyle, S., and Smith, R. (2004). Modelling and Optimization of Utility Systems. *Chemical Engineering Research and Design*, 82(5):561–578.

Varbanov, P. S. and Klemeš, J. J. (2011). Integration and management of renewables into Total Sites with variable supply and demand. *Computers & Chemical Engineering*, 35(9):1815–1826.

Velasco-Garcia, P., Varbanov, P. S., Arellano-Garcia, H., and Wozny, G. (2011). Utility systems operation: Optimisation-based decision making. *Applied Thermal Engineering*, 31(16):3196–3205.

Victoria, M., Zhu, K., Brown, T., Andresen, G., and Greiner, M. (2019). The role of storage technologies throughout the decarbonisation of the sector-coupled European energy system. *Energy Conversion and Management*, 201:111977.

Volkart, K., Mutel, C. L., and Panos, E. (2018). Integrating life cycle assessment and energy system modelling: Methodology and application to the world energy scenarios. *Sustainable Production and Consumption*, 16:121–133.

Volkart, K., Weidmann, N., Bauer, C., and Hirschberg, S. (2017). Multi-criteria decision analysis of energy system transformation pathways: A case study for Switzerland. *Energy Policy*, 106:155–168.

Voll, P. (2014). *Automated optimization-based synthesis of distributed energy supply systems: Zugl.: Aachen, Techn. Hochsch., PhD thesis., 2013*, volume 1 of *Aachener Beiträge zur Technischen Thermodynamik*. Wissenschaftsverl. Mainz, Aachen.

Voll, P., Jennings, M., Hennen, M., Shah, N., and Bardow, A. (2015). The optimum is not enough: A near-optimal solution paradigm for energy systems synthesis. *Energy*, 82:446–456.

Voll, P., Klaffke, C., Hennen, M., and Bardow, A. (2013). Automated superstructure-based synthesis and optimization of distributed energy supply systems. *Energy*, 50:374–388.

Voorspools, K. R. and D'haeseleer, W. D. (2000). An evaluation method for calculating the emission responsibility of specific electric applications. *Energy Policy*, 28:967–980.

Wallerand, A. S., Kermani, M., Voillat, R., Kantor, I., and Maréchal, F. (2018). Optimal design of solar-assisted industrial processes considering heat pumping: Case study of a dairy. *Renewable Energy*, 128:565–585.

Wallerand, A. S., Selviaridis, A., Ashouri, A., and Maréchal, F. (2016). Targeting Optimal Design and Operation of Solar Heated Industrial Processes: A MILP Formulation. *Energy Procedia*, 91:668–680.

Wang, Q., McCalley, J., Zheng, T., and Litvinov, E. (2016). Solving corrective risk-based security-constrained optimal power flow with Lagrangian relaxation and Benders decomposition. *International Journal of Electrical Power & Energy Systems*, 75:255–264.

Weitemeyer, S., Kleinhans, D., Wienholt, L., Vogt, T., and Agert, C. (2016). A European Perspective: Potential of Grid and Storage for Balancing Renewable Power Systems. *Energy Technology*, 4(1):114–122.

Wernet, G., Bauer, C., Steubing, B., Reinhard, J., Moreno-Ruiz, E., and Weidema, B. (2016). The ecoinvent database version 3 (part I): overview and methodology. *The International Journal of Life Cycle Assessment*, 21(9):1218–1230.

Wu, Q., Ren, H., Gao, W., and Ren, J. (2016). Multi-objective optimization of a distributed energy network integrated with heating interchange. *Energy*, 109:353–364.

Yang, Y., Zhang, S., and Xiao, Y. (2015a). An MILP (mixed integer linear programming) model for optimal design of district-scale distributed energy resource systems. *Energy*, 90:1901–1915.

Yang, Y., Zhang, S., and Xiao, Y. (2015b). Optimal design of distributed energy resource systems coupled with energy distribution networks. *Energy*, 85:433–448.

Yokoyama, R. (2013). Optimal Operation of a Gas Turbine Cogeneration Plant in Consideration of Equipment Minimum Up and Down Times. *Journal of Engineering for Gas Turbines and Power*, 135(7):122–130.

Yokoyama, R. and Ito, K. (1996). A Revised Decomposition Method for MILP Problems and its Application to Operational Planning of Thermal Storage Systems. *Journal of Energy Resources Technology*, 118:277–284.

Yokoyama, R., Shinano, Y., Taniguchi, S., Ohkura, M., and Wakui, T. (2015). Optimization of energy supply systems by MILP branch and bound method in consideration of hierarchical relationship between design and operation. *Energy Conversion and Management*, 92:92–104.

Yokoyama, R., Shinano, Y., Taniguchi, S., and Wakui, T. (2018a). Search for K-best solutions in optimal design of energy supply systems by an extended MILP hierarchical branch and bound method. *Energy*, 184:45–57.

Yokoyama, R., Shinano, Y., Wakayama, Y., and Wakui, T. (2018b). Optimal Design of a Gas Turbine Cogeneration Plant by a Hierarchical Optimization Method With Parallel Computing. In *Proceedings of the ASME Turbo Expo: Turbomachinery Technical Conference and Exposition–2018*. ASME.

Yokoyama, R., Shinano, Y., Wakayama, Y., and Wakui, T. (2019). Model reduction by time aggregation for optimal design of energy supply systems by an MILP hierarchical branch and bound method. *Energy*, 181:782–792.

Yokoyama, R., Tokunaga, A., and Wakui, T. (2018c). Robust optimal design of energy supply systems under uncertain energy demands based on a mixed-integer linear model. *Energy*, 153:159–169.

Zatti, M., Gabba, M., Freschini, M., Rossi, M., Gambarotta, A., Morini, M., and Martelli, E. (2019). k-MILP: A novel clustering approach to select typical and extreme days for multi-energy systems design optimization. *Energy*, 181:1051–1063.

Zentrum für Energieforschung (2012). Stromspeicherpotenziale für Deutschland: (German).

Zerrahn, A., Schill, W.-P., and Kemfert, C. (2018). On the economics of electrical storage for variable renewable energy sources. *European Economic Review*, 108:259–279.

Zhou, Z., Liu, P., Li, Z., Pistikopoulos, E. N., and Georgiadis, M. C. (2013). Impacts of equipment off-design characteristics on the optimal design and operation of combined cooling, heating and power systems. *Computers & Chemical Engineering*, 48:40–47.

Zhu, Q., Luo, X., Zhang, B., and Chen, Y. (2017). Mathematical modelling and optimization of a large-scale combined cooling, heat, and power system that incorporates unit changeover and time-of-use electricity price. *Energy Conversion and Management*, 133:385–398.

ZIV (2017a). Anzahl der Gasheizungen in Deutschland nach Leistungsklasse im Jahr 2016 (in 1.000 Stück): (German). Online available at `https://de.statista.com/statistik/daten/studie/380951/umfrage/anzahl-der-gasheizungen-in-deutschland-nach-leistungsklasse`.

ZIV (2017b). Anzahl der Ölheizungen in Deutschland nach Leistungsklasse im Jahr 2016 (in 1.000 Stück): (German). Online available at https://de.statista.com/statistik/daten/studie/380931/umfrage/anzahl-der-oelheizungen-in-deutschland-nach-leistungsklasse.

Zou, H., Gratz, E., Apelian, D., and Wang, Y. (2013). A novel method to recycle mixed cathode materials for lithium ion batteries. *Green Chemistry*, 15:1183–1191.

Aachener Beiträge zur Technischen Thermodynamik

ABTT 1
Philip Voll
Automated Optimization-Based Synthesis of Distributed Energy Supply Systems
1. Auflage 2014
ISBN 978-3-86130-474-6

ABTT 2
Johannes Jung
Comparative Life Cycle Assessment of Industrial Multi-Product Processes
1. Auflage 2014
ISBN 978-3-86130-471-5

ABTT 3
Franz Lanzerath
Modellgestützte Entwicklung von Adsorptionswärmepumpen
1. Auflage 2014
ISBN 978-3-86130-472-2

ABTT 4
Thorsten Brands
Einfluss der Gemischzusammensetzung auf die Verbrennung im Diesel- und GCAI-Motor
1. Auflage 2014
ISBN 978-3-95886-006-3

ABTT 5
Dominique Dechambre
Efficient Measurement of Liquid-Liquid Equilibria using Automation and Optimal Experimental Design
1. Auflage 2016
ISBN 978-395886-077-3

ABTT 6
Niklas von der Aßen
From Life-Cycle Assesement towards life-Cycle Design of Carbon Dioxide Capture and Utilization
1. Auflage 2016
ISBN 978-3-95886-080-3

ABTT 7
Matthias Lampe
Integrated Process and Organic Rankine Cycle Working Fluid Design in the Continuous-Molecular Targeting Framework
1. Auflage 2016
ISBN 978-3-95886-086-5

ABTT 8
Thomas Hülser
Optische Untersuchung der Zündvorgänge und deren Auswirkung auf die Verbrennung in PKW-Motoren
1. Auflage 2016
ISBN 978-3-95886-090-2

Aachener Beiträge zur Technischen Thermodynamik

ABTT 9
Malte Döntgen
Reaction Models from Reactive Molecular Dynamics and High-Level Kinetics Predictions
1. Auflage 2016
ISBN 978-3-95886-156-5

ABTT 10
Heike Schreiber
Experiments and Validated Models for Adsorption Thermal Energy Storage in Industrial and Residential Application
1. Auflage 2017
ISBN 978-3-95886-178-7

ABTT 11
André Dirk Sternberg
System-Wide Perspective for Life Cycle Assesment of CO_2-based C1-Chemicals
1. Auflage 2017
ISBN 978-3-95886-193-0

ABTT 12
Uwe Bau
From Dynamic Simulation to Optimal Design and Control of Adsorption Energy Systems
1. Auflage 2018
ISBN 978-3-95886-216-6

ABTT 13
Christian Jens
Modellbasiertes Design von Produkt, Lösungsmittel und Prozess für die Ameisensäuresynthese aus CO_2 und H_2
1. Auflage 2018
ISBN 978-3-95886-231-9

ABTT 14
Jan David Scheffczyk
Integrated Computer-Aided Design of Molecules and Processes using COSMO-RS
1. Auflage 2018
ISBN 978-3-95886-236-4

ABTT 15
Björn Bahl
Optimization-Based Synthesis of Large-Scale Energy Systems by Time-Series Aggregation
1. Auflage 2018
ISBN 978-3-95886-240-1

Aachener Beiträge zur Technischen Thermodynamik

ABTT 16
Bastian Liebergesell
A Milliliter-Scale Setup for the Efficient Characterization of Multicomponent Vapor-Liquid Equilibria Using Raman Spectroscopy
1. Auflage 2018
ISBN 978-3-95886-247-0

ABTT 17
Stefan Wilhelm Graf
A Design Approach for Adsorption Energy Systems Integrating Dynamic Modeling with Small-Scale Experiments
1. Auflage 2018
ISBN 978-3-95886-258-6

ABTT 18
Sebastian Kaminski
Quantum-Mechanics-Based Prediction of SAFT Parameters for Non-Associating and Associating Molecules Containing Carbon, Hydrogen, Oxygen and Nitrogen
1. Auflage 2019
ISBN 978-3-95886-270-8

ABTT 19
Maike Renate Hennen
Decision Support for the Synthesis of Energy Systems by Analysis of the Near-Optimal Solution Space
1. Auflage 2019
ISBN 978-3-95886-277-7

ABTT 20
Peyman Yamin
COSMO-RS-Based Methods for Improved Modelling of Complex Chemical Systems
1. Auflage 2019
ISBN 978-3-95886-288-3

ABTT 21
Meltem Erdogan
Assessement of Adsorbents for Drying by Experiments and Dynamic Simulations
1. Auflage 2019
ISBN 978-3-95886-303-3

ABTT 22
Christian Schulz
SRS/LIF-Messungen zur Charakterisierung rußarmer dieselähnlicher Flammen von alternativen Kraftstoffen und n-Heptan
1. Auflage 2019
ISBN 978-3-91886-310-1

Aachener Beiträge zur Technischen Thermodynamik

ABTT 23
Peter Beumers
Physically-Based Models for the Analysis of Raman Spectra
1. Auflage 2019
ISBN 978-3-95886-319-4

ABTT 24
Arne Kätelhön
Technology Choice Model for Consequential Life Cycle Assessment
1. Auflage 2019
ISBN 978-3-95886-324-8

ABTT 25
Christine Peters
Measurement of Multicomponent Diffusion in Liquids Using Raman Microspectroscopy and Microfluidics
1. Auflage 2020
ISBN 978-3-95886-337-8

ABTT 26
Dinah Elena Hollermann
Reliable and Robust Optimal Design of Sustainable Energy Systems
1. Auflage 2020
ISBN 978-3-95886-346-0

ABTT 27
Thomas Raffius
Laserspektroskopische Analyse von selbstzündenden motorischen Einspritzstrahlen alternativer Biokraftstoffe
1. Auflage 2020
ISBN 978-3-95886-358-3

ABTT 28
Johannes Schilling
Integrated Thermo-Economic Design of Processes and Molecules Using PC-SAFT
1. Auflage 2020
ISBN 978-3-95886-368-2

ABTT 29
Nils Julius Baumgärtner
Optimization of Low-Carbon Energy Systems from Industrial to National Scale
1. Auflage 2020
ISBN 978-3-95886-385-9